Operations Research Analysis in Test and Evaluation

Donald L. Giadrosich
Former Director of Operations Analysis,
U.S. Air Forces in Europe
and Chief Scientist, U.S. Air Force Air Warfare Center,
Eglin Air Force Base, Florida

EDUCATION SERIES
J. S. Przemieniecki
Series Editor-in-Chief
Air Force Institute of Technology
Wright-Patterson Air Force Base, Ohio

Published by
American Institute of Aeronautics and Astronautics, Inc.
370 L'Enfant Promenade, SW, Washington, DC 20024-2518

American Institute of Aeronautics and Astronautics, Inc., Washington, DC
370 L'Enfant Promenade, SW, Washington, DC 20024-2518

Library of Congress Cataloging-in-Publication Data

Giadrosich, Donald L.
    Operations research analysis in test and evaluation / Donald L. Giadrosich.
       p.   cm.—(AIAA education series)
    Includes bibliographical references and index.
    1. Weapons industry—Quality control.   2. Operations research.
  I. Title.   II. Series.
UF530.G53   1995   623.4'011—dc20   94-48188
ISBN 1-56347-112-4

Copyright © 1995 by the American Institute of Aeronautics and Astronautics, Inc. All rights reserved. Printed in the United States of America. No part of this book may be reproduced or transmitted in any form or by any means, or stored in a database or retrieval system, without the prior written permission of the publisher.

Data and information appearing in this book are for informational purposes only. AIAA is not responsible for any injury or damage resulting from use or reliance, nor does AIAA warrant that use or reliance will be free from privately owned rights.

Texts Published in the AIAA Education Series

Flight Performance of Aircraft
   S. K. Ojha, 1995
Operations Research Analysis in Test and Evaluation
   Donald L. Giadrosich, 1995
Radar and Laser Cross Section Engineering
   David C. Jenn, 1995
Introduction to the Control of Dynamic Systems
   Frederick O. Smetana, 1994
Tailless Aircraft in Theory and Practice
   Karl Nickel and Michael Wohlfahrt, 1994
Mathematical Methods in Defense Analyses
   Second Edition
   J. S. Przemieniecki, 1994
Hypersonic Aerothermodynamics
   John J. Bertin, 1994
Hypersonic Airbreathing Propulsion
   William H. Heiser and David T. Pratt, 1994
Practical Intake Aerodynamic Design
   E. L. Goldsmith and J. Seddon, Editors, 1993
Acquisition of Defense Systems
   J. S. Przemieniecki, Editor, 1993
Dynamics of Atmospheric Re-Entry
   Frank J. Regan and Satya M. Anandakrishnan, 1993
Introduction to Dynamics and Control of Flexible Structures
   John L. Junkins and Youdan Kim, 1993
Spacecraft Mission Design
   Charles D. Brown, 1992
Rotary Wing Structural Dynamics and Aeroelasticity
   Richard L. Bielawa, 1992
Aircraft Design: A Conceptual Approach
   Second Edition
   Daniel P. Raymer, 1992
Optimization of Observation and Control Processes
   Veniamin V. Malyshev, Mihkail N. Krasilshikov, and Valeri I. Karlov, 1992
Nonlinear Analysis of Shell Structures
   Anthony N. Palazotto and Scott T. Dennis, 1992
Orbital Mechanics
   Vladimir A. Chobotov, 1991
Critical Technologies for National Defense
   Air Force Institute of Technology, 1991
Defense Analyses Software
   J. S. Przemieniecki, 1991
Inlets for Supersonic Missiles
   John J. Mahoney, 1991
Space Vehicle Design
   Michael D. Griffin and James R. French, 1991
Introduction to Mathematical Methods in Defense Analyses
   J. S. Przemieniecki, 1990
Basic Helicopter Aerodynamics
   J. Seddon, 1990
Aircraft Propulsion Systems Technology and Design
   Gordon C. Oates, Editor, 1989

(Continued on next page.)

Texts Published in the AIAA Education Series (continued)

Boundary Layers
   A. D. Young, 1989
Aircraft Design: A Conceptual Approach
   Daniel P. Raymer, 1989
Gust Loads on Aircraft: Concepts and Applications
   Frederic M. Hoblit, 1988
Aircraft Landing Gear Design: Principles and Practices
   Norman S. Currey, 1988
Mechanical Reliability: Theory, Models and Applications
   B. S. Dhillon, 1988
Re-Entry Aerodynamics
   Wilbur L. Hankey, 1988
Aerothermodynamics of Gas Turbine and Rocket Propulsion,
Revised and Enlarged
   Gordon C. Oates, 1988
Advanced Classical Thermodynamics
   George Emanuel, 1988
Radar Electronic Warfare
   August Golden Jr., 1988
An Introduction to the Mathematics and Methods of Astrodynamics
   Richard H. Battin, 1987
Aircraft Engine Design
   Jack D. Mattingly, William H. Heiser, and Daniel H. Daley, 1987
Gasdynamics: Theory and Applications
   George Emanuel, 1986
Composite Materials for Aircraft Structures
   Brian C. Hoskins and Alan A. Baker, Editors, 1986
Intake Aerodynamics
   J. Seddon and E. L. Goldsmith, 1985
Fundamentals of Aircraft Combat Survivability Analysis and Design
   Robert E. Ball, 1985
Aerothermodynamics of Aircraft Engine Components
   Gordon C. Oates, Editor, 1985
Aerothermodynamics of Gas Turbine and Rocket Propulsion
   Gordon C. Oates, 1984
Re-Entry Vehicle Dynamics
   Frank J. Regan, 1984

Published by
American Institute of Aeronautics and Astronautics, Inc., Washington, DC

Foreword

Operations Research Analysis in Test and Evaluation by Donald L. Giadrosich represents a significant addition to the technical literature in this field. Although the text focuses on processes and methodologies that are vital to the military, these techniques are also applicable to civilian problems.

The text provides an insight into the importance of operations research analysis, test and evaluation (T&E), and how they both affect weapons systems at critical decision points in their life cycle. Risk reduction in the decision process through operations research analysis and T&E becomes increasingly important as budgetary pressures impact the acquisition process and the nation moves toward a smaller defense force. The application of advanced technologies in computers; distributed and interactive simulation; information processing, synthesis, and display; virtual and hybrid reality; global positioning; automated systems and diagnostics; and other areas cited in this text are extremely important in reducing both the costs and risks associated with future weapons systems.

This text is recommended reading for both civilian and military managers and officials involved in making critical decisions requiring timely analysis. The text should also be of great value to those who would like a better perspective into the advanced aspects of applied operations research analysis and T&E.

General Robert D. Russ
United States Air Force, Retired
Former Commander of the Tactical Air Command

Preface

Critical decisions are made daily throughout commercial industry and government based on the results of operations research analysis and test and evaluation (T&E). A wide variety of questions must be answered, such as which products to develop and buy, how much is enough, and how to efficiently apply those products that are available. Incorrect decisions can lead to increases in costs and losses in capability, profit, and life. Operations research analysis and T&E results help the decision maker by reducing risks and/or bounding the unknowns, and thus risks associated with decisions. Often, when the wrong decision is made, it can be traced to inadequate or improper analysis and T&E, and can even be traced at times to someone's having ignored the results of analysis and T&E.

Because the consequences of T&E can be extremely expensive and sometimes catastrophic, scientific and multidisciplined methods to include operations research analysis have found increased use and importance, especially during the last 30 years. The origins of many of the technical and analytical methods applied find their basis in many of disciplines including engineering, physics, mathematics, economics, probability, statistics, human factors, and a wide variety of the social sciences. The analytical power and synergism of these methods continue to increase the utility and influence of T&E.

The nature and environment of T&E is changing at a rapid pace. New technologies in the internal functioning of systems, measurement facilities, integration laboratories, hardware-in-the-loop test facilities, computer modeling and simulation, open air test range instrumentation, and the electronic linking of simulations and open air test ranges are all in a revolutionary state of improvement. The changing world situation has also resulted in the need for major reductions in the support infrastructure of the Department of Defense. This has led to significant efforts in downsizing, consolidation, and the elimination of some analysis and T&E capabilities. Numerous studies have recently taken place and ongoing efforts continue to address how best to provide the essential facilities and capabilities in the most efficient manner. Although some of the organizational information contained in this text may change due to these ongoing studies, the concepts and functions described remain representative of the essential types of facilities required and available.

Opportunities exist today through new techniques and environments to significantly improve T&E as well as reduce its costs. The advent of advanced computer technology, advanced distributed simulation, distributed interactive simulation, advanced readout and display technologies that include virtual and hybrid reality, automated diagnostics, and the global positioning system all have great potential to improve T&E. The move toward being able to capitalize on these and other technologies requires significant change, and has met the normal political and institutional resistance to change. Significant change also requires improved methods to validate the technologies and to increase the confidence of decision makers in the results of T&E from these advanced capabilities.

This book treats a wide range of subjects important in operations research analysis and T&E. It is unique in its attempt to focus on the importance of T&E and demonstrate how T&E relates to and affects the acquisition and operational implementation of systems. It is designed to provide a global presentation of applied operations research analysis in test and evaluation, yet provide some new in-depth work in selected subjects. The intent has been to introduce and provide insight into the major factors and subject areas important in T&E. At the same time, a goal has been to provide selected example problems, techniques, and applications that give the reader a feel for the complexity, characteristics, and benefits of operations research analysis in properly designed and conducted T&E.

The successful completion of a project such as this book requires the help and coordinated efforts of many individuals and organizations. I would like to thank all of the many people and organizations where I have worked in the past and from which I have learned and practiced operations research analysis in T&E. In particular, I would like to acknowledge my colleague and friend, Mr. Hugh Curtis, who over the years worked with me in operations research analysis and T&E. Hugh and I collaborated on several papers that helped formulate some of my ideas and work in this area. It has been through the exchange and interactions with many people like Hugh that I have reached this point. Organizations that contributed greatly to my knowledge and experience include Hughes Aircraft Company, the U.S. Navy, the Joint Chiefs of Staff, and the U.S. Air Force.

I would like to thank Dr. John Przemieniecki for helping me get this book published. I would also like to thank Colonel Anthony Gardecki (USAF Retired), Ms. Geri Lentz, and Dr. Rita Gregory for reviewing and commenting on the initial manuscript, and Ms. Sherry Griffith and Ms. Ingrid Chavez for their outstanding support on parts of the initial draft. I also owe thanks to Mr. Randy Bryant for review and comments on the example problems in Appendix 1 and help in computer generation of finished graphics. The contribution and assistance of all of these individuals was invaluable.

I owe a special thanks to my children, Kirk, Dana, and Keith, for putting up with a fanatic in operations research analysis and T&E. I also thank my daughter, Dana, for her moral support and help with proofreading. I would like to thank my parents, Edward and May Giadrosich, for getting me started in a way that I could eventually write this book. I would like to acknowledge the help and encouragement throughout the years of my father-in-law and mother-in-law, Riley and Nella Davidson, and Aunt Fay McDonald.

Finally, and most importantly, I wish to thank my wife, Diana, for her support and for always being there. Without her patience, concern, and support, this book would not have been possible. Her health during this period has not been the best, but often she has taken on some of my responsibilities to allow me to complete this effort. It is to all of my family that this book is dedicated. I know they have made great sacrifices for me and hope in some way I can repay them for all that they have done.

Donald L. Giadrosich
June 1995

Table of Contents

Foreword

Preface

Chapter 1. Introduction 1
  1.1 History of Test and Evaluation. 1
  1.2 General. 3
      1.2.1   Test. 3
      1.2.2   Evaluation 4
  1.3 Quality Measurement and Improvement 4
      1.3.1   The Customer. 4
      1.3.2   The Process 4
      1.3.3   The Product 5
  1.4 Establishing the Need for New or Improved Systems. 5
  1.5 Acquiring Systems. 5
  1.6 Assessing and Validating Requirements 7
  1.7 Program Flow and Decision Points 9
  1.8 Reducing Risks Through Test and Evaluation. 10
      1.8.1   Development Test and Evaluation 10
      1.8.2   Operational Test and Evaluation 11
      1.8.3   Production Acceptance Test and Evaluation 13
      1.8.4   Joint Service Test and Evaluation 13
  1.9 Scientific Approach 13
      1.9.1   What To Do 14
      1.9.2   How To Do It. 14
  1.10 Ideals of Test and Evaluation. 14
  References. 15

Chapter 2. Cost and Operational Effectiveness Analysis 17
  2.1 General. 17
  2.2 Approach 18
  2.3 Acquisition Issues 18

TABLE OF CONTENTS

      2.3.1 Threat 18
      2.3.2 Need 20
      2.3.3 Environment 20
      2.3.4 Constraints 21
      2.3.5 Operational Concept 21
2.4 Alternatives 21
      2.4.1 Performance Objectives 21
      2.4.2 Description of Alternatives 22
2.5 Analysis of Alternatives 23
      2.5.1 Models 23
      2.5.2 Measures of Effectiveness 24
      2.5.3 Costs 24
      2.5.4 Trade-Off Analyses 26
      2.5.5 Decision Criteria 27
2.6 Scope by Milestone 28
2.7 Cost and Effectiveness Comparisons 29
      2.7.1 System Concept Level Comparisons 29
      2.7.2 System Design Level Comparisons 30
      2.7.3 System Comparison Level 32
      2.7.4 Force Level Comparisons 32
2.8 Addressing Risks 36
      2.8.1 Uncertainty in Estimates 36
      2.8.2 Risks in Terms of Probability 38
2.9 Other Cost and Effectiveness Considerations 38
      2.9.1 Costs and Effectiveness Over Time 40
2.10 Optimization and Partial Analysis 45
      2.10.1 Lagrangian Function (L) 46
      2.10.2 Partial Analysis 47
2.11 Affordability Assessments 49
References 50

Chapter 3. Basic Principles 53
3.1 General 53
      3.1.1 Effectiveness 53
      3.1.2 Suitability 53
      3.1.3 Tactics, Procedures, and Training 53
3.2 Defining the Test and Evaluation Problem 54
3.3 Establishing an Overall Test and Evaluation Model 55
3.4 Measurement 57
      3.4.1 Physical Measurements 58
      3.4.2 Measurement Process 59
      3.4.3 Validity of Measurements 59
3.5 Statistical Nature of Test and Evaluation Data 65
3.6 Typical Range System Measurement Capabilities 66

TABLE OF CONTENTS

    3.6.1 Time-Space-Position Information 66
    3.6.2 Electromagnetic Environment Testing.................. 68
    3.6.3 Engineering Sequential Photography................... 69
    3.6.4 Range-Timing Systems 69
    3.6.5 Aerospace Environmental Support 69
    3.6.6 Calibration and Alignment........................... 69
References.. 70

Chapter 4. Modeling and Simulation Approach 73
4.1 General.. 73
4.2 Model Concept... 73
    4.2.1 Decomposition of a Model........................... 76
    4.2.2 Applications....................................... 78
4.3 Verification, Validation, and Accreditation..................... 78
    4.3.1 Verification....................................... 81
    4.3.2 Validation.. 81
    4.3.3 Sources of Information.............................. 85
    4.3.4 Sensitivity Analysis 85
    4.3.5 Tasks for Model Verification, Validation, and
            Accreditation 86
    4.3.6 Stakeholders in Model Verification, Validation, and
            Accreditation 86
    4.3.7 Model Verification, Validation, and Accreditation Plan...... 88
    4.3.8 Documentation of the Model Verification, Validation, and
            Accreditation Efforts 89
    4.3.9 Special Considerations When Verifying, Validating, and
            Accrediting Models 90
    4.3.10 Summary 91
4.4 Establishing an Overall Test and Evaluation Model 92
    4.4.1 System Test and Evaluation Environments 92
    4.4.2 Other Modeling and Simulation....................... 94
4.5 Future Challenges.. 95
References.. 95

Chapter 5. Test and Evaluation Concept 99
5.1 General... 99
5.2 Focusing on the Test and Evaluation Issues.................... 101
5.3 Supporting Data and System Documentation 102
    5.3.1 Design Concept 103
    5.3.2 Operations Concept 103
    5.3.3 Maintenance and Support Concept 104
5.4 Critical Issues and Test Objectives........................... 104

TABLE OF CONTENTS

      5.4.1   Relating Critical Issues and Objectives to the Military Mission. 104
      5.4.2   Researching the Critical Issues and Objectives 105
5.5   Measures of Effectiveness, Measures of Performance, and Criteria 105
5.6   Data Analysis and Evaluation 106
5.7   Scenarios 106
5.8   Requirements for Test and Evaluation Facilities 107
5.9   Scope and Overall Test and Evaluation Approach 107
5.10 Non-Real Time Kill Removal Versus Real Time Kill Removal..... 107
      5.10.1 Non-Real Time Kill Removal Testing 108
      5.10.2 Real Time Kill Removal Testing 108
5.11 Use of Surrogates 108
5.12 Managing Change 109
References... 110

Chapter 6. Test and Evaluation Design........................... 113
6.1   General 113
6.2   Test and Evaluation Design Process........................... 114
6.3   Procedural Test and Evaluation Design 114
      6.3.1   Critical Issues and Test Objectives..................... 114
      6.3.2   Test and Evaluation Constraints....................... 116
      6.3.3   Test and Evaluation Method. 116
      6.3.4   Measures of Effectiveness and Measures of Performance... 118
      6.3.5   Measurements and Instrumentation 119
      6.3.6   Selecting the Size of the Sample 121
      6.3.7   Data Management and Computer-Assisted Analysis........ 126
      6.3.8   Decision Criteria 127
      6.3.9   Limitations and Assumptions.......................... 128
6.4   Experimental Test and Evaluation Design 128
      6.4.1   Test Variables... 128
      6.4.2   Sensitivity of Variables 129
      6.4.3   Operations Tasks 129
      6.4.4   Analysis 130
      6.4.5   Factor Categories 130
      6.4.6   Design Considerations 131
      6.4.7   Example of Categorizing Factors for a Multiple-Factor Test. 131
      6.4.8   Analysis of Variance 134
6.5   Sequential Methods 141
6.6   Nonparametric Example.. 141
References... 143

TABLE OF CONTENTS

Chapter 7. Test and Evaluation Planning 145
  7.1   General 145
  7.2   Advanced Planning 146
  7.3   Developing the Test and Evaluation Plan 147
       7.3.1   Introduction 147
       7.3.2   Test and Evaluation Concept 149
       7.3.3   Methodology 154
       7.3.4   Administration 157
       7.3.5   Reporting 161
       7.3.6   Test and Evaluation Plan Supplements 162
  References 162

Chapter 8. Test and Evaluation Conduct, Analysis, and Reporting 165
  8.1   General 165
  8.2   Test Conduct 165
       8.2.1   Professional Leadership and Test Team Performance 165
       8.2.2   Systematic Approach 166
  8.3   Analysis 170
  8.4   Reporting 171
       8.4.1   General 171
       8.4.2   Planning 172
       8.4.3   Gathering Data 172
       8.4.4   Organizing 173
       8.4.5   Outlining 173
       8.4.6   Writing 174
       8.4.7   Editing 175
  References 176

Chapter 9. Software Test and Evaluation 179
  9.1   General 179
  9.2   Important Terms and Concepts 179
  9.3   Software Maintenance 180
  9.4   Modifying Software 180
       9.4.1   Reducing Debugging Problems 182
       9.4.2   Software Errors 182
       9.4.3   Top-Down Programming 183
       9.4.4   Top-Down Testing 184
       9.4.5   Bottom-Up Testing 184
  9.5   Assessment of Software 184
       9.5.1   Design Concepts and Attributes 185
       9.5.2   Performance Measurement 186
       9.5.3   Suitability Measurement 186

TABLE OF CONTENTS

9.6 Test Limitations 187
References 187

Chapter 10. Human Factors Evaluations **191**
  10.1 General 191
  10.2 System Analysis and Evaluation 192
      10.2.1 System Analysis 192
      10.2.2 System Evaluation 199
      10.2.3 Questionnaires and Subjective Assessments 201
References 205

Chapter 11. Reliability, Maintainability, Logistics Supportability, and Availability **209**
  11.1 General 209
  11.2 System Cycles 212
      11.2.1 System Daily Time Cycle 212
      11.2.2 System Life Cycle 213
  11.3 System Degradation 215
  11.4 Reliability 215
      11.4.1 Combining Components 216
      11.4.2 Example Complex System Reliability Program 217
  11.5 Maintainability 218
      11.5.1 Maintenance Concept 219
      11.5.2 Repairability and Serviceability 221
      11.5.3 Maintenance Personnel and Training 222
  11.6 Logistics Supportability 222
  11.7 Analysis and Predictions 222
      11.7.1 Mean Time Between Failure 223
      11.7.2 Mean Time To Repair 223
      11.7.3 Mean Downtime 223
      11.7.4 Availability 224
  11.8 Fault-Tolerant Systems 226
References 227

Chapter 12. Test and Evaluation of Integrated Weapons Systems **229**
  12.1 General 229
  12.2 Integration Trend 230
      12.2.1 Segregated Systems 230
      12.2.2 Federated Systems 230
      12.2.3 Integrated Systems 230
  12.3 System Test and Evaluation 230
      12.3.1 Engineering Level Test and Evaluation 231

TABLE OF CONTENTS

	12.3.2	Operational Level Test and Evaluation 232
	12.3.3	Testability in Systems............................. 232
	12.3.4	Emphasizing Designed-In Capabilities 233
	12.3.5	Real-World Threats and Surrogates 235
	12.3.6	Global Positioning System as a Precision Reference..... 236
	12.3.7	Shared Data Bases 236
	12.3.8	Summary 237
12.4	Automatic Diagnostic Systems 237
	12.4.1	Purpose of Automatic Diagnostic Systems 238
	12.4.2	Probabilistic Description of Automatic Diagnostic Systems 239
	12.4.3	Estimating Automatic Diagnostic System Probabilities... 242
	12.4.4	Automatic Diagnostic System Measures of Effectiveness 246
	12.4.5	Summary 250
References.. 250

Chapter 13. Measures of Effectiveness and Measures of Performance... 253
- 13.1 General.. 253
- 13.2 Systems Engineering Approach 255
- 13.3 Operational Tasks/Mission Approach 258
- 13.4 Example Measures of Performance 262
- 13.5 Example Measures of Effectiveness........................ 262
- 13.6 Example Measures of Effectiveness for Electronic Combat Systems.. 263
- 13.7 Example Measures of Effectiveness for Operational Applications, Concepts, and Tactics Development........................ 268
- References.. 268

Chapter 14. Measurement of Training 271
- 14.1 General.. 271
- 14.2 Learning by Humans 272
- 14.3 Transfer of Training 272
- 14.4 Example Complex Training System......................... 273
  - 14.4.1 Tracking Instrumentation Subsystem.................. 273
  - 14.4.2 Aircraft Instrumentation Subsystem................... 275
  - 14.4.3 Control and Computational Subsystem 275
  - 14.4.4 Display and Debriefing Subsystem 275
  - 14.4.5 ACMI System Training Attributes 275
- 14.5 Example Measures for Training Activities.................... 276
- References.. 277

TABLE OF CONTENTS

Chapter 15. Joint Test and Evaluation 301
  15.1   General 301
  15.2   Joint Test and Evaluation Program 301
        15.2.1   Joint Test and Evaluation Nomination and Joint Feasibility Study Approval 302
        15.2.2   Joint Feasibility Study Review and Chartering of the Joint Test Force 305
  15.3   Activating and Managing the Joint Test Force 305
        15.3.1   Command Relationship 306
        15.3.2   Example Joint Test Force Organization 308
  15.4   Developing the Joint Test and Evaluation Resource Requirements 308
  15.5   Preparing the Test Design 308
  15.6   Preparing the Test Plan 310
  15.7   Conducting the Joint Test and Evaluation 310
  15.8   Joint Test and Evaluation Analysis and Reporting 311
  15.9   Disposition of Joint Test and Evaluation Resources 313
  15.10 Example Joint Test and Evaluation Programs 313
        15.10.1 Joint Test and Evaluation of Combat Air Support Target Aquisition (SEEKVAL) 313
        15.10.2 Joint Test and Evaluation for Validation of Simulation Models (HITVAL) 313
        15.10.3 Joint Test and Evaluation for Electronic Warfare Joint Test (EWJT) 313
        15.10.4 Joint Test for A-7/A-10 Fly-Off 314
        15.10.5 Joint Test and Evaluation of Tactical Aircraft Effectiveness and Survivability in Close Air Support Antiarmor Operations (TASVAL) 314
        15.10.6 Joint Test and Evaluation for Data Link Vulnerability Analysis (DVAL) 314
        15.10.7 Joint Test and Evaluation for Identification Friend, Foe, or Neutral (IFFN) 314
        15.10.8 Joint Test and Evaluation for Forward Area Air Defense (JFAAD) 314
        15.10.9 Joint Test and Evaluation for Command, Control, and Communications Countermeasures (C^3CM) 315
        15.10.10 Joint Test and Evaluation for Electromagnetic Interference (JEMI) 315
        15.10.11 Joint Test and Evaluation for Over-the-Horizon Targeting (JOTH-T) 315
        15.10.12 Joint Test and Evaluation of Camouflage, Concealment, and Deception (JCCD) 315
        15.10.13 Joint Test and Evaluation for Air Defense Operations/Joint Engagement Zone (JADO/JEZ) 316

TABLE OF CONTENTS

    15.10.14 Joint Test and Evaluation for Logistics over the Shore (JLOTS) 316
    15.10.15 Joint Test and Evaluation for Band IV (Infrared Countermeasures) 316
    15.10.16 Joint Test and Evaluation for Tactical Missile Signatures (JTAMS) 316
    15.10.17 Joint Crisis Action Test and Evaluation (JCATE) 316
    15.10.18 Joint Test and Evaluation for Smart Weapons Operability Enhancement (SWOE) Process 317
    15.10.19 Joint Advanced Distributed Simulation (JADS) 317
    15.10.20 Joint Theater Missile Defense (JTMD) 317
References 317

Appendix 1. Example Test and Evaluation Quantification Problems and Solutions 321

Appendix 2. Example Department of Defense Test and Evaluation Facilities 339

Appendix 3. Glossary of Abbreviations and Acronyms 345

Selected Bibliography 355

Subject Index 363

1
Introduction

1.1 History of Test and Evaluation

There is little doubt that the history of mankind has probably always included some form of trial and error, experimentation, evaluation, and learning processes. However, the name of the first test-and-evaluator has not been handed down for posterity. He/she was probably a person of prehistoric time who, through both curiosity and necessity, needed to determine if something worked, and then refined his/her effort based on the results of his/her inquiry.

During the first three or four decades of this century, the early engineers in industry carried out many forms of methodical inquiry that could certainly be classified as Test and Evaluation (T&E). Most of these early testers appeared to be people who were a combination of researcher, scientist, innovator, and sometimes daredevil. In the field of aviation, Orville Wright flew his "Wright Flyer" on December 17, 1903, and Wilbur Wright delivered what could be called an early T&E report, "Experiments and Observations in Soaring Flight," to the Western Society of Engineers in 1903 (Reference 1). The first test flight was made from a ship by Eugene Ely on November 14, 1910, and he also made the first landing on a ship in January 1911 (Reference 2). Early forms of experimental and demonstration testing were followed by testing with broader themes. For example, after the Army had determined that the Wrights' aircraft met the required standards, the military Services then had to figure out where and how to use flying machines to gain the military advantage in combat. This type of military testing has had many titles over the years; however, it has been most commonly referred to as military operational suitability testing. Today, it is described in terms of development test and evaluation, operational test and evaluation, and tactics development and evaluation. These areas will be described in more detail later in this chapter and addressed throughout the book.

Although the information presented in this book is focused on operations research analysis in the T&E of military weapons systems, the fundamental principles and concepts also apply to commercial products and systems. Both government and industry equally face the challenge of producing high quality products and systems within limited resources. The commercial customer and the military combat user both demand high quality products and systems that meet needs and that are continually improved. Good T&E measurement examines the system environment in total, along with the person-machine interfaces, and provides a meaningful basis for objective feedback on performance and sound acquisition and product improvement decisions.

When T&E is avoided, or improperly performed, it can lead to costly and even disastrous results. History has shown that inadequate T&E can result in aircraft that do not fly safely, weapons systems than can not fire missiles, and even systems that falsely identify friends as enemies and vice versa. For example, the U.S.

Navy spent over $1.5 billion developing an electronic warfare jamming system over an extended period of time, during which T&E was supposedly successful. However, the system was subsequently rejected in total prior to any significant level of production. The U.S. Army managed a similar situation with an expensive anti-aircraft gun system, and the U.S. Air Force experienced the same type of problem with expensive flightline test equipment for electronic warfare systems. These systems and their costs to taxpayers are a few of the many examples that cause concern among responsible leaders. These examples demonstrate to those in the T&E business the critical importance of methodically conducting effective and adequate T&E.

Inadequate T&E and incorrect analysis of results have to some degree contributed to extremely catastrophic situations. In some cases, tests of systems that are less than full have resulted in the inability of weapons systems to fire missiles that guide successfully and achieve the expected levels of kill. Even critical space systems such as the spacecraft *Challenger* have not been able to escape the spotlight and second-guessing of whether and how T&E was accomplished.

Various forms of testing in the commercial industry have had their share of difficulties, but have also contributed significantly to the knowledge base for T&E. Management in practically every industry has utilized some form of T&E to implement policies of production control and to assure product quality in the competitive market. Design engineers, metallurgists, materials engineers, and operations researchers have applied T&E techniques to attain better use of engineering materials. Production executives and engineers have used T&E to decrease losses of production time, eliminate material wastage, and maintain high industrial productivity. Since World War II, the Japanese have made a major impact on the quality of products produced for consumption through the application of W. Edwards Deming's method for total quality management (Reference 3). This method is highly dependent upon some form of T&E to quantitatively establish the effectiveness of production processes in order that they may be continuously improved. The theory here is that one must never be satisfied with the level of quality produced—it must always be improved in subsequent production runs and applications. Maintenance experts have also used T&E to improve the performance of systems in service, to decrease maintenance time and costs, and to increase the operational life of systems. Reliability, maintainability, and availability (RM&A) experts have found T&E to be a highly valuable tool to help measure and achieve increased product integrity and serviceability (Reference 4). Applied in such a variety of ways, T&E has contributed, and is likely to continue to contribute, to increased productivity, lower production and operating costs, and greater profits in industry.

During World War II, military T&E made great strides and had to rapidly expand to address the many pressing complex developments, acquisitions, and operational problems. Throughout and subsequent to this period, there were numerous changes in military T&E organizations and in testing procedures. These changes primarily attempted to make the T&E processes and inquiries more realistic, accurate, responsive, objective, and effective. Although competent individuals have differed, and will probably continue to do so, in their views on the nature of T&E, there is little doubt regarding the importance of its role and how it has affected the capability of the U.S. military and commercial industry as well as the lives of most people. This role ranges all the way from the examination of consumer products to determine their utility and the preferences of customers, to the identification of which military systems the U.S. Congress will support for procurement.

During modern times, the failure to properly use T&E has often led to the loss of markets and reputations when governments and industries sought to develop new fields of application for their products. Disaster and loss of life or loss of costly investments in complex developments can result from the omission of necessary T&E on components and systems in such fields as high performance aircraft, missiles, and spacecraft, as well as nuclear systems (Reference 5). As such systems have become more complex and environmental restrictions and applications more severe, the cost of such failures have risen dramatically. Errors in judgment, which result in the failure to carry out effective T&E to ensure system integrity and reliability, cannot be tolerated in the modern development, production, and management environments.

1.2 General

The proof or acceptance of a component, system, concept, or technical approach is heavily dependent upon the ability to provide convincing evidence that compels acceptance by the community of operators, supporters, and other experts involved in its development and application. Historically, both operations research analysis and T&E have played a vital role in the acceptance (or rejection) process by creating environments and situations as well as objective measurement data that lead to proof or disproof. T&E in the simplest terms can be defined in its two parts: 1) test and 2) evaluation. Test or testing is a systematic means of collecting data which can be analyzed and used to formulate statements regarding the performance of a component, system, or concept that is within certain limits of error. Operations Research Analysis is a scientific method using a formalized structured approach with quantative models and statistical methods that help evaluate the test results. Evaluation is the process by which one examines the test data and statements, as well as any other influencing factors, to arrive at a judgment of the significance or worth of the component, system, or concept. Operations Research Analysis and T&E are vital to the process of successfully acquiring and operating commercial and military systems.

1.2.1 Test

Some degree of risk and uncertainty is almost always inherent in important real world situations. Formulating statements about the risks and uncertainty that are within certain limits of error is critical to the conduct of effective testing. Consequently, the risk and uncertainty must be characterized and quantified to the degree practical during the T&E process. Test or testing involves the physical exercising (i.e., a trial use or examination) of a component, system, concept, or approach for the sole purpose of gathering data and information regarding the item under test. These data and this information may cover a broad range of operating and environmental conditions and can be both descriptive and inferential. To ensure credibility in the results, the test must be objective, unbiased, and operationally and statistically significant, as well as operationally realistic. These important characteristics will be further defined and addressed in considerable detail throughout this text. The test must be tailored to provide information about the phenomenon under investigation, and may also contribute to the science and methodology of the test investigative processes themselves. Subsequent chapters of this book will address the fundamental principles and concepts that are dedicated to designing, planning, conducting, analyzing, and reporting tests that can achieve these important characteristics.

1.2.2 Evaluation

Evaluation is the process of establishing the value or worth of a component, system, concept, or approach. When used in the context of "test and evaluation," there is the implied premise that the "evaluation" will be based on the data and information derived from the "test." In most cases, the evaluation involves an inferential process where one extrapolates from the limited test results in some manner to real world problems of interest. Often this extrapolation is done with the aid of mathematical modeling and simulation of the real world situation. Additionally, the evaluation requires some form of judgmental basis (i.e., criteria) for the conclusions deduced. These criteria, normally established in advance of the test, specify the level at which the activity must perform to be judged as successful, or acceptable. In a broader sense, evaluations may be and usually are based on a wider range of data and information than that derived from a single physical test. For example, historical data and information regarding the item (or similar items) may be available from previous experience, and/or generated by analytical modeling and simulation. Because of environmental and safety restrictions, the data and information are sometimes generated in one environment or set of conditions and analytically projected or extrapolated to others.

1.3 Quality Measurement and Improvement
(See References 6–10)

The philosophical concept of perpetual quality improvement indeed plays an important role in T&E. This concept is based on meeting the needs and expectations of the customer and a continuing effort to improve product quality throughout the life cycle. This requires a clear understanding of who the real customers are, their expectations and requirements, planning and conducting effective T&E with the potential to meet those expectations and requirements, and maintaining the utmost integrity in the analysis and reporting of results.

1.3.1 The Customer

Although there are many layers of customers in both the civilian and military T&E processes to include the testers themselves, the real customers are those who will eventually receive and operate the systems and concepts in the civilian world or in combat. Working only to satisfy the expectations and requirements of customers short of the real end users can lead to waste and, in most cases, solutions that fall short of the desired end results.

1.3.2 The Process

During the acquisition of a weapons system, it is extremely important to cor-

INTRODUCTION

rectly determine the combat needs (i.e., the requirement), to correctly state the contractual specifications for design and procurement of the system, and to conduct sufficient and effective T&E to minimize the risk associated with acquiring the system. Once acquired, effective T&E is essential to maximize the potential and implementation of the system throughout its life cycle.

The T&E process described throughout this book is designed to improve both government and industry system acquisition and operational excellence. It is dedicated to objective and accurate quantification of those things that truly make a difference in a realistic system acquisition and operational environment. In today's modern complex world, this can best be achieved by a multidisciplined group of expert operators, developers, maintainers, operations research analysts, intelligence officers, logisticians, testers, and other supporting personnel. This group, working as a team, must apply a broad spectrum of T&E tools to include operations research analysis, analytical modeling and simulation, hybrid simulators, laboratories, ground test facilities driven by computer simulations, and flight/field testing environments.

1.3.3 The Product

The product of T&E is to ensure that the concepts and systems judged qualified for operational service indeed meet the needs and expectations of the civilian customer or the military operational combat user. This requires an extraordinary amount of teamwork and a close interrelationship among the decision makers, operators, developers, analysts, supporters, and testers. Full application of the quality culture and philosophy in the T&E acquisition process can lead to a significant and continuing improvement in the T&E product.

1.4 Establishing the Need for New or Improved Systems (See References 11–19)

Military operators, as well as commercial customers, must constantly seek ways to obtain and maintain improved efficiency, reduced costs and, in general, the competitive edge in mission accomplishment. To achieve this, the Department of Defense (DoD) has established and maintains a formal process for requirements definition and acquisition of military systems. The process starts at the highest levels of government with the formulation of a National Military Strategy. This strategy provides information essential to support more detailed defense planning at the war fighting and supporting command levels. The capabilities evolving from this strategy must cover the full spectrum of military doctrine to include Strategic Deterrence, and Defense; Forward Presence; Crisis Response; and Reconstitution.

1.5 Acquiring Systems (See References 20–22)

Acquisition programs within DoD are directed, funded efforts designed to provide new or improved material capabilities in response to validated needs of the Services. Information regarding costs, categories, and approval officials for these programs has been extracted from the cited references. DoD acquisition programs can be described as major programs, nonmajor programs, or highly sensitive classified programs. Highly sensitive classified programs are established in accordance with DoD 5200.1-R and are managed in accordance with DoD Directive 0-5205.7, "Special Access Program Policy."

Major programs are based on criteria established in Title 10, United States Code, Section 2430, "Major defense acquisition programs defined" and reflects authorities delegated in DoD Directive 5134.1, "Under Secretary of Defense for Acquisition." A major program must be designated by the Under Secretary of Defense for Acquisition [USD(A)] and be estimated to require the following:

1) An eventual total expenditure for research, development, test, and evaluation of more than $200 million in fiscal year 1980 constant dollars (approximately $300 million in fiscal year 1990 constant dollars), or

2) An eventual total expenditure for procurement of more than $1 billion in fiscal year 1980 constant dollars (approximately $1.8 billion in fiscal year 1990 constant dollars).

A major system is defined based on criteria established in Title 10, United States Code, Section 2302(5), "Definitions: Major System." It is defined to be a combination of elements that function together to produce the capabilities required to fulfill a mission need, including hardware, equipment, software, or any combination thereof, but excluding construction of other improvements to real property.

A system is considered major if it is estimated by the USD(A) to require the following:

1) An eventual total expenditure for research, development, test, and evaluation of more than $75 million in fiscal year 1980 constant dollars (approximately $115 million in fiscal year 1990 constant dollars), or

2) An eventual total expenditure for procurement of more than $300 million in fiscal year 1980 constant dollars (approximately $540 million in fiscal year 1990 constant dollars).

A nonmajor defense acquisition program is a program designated to be other than a major program, or a highly sensitive classified program. Defense programs are also placed in acquisition categories based on costs and level of program decision authority. These categories are depicted in Table 1-1.

Table 1-1. Defense Acquisition Program Categories

Category	Criteria	Decision Authority
ID	$300M Research, Development, and Test (1990) $1.8B Procurement	Under Secretary of Defense (Acquisition)
IC	$300M Research, Development, and Test (1990) $1.8B Procurement	DoD Component Head or delegated Component Acquisition Executive
II	$115M Research, Development, and Test (1990) $540M Procurement	DoD Component Head or delegated Component Acquisition Executive
III	Less than Category I or II	Lowest level deemed appropriate by DoD Component Acquisition Executive
IV	Less than Category III	Lowest level deemed appropriate by DoD Component Acquisition Executive

INTRODUCTION

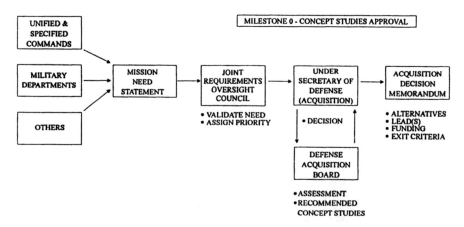

Figure 1-1. Military Requirements Process

1.6 Assessing and Validating Requirements

The assessment and validation of mission requirements are conducted by the Unified or Specified Commands, the Military Departments, the Office of the Secretary of Defense, and the Chairman of the Joint Chiefs of Staff. Figure 1-1 provides a general flowchart illustrating this process. Each Service has military roles and missions assigned that it must be capable of performing. Continuous monitoring and review of Service capabilities for these roles and missions in light of the projected threat/defense policy are conducted through a process of military mission area assessments. Campaign analysis tools are applied to assess each combatant command's area of responsibility in terms of identifying the operational effectiveness of alternative means of completing campaign objectives. When these assessments reveal shortfalls in capability, they must be corrected through either nonmaterial or material solutions. Operational needs are based on both short- and long-term capability objectives and can also result from changes in policy and cost reduction, as well as opportunities created by the development of advanced technologies. Nonmaterial solutions include changes in doctrine, operational concepts, tactics, training, or organization. When a need cannot be met by nonmaterial changes, a broad statement of mission need (expressed in terms of an operational capability, not a system-specific solution) is identified in a Mission Need Statement (MNS). The MNS also identifies the threat to be countered and the projected threat environment. The mission need is normally prioritized relative to other documented needs. The MNSs, when validated by individual Services, are forwarded to the Joint Requirements Oversight Council (JROC) if it appears that the potential solutions will be of sufficient magnitude to warrant a new major program. The JROC is chaired by the Vice Chairman of the Joint Chiefs of Staff, and consists of the Vice Chiefs of the Army, Navy, and Air Force and the Commandant of the Marine Corps. The JROC validates the need, considers how joint Service requirements can be met, and assigns the proposed acquisition a priority. The JROC-approved MNSs are, in turn, evaluated by the Defense Acquisition Board (DAB) about once a year. Approval by the DAB constitutes the start of the "Mile-

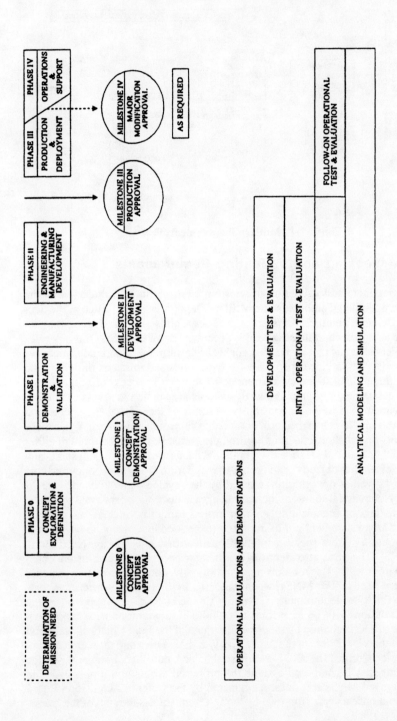

Figure 1-2. Acquisition Milestones and Phases

INTRODUCTION

stone 0" Concept Studies Approval decision process. As will be discussed later, a favorable review by the DAB at Milestone 0 represents the determination that a study effort of alternative concepts to meet the need is warranted. The DAB establishes whether the need is based on a validated threat, if it must be met by a material solution, and if it is sufficiently important to warrant funding.

If, after review of the requirement, the DAB and the USD(A) concur in its validity, an Acquisition Decision Memorandum (ADM) is issued by the USD(A) to the Service responsible for acquisition. Funding of Milestone 0 activities is provided by the submitting major command or a central studies fund controlled by USD(A), or both. This same overall review concept is followed for lesser nonmajor programs, however, the approval authority is at a lower level in the command structure depending on the acquisition importance of the program.

1.7 Program Flow and Decision Points

The five major milestones and phases of an acquisition program are depicted in Figure 1-2.

The initial formal interface of the requirements generation process and the acquisition management system occurs at Milestone 0. Approval at Milestone 0 allows studies and analyses of alternative material concepts to be conducted to identify the most promising potential solution(s) to validated user needs. Critical information regarding these competing concepts can be derived through cost and operational effectiveness analyses (COEAs), analytical modeling and simulation, operational evaluations, and operational demonstrations. Milestone I, concept demonstration approval, marks the official start of a new acquisition program. For program approval, it is essential that the products of the requirements generation process, the acquisition management system, and the planning, programming, and budgeting system be effectively integrated. The results of Phase 0 studies are evaluated and the acquisition strategy and proposed concept with cost, schedule, and performance objectives are assessed in light of affordability constraints. "Performance" is defined as "those operational and support characteristics of the system which allow it to effectively and efficiently perform its assigned mission over time. The support characteristics of the system include both supportability aspects of the design and the support elements necessary for system operation."

Phases and milestones after Phase I are designed to facilitate the orderly translation of broadly stated mission needs into system-specific performance requirements and a stable design that can be efficiently produced.

At each milestone decision point, assessments are made of the program status, the plans for the next phase and the remainder of the program. Both analytical modeling and simulation and the various types of T&E play a major role in providing information to support the various decision points. The risks associated with the program and the adequacy of risk management should be explicitly addressed in T&E. Additionally, program-specific results to be required in the next phase, called exit criteria, should be established. These criteria are based on such things as achieving a specified level of performance in testing, passing a critical design review prior to committing funds for long lead time item procurement, and demonstrating the adequacy of a new manufacturing process prior to entry into low small-scale initial production. Such criteria can be viewed as gates through which a program must pass before exiting one phase and entering another.

1.8 Reducing Risks Through Test and Evaluation

The primary purpose of T&E during the acquisition process is to allow the reduction of risk, either the risk that the system or equipment will not meet performance specifications, or the risk that the system or equipment cannot be effectively employed in its intended combat environment. The lower portion of Figure 1-2 depicts the important relationship among program milestones and phases, analytical modeling and simulation, operational evaluations and demonstrations, and T&E.

T&E is the principal means of demonstrating that program objectives can be met. The two major types of testing associated with acquisition programs are Development Test and Evaluation (DT&E) and Operational Test and Evaluation (OT&E). OT&E is further differentiated by identifying OT&E accomplished prior to the production decision (i.e., Milestone III) as Initial Operational Test and Evaluation (IOT&E) and OT&E accomplished after the production decision as Follow-on Operational Test and Evaluation (FOT&E).

1.8.1 Development Test and Evaluation

DT&E is conducted to assist the engineering design and development process, to demonstrate that design risks are minimized, and to verify the attainment of technical performance specifications and objectives. DT&E is essentially a detailed engineering analysis of a system's performance beginning with individual subsystems and progressing through a complete system, where system design is tested and evaluated against engineering and performance criteria by the implementing command. DT&E is a natural part of the contractor development process, and is initiated as early in the development cycle as possible. It includes the testing of component(s), subsystem(s), and prototype or preproduction model(s) of the entire system.

The implementing command is responsible for DT&E program management. Other commands support DT&E as specified in the program directives and in the coordinated planning documents. The contractor, under the direction of the implementing command, usually conducts the early part of DT&E, which includes the preproduction qualification tests.

DT&E is conducted to assess the critical issues of the system and carry out the development objectives specified in the program management documents. DT&E should accomplish the following items:

1) Provide data to assist in the system's development process.

2) Identify deficiencies in the system and its performance specifications, and determine the degree to which those specifications have been met.

3) Ensure the compatibility and performance of the support items (e.g., simulators, life support systems, support equipment, computer resources, technical manuals, and other data).

4) Provide estimates of system reliability and maintainability to be expected when deployed.

5) Determine whether the system is safe and ready for operational testing.

6) Accumulate and provide data for the estimation of survivability, vulnerability, and logistics supportability of the system.

7) Provide data for the compatibility and interoperability of the new system/

equipment.

8) Provide data for refining estimates of requirements for training programs and training equipment.

9) Provide information on environmental issues to be used in preparing impact assessments.

10) Ensure design integrity over the specified operational and environmental range by conducting preproduction qualification tests.

11) Validate and verify technical orders and computer software.

Qualification Test and Evaluation (QT&E) is conducted in lieu of DT&E for systems which require no development or Research, Development, Test, and Evaluation (RDT&E) funding. It is conducted after the first major production decision but prior to acceptance of the first article. QT&E is conducted to demonstrate that engineering design is complete, that design and production risks are minimized, and that the items fulfill the requirements and specifications of the procuring contract or agreement. QT&E is managed and conducted similar to DT&E with a few minor exceptions in test objectives and the documentation process.

1.8.2 Operational Test and Evaluation

OT&E is conducted to estimate a system's operational effectiveness and operational suitability, and to identify needed modifications. It also includes tests for compatibility, interoperability, reliability, maintainability, logistics supportability, software supportability, and training requirements. In addition, OT&E provides information on organization, personnel requirements, doctrine, and tactics, and may result in changes to operation employment and maintenance concepts. It should provide data to support or verify data in operating instructions, publications, and handbooks. OT&E is accomplished by operational and support personnel of the type and qualifications of those expected to use and maintain the system when deployed in combat. It is conducted in phases, each keyed to an appropriate decision point. OT&E should be continued as necessary during and after the production period to refine estimates, to evaluate changes, and to reevaluate the system to ensure that it continues to meet operational needs and retain its effectiveness in a new environment or against a new threat.

Each Service has a designated organization to manage the conduct of OT&E independently of the developing and using commands. For example, the Air Force Operational Test and Evaluation Center (AFOTEC) is responsible for the management of Air Force independent OT&E. AFOTEC manages OT&E on all major systems plus some nonmajor systems which are designated by HQ U.S. Air Force. On these OT&E programs, the AFOTEC Commander appoints the OT&E test director from the AFOTEC staff or from the appropriate major command. The test director exercises operational control over the OT&E team. The major command conducts OT&E on all other nonmajor systems. AFOTEC monitors IOT&E programs that are managed by major commands and selected FOT&E programs. The Operational Test and Evaluation Command (OPTEC) conducts similar independent operational testing in the U.S. Army and the Operational Test and Evaluation Force (OPTEVFOR) conducts it for the U.S. Navy.

1.8.2.1 Initial operational test and evaluation IOT&E is accomplished prior to the first major production decision and provides critical data to support the decision. Planning of IOT&E begins as early as practical in the acquisition process. IOT&E is conducted using preproduction items, prototypes, or pilot production items. These items must be sufficiently representative of the production article to provide a valid estimate of the operational effectiveness and suitability of the production system. During IOT&E, operational deficiency and proposed configuration changes are identified as early as possible to the implementing commands. It is especially important to provide realistic operational environments for IOT&E in order to assure that performance, safety, maintainability, reliability, human factors, and logistics supportability criteria can be evaluated under conditions similar to those that will exist when the system is employed in combat.

Operational testing should be separate from development testing. However, DT&E and IOT&E are often combined because separation would cause an unacceptable delay or would cause an increase in the acquisition cost of the system. When combined testing is conducted, the necessary test conditions and test data required by both DT&E and OT&E organizations must be obtained. Chapter 12 provides some recommendations and discussion on how the cost effectiveness of this process might be improved through the use of a common data base that is both certified and protected. The OT&E agency and implementing command must ensure that the combined test is planned and executed to provide the necessary operational test information and the OT&E agency participated actively in the test and provides an independent evaluation of the resultant operational test. IOT&E is budgeted for and funded by the implementing command using 3600 RDT&E funds.

Qualification Operational Test and Evaluation (QOT&E) is conducted in lieu of IOT&E for systems where there is no RDT&E funding required. It is conducted after the first major production decision, but prior to acceptance of the first article. Normally, this category of tests includes Class IV modifications, logistic service tests, Class V modifications, simulators, one-of-a-kind systems, and off-the-shelf items. QOT&E is managed and conducted in a manner similar to IOT&E with some exceptions in the documentation process.

1.8.2.2 Follow-on operational test and evaluation. FOT&E is conducted after a production decision is made on a system. Within the Air Force, for most major programs and nonmajor programs assigned to AFOTEC, FOT&E may be conducted in phases. AFOTEC conducts the first phase through first production article testing to refine initial estimates made during IOT&E. The second phase is conducted by the major command scheduled to receive the system to refine tactics and techniques and training programs for the system. Transition of FOT&E from Phase I to Phase II is normally determined on a program-to-program basis. AFOTEC closely monitors the Phase II effort to ensure that problems it identified in earlier testing were corrected. Major commands perform FOT&E on all other systems after the production decision.

FOT&E is conducted to refine the initial estimates made during IOT&E and to ensure that production article performance and operational effectiveness and suitability are equal to or greater than the preproduction article. FOT&E is used to verify that deficiencies previously identified have been remedied and that new deficiencies are identified and corrected. FOT&E also introduces an evaluation of the organization, personnel, doctrine, and tactics for operational employment of the system.

1.8.3 Production Acceptance Test and Evaluation

Production Acceptance Test and Evaluation (PAT&E) is conducted on production items to demonstrate that the items procured fulfill the requirements and specifications of the procuring contract or agreements. It is normally the responsibility of the government plant representative to accomplish the necessary PAT&E throughout the production phase of the acquisition process. Acceptance test plans normally are procured by the development agency from the prime items contractor as part of the production specification.

1.8.4 Joint Service Test and Evaluation

The Deputy Director of T&E of the Office of the Under Secretary of Defense (acquisition and test) directs joint tests by the Services. These tests are normally conceptually oriented and designed to use existing equipment and Service procedures to gather baseline data for future application to concepts and hardware design. When the Air Force is the lead Service in a joint Service acquisition program, JT&E is conducted as outlined by Air Force Directives. If another Service is selected to serve as executive agent for testing sponsored by the Office of the Secretary of Defense (OSD), JT&E is conducted in accordance with agreements between OSD, the Air Force, and other Services. The agency selected to conduct joint testing normally forms a Joint Test Force (JTF) of the participating Services as agreed to in a charter approved by OSD. The JTF drafts a detailed test plan for conducting the JTF. The planning, conduct, and reporting of the joint test are closely monitored by OSD and the headquarters of each participating Service.

Data analysis and progress reporting are accomplished as specified in the test plan. Upon completion of the joint test, the JT&E prepares, publishes, and coordinates the final test report. Sometimes, in addition to the JTF report, independent comments and reports are provided by the participating Services. The realities of joint testing, as with all T&E, make adherence to rigidly structured procedural steps, operational control lines, and interservice liaison difficult.

1.9 Scientific Approach (See References 23–34)

The methods of T&E presented in this book are indeed based on the principles of operations research analysis and the scientific approach. These methods also depend heavily upon the experience of experts, common sense, and continuous quality improvement. The recommended approach centers around the basic operations research principle of defining the right T&E problem, establishing an overall T&E model, making assumptions regarding the model, testing the model, and deriving a solution from the model for real world problems. Problem definition consists of specifying critical issues and objectives in sufficient detail to allow T&E to quantify their value and to seek methods of improvement. In military T&E, the model is almost always abstract to some degree and is rarely totally representative of the real world combat situation. It is the responsibility and challenge of the operational user to interpret and project the model results to the real world situations of interest. It is also the responsibility of the operational user to continually seek methods to optimize the military combat capability within the budget made available by the U.S. taxpayers. Often it is necessary to apply an iterative "model, test, fix" approach when addressing and solving problems associated with complex weapons systems. Although it may appear in a variety of formats, the scientific approach is advocated and applied in the methodology throughout this text. This book addresses both the "what to do" and the "how to do it" aspects of operations research analysis in T&E.

1.9.1 What To Do

The chapters of this book essentially follow the flow for accomplishing operations research analysis in quality T&E. Chapter 2 addresses the Cost and Operational Effectiveness Analysis (COEA), which is the analytical backbone of all good defense acquisition programs. The COEA provides the basis for a close linkage between research, development, design, production, and T&E. Chapters 3, 4, and 5 are dedicated to the basic principles, the modeling and simulation approach, and the concepts of T&E. These include analytical modeling and simulation as well as the typical range system measurement capabilities available for military T&E. All of these subjects are essential to understanding "what to do" in T&E. They are also extremely important to the operations research analysis process of meaningful problem definition. The knowledge and experience of highly qualified operators, operations research analysts, and other T&E team experts are essential to carrying out these multidisciplined processes. Making sure that the correct problems are being addressed by the testing, that the essential resources are being applied, and that the right judgments are being rendered during evaluations is fundamental to the success of quality T&E.

1.9.2 How To Do It

The remaining chapters of the book are more concerned with "how to do it." A number of example problems and solutions are spread throughout the text. Chapter 6 addresses T&E Design which provides greater detail on the T&E analysis model and the impact of appropriate assumptions regarding the model. Chapter 7 (T&E Planning) addresses the importance of formally documenting the overall T&E problem, the T&E design, and the detailed procedures for getting the T&E accomplished. Chapter 8 addresses the important tasks of T&E Conduct, Analysis, and Reporting. Chapters 9, 10, and 11 address some special areas of T&E that require dedicated emphasis. These include software; human factors; and reliability, maintainability, logistics supportability, and availability. Chapter 12 addresses the T&E of integrated weapons systems and Chapter 13 explains measures of effectiveness and measures of performance that are vital to the T&E process. Chapter 14 describes the important subject of training measurement and Chapter 15 addresses joint testing and evaluation. Finally, Appendix 1 presents a wide variety of example T&E quantification problems and solutions, Appendix 2 addresses some example government laboratories and test range facilities, and Appendix 3 provides a glossary of abbreviations and acronyms used in the text.

1.10 Ideals of Test and Evaluation

T&E results are relied upon heavily by acquisition decision makers and system operators. Congressional officials and other high level government executives are highly dependent upon T&E to provide timely, unbiased quantitative input to decisions. Combat users must have essential system information that can be derived in no other manner. T&E results represent the most credible source of system information available. Consequently, it is imperative that those who are charged with the conduct of T&E maintain the highest ideals and standards: integrity, objectivity, operational realism, and scientific validity. This book is dedicated to helping achieve and maintain those ideals.

References

[1] Mondey (ed.), *The International Encyclopedia of Aviation*. New York: Crown Publishers, Inc., 1977.

[2] Swanborough, G., and Bowers, P.M., *United States Navy Aircraft Since 1911* (2nd ed.). Annapolis, Maryland: Naval Institute Press, 1976.

[3] Walton, M., *The Deming Management Method*. New York: The Putnam Publishing Group, 1986.

[4] Jones, J.V., *Engineering Design, Reliability, Maintainability, and Testability*. Blue Ridge Summit, Pennsylvania: Tab Books, Inc., 1988.

[5] Shayler, D., *Shuttle Challenger*. Englewood Cliffs, New Jersey: Prentice-Hall, 1987.

[6] Peters, T., and Austin, N.K., *A Passion for Excellence: The Leadership Difference*. New York: Random House, 1985.

[7] Deming, W.E., *Out of Crisis*. Cambridge, Massachusetts: MIT Center for Advanced Engineering Study, 1986.

[8] Juran, J.M., *Juran on Planning for Quality*. New York: Free Press, 1988.

[9] Ross, P.J., *Taguchi Techniques for Quality Engineering*. New York: McGraw-Hill, 1988.

[10] Crosby, P.B., *Quality Is Free: The Art of Making Quality Certain*. New York: McGraw-Hill, 1979.

[11] White House, "National Security Strategy of the United States." Washington, D.C.: U.S. Government Printing Office, May 1992.

[12] Department of Defense, The Pentagon, "Defense Strategy for The 1990s: The Regional Defense Strategy." Washington, D.C.: U.S. Government Printing Office, January 1993.

[13] Department of the Air Force, The Pentagon, "Basic Aerospace Doctrine of the United States Air Force," AFM 1-1. Washington, D.C.: U.S. Government Printing Office, March 1992.

[14] Department of the Army, The Pentagon, "The Army," Publication FM 100-1. Washington, D.C.: U.S. Government Printing Office, 10 December 1991.

[15] Department of the Navy, The Pentagon, "From the Sea: Preparing the Naval Service for the 21st Century," MNS 104, 30 September 1992.

[16] Loh, Gen. J.M., USAF, "Advocating Mission Needs in Tomorrow's World." *Airpower Journal*, Vol. 6, No. 1, Spring 1992.

[17] Schelling, T.C., *Arms and Influence*. New Haven, Connecticut: Yale University Press, 1966.

[18] Chandler, D., *The Campaigns of Napoleon*. New York: The Macmillan Company, 1966.

[19] Enthoven, A.C., and Smith, K.W., *How Much Is Enough?* New York: Harper and Row, 1971.

[20]DoD Directive 5000.1, "Defense Acquisition," Office of the Secretary of Defense, Under Secretary of Defense for Acquisition, 23 February 1991.

[21]DoD Instruction 5000.2, "Defense Acquisition Management Policy and Procedures," Office of the Secretary of Defense, Under Secretary of Defense for Acquisition, 23 February 1991.

[22]Przemieniecki, J.S. (ed.), *Acquisition of Defense Systems*. AIAA Education Series. Washington, D.C.: American Institute of Aeronautics and Astronautics, 1993.

[23]Couger, J.D., and Knapp, R.W., *Systems Analysis Techniques*. New York: John Wiley and Sons, Inc., 1974.

[24]Ackoff, R.L., *Scientific Method: Optimizing Applied Research Decisions*. New York and London, England: John Wiley and Sons, Inc., 1962.

[25]Blackwell, D., and Girshick, M.A., *Theory of Games and Statistical Decisions*. New York: John Wiley and Sons, Inc., 1954.

[26]Cochrane, J.L., and Zeleny (eds.), *Multiple Criteria Decision Making*. Columbia, South Carolina: University of South Carolina Press.

[27]Rudwick, B.H., *Systems Analysis for Effective Planning: Principles and Cases*. New York: John Wiley and Sons, Inc., 1969.

[28]Arrow, K.J., *Social Choice and Individual Values* (2nd ed.). New York: John Wiley and Sons, Inc., 1963.

[29]Brown, R.V., Kahr, A.S., and Peterson, C., *Decision Analysis for the Manager*. New York: Holt, Rinehart, and Winston, 1974.

[30]Raiffa, H., *Decision Analysis: Introductory Lectures on Choices Under Uncertainty*. Reading, Massachusetts: Addison-Wesley Publishing Company, 1968.

[31]Hitch, C.J., *Decision Making for Defense*. Berkeley, California: University of California Press, 1965.

[32]Kahn, H., and Mann, I., *Techniques of Systems Analysis*. The RAND Corporation, RM-1829-1, 1957.

[33]Quade, E.S. (ed.), *Analysis for Military Decisions*. The RAND Corporation, 1964.

[34]Wilde, D.J., *Optimum Seeking Methods*. Englewood Cliffs, New Jersey: Prentice-Hall, 1964.

2
Cost and Operational Effectiveness Analysis

2.1 General (See References 1–5)

A Cost and Operational Effectiveness Analysis (COEA) is a formal analysis conducted within the Department of Defense (DoD) to aid responsible officials in making well-informed decisions regarding an acquisition program. The analysis is presented in terms of a problem of choice among competing alternatives based on effectiveness and costs, while considering risk and uncertainty in the process. The COEA should attempt to clearly define the program issues and alternatives, provide a full, accurate, and meaningful summary of as many of the relevant facts as possible, and provide the analytical underpinning or rationale for important program decisions. It should address the relative advantages and disadvantages of the alternatives being considered and show the sensitivity of each alternative to possible changes in key assumptions or variables (e.g., the threat, selected performance, capabilities).

The COEA provides a comparison of the alternative courses of action in terms of their costs and effectiveness in attaining specified objectives. Usually this comparison is made by attempting to designate an alternative that will minimize costs, subject to some fixed performance requirement or, conversely, provide maximum performance at some fixed level of costs. To be most beneficial, the analysis requires the specification of sensible objectives, the establishment of a meaningful way to measure performance, the illumination of considerations that cannot be quantified and, most importantly, the identification of the best alternatives for consideration.

The COEA should aid the decision makers in judging whether or not any of the proposed alternatives offer sufficient military advantage over the current program. That is, does the best alternative provide enough military worth to warrant the change in operations, support, and costs? It should also serve as a vehicle to facilitate communications and debate regarding issues and alternatives among decision makers and staffs at all levels. Both the strengths and weaknesses of each alternative should be highlighted to make them open and specific to reviewing officials and decision makers. These activities are designed to promote open discussion and debate on all issues and alternatives and to allow illumination at the program critical decision points. The analysis should be objective and as quantitative as practical, and promote full understanding and debate on key issues affecting decisions. The COEA is essentially a program foundation type of document with significant impact beginning at Milestone I (Concept Demonstration Approval) and continuing throughout a program. As such, it helps tie together a number of important analytical and managerial aspects of a program. For example, the measures of effectiveness (MOEs) addressed in the COEA to justify a concept or system in terms of its military worth are tracked throughout the acquisition cycle. These measures (or factors that can be directly related to them to include development measures of performance) are monitored later during the test and evaluation (T&E) of a developed system to ensure that it indeed achieves its programmed level of effectiveness. Any differences in what a system was programmed to achieve and what is demonstrated during T&E should be documented and reconciled prior to commitment to large-scale production. The COEA should provide a historical record of the alternatives considered at each milestone decision point.

2.2 Approach (See References 1–16)

The COEA's primary role is to serve as an aid to the decision making process. The focus of the analysis at any point in time is tied to the important decisions being made at program milestones. Figure 2-1 (extracted from DoD Instruction 5000.2, 23 February 1991) depicts the relationship of important acquisition program elements to the milestone decision process. The COEA should be structured to clearly define and address acquisition issues and the alternative solutions. It should provide an analysis of the alternatives and a summary of results (see Figure 2-2 for an example COEA format). The primary purpose of the COEA is to provide facts and objective analysis that aid in the decisions that must be made during acquisition and operational implementation.

The acquisition issues address the need, threat, environment, constraints, and operational concept. The alternatives are addressed in terms of performance objectives along with a comprehensive description of what is being investigated in the analysis. The analysis identifies and describes the models and simulations used, MOEs, costs, trade-off analyses, and the decision criteria. The summary of results highlights the major findings of the analysis and the factors affecting the acceptability and affordability of the alternatives, both in absolute and in relative terms.

2.3 Acquisition Issues

2.3.1 Threat

The threat analysis examines those elements against which a system might be used and the forces and capabilities that could be used against that system. It should include broad considerations (such as the nature and size of opposing forces or conventional versus nuclear weapons) as well as the essential detailed information (e.g., the strength of projectile attacks, electronic warfare devices). A range of plausible threats should be examined to allow for the uncertainty inherent in threat projections and enemy countermeasures. Both enemy and friendly objectives should be carefully analyzed in terms of constraints, along with logistics, personnel, infrastructure, and other factors that could affect the performance of the weapons systems. The analysis should address postulated countermeasures and enemy responses. It is important not to grossly overestimate or underestimate the threat. Overestimating enemy force size or capability could invite consideration of an unachievable or prohibitively expensive solution, while underestimating enemy projected capabilities might lead to inadequate provision for future product improvements.

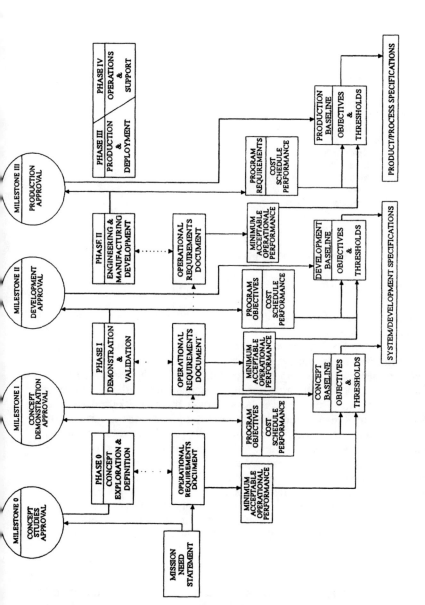

Figure 2-1. Important Acquisition Program Elements and Decision Points

COEA
For
Program Title

1. The Acquisition Issue
   a. Threat
   b. Need
   c. Environment
   d. Constraints
   e. Operational Concept
2. Alternatives
   a. Performance Objectives
   b. Description of Alternatives
3. Analysis of Alternatives
   a. Models
   b. MOEs
   c. Costs
   d. Trade-off Analyses
   e. Decision Criteria
4. Summary of Results

Figure 2-2. Example Format for COEA

2.3.2 Need

A desired prerequisite to the identification and documentation of a mission need is a comprehensive mission area assessment. This analysis, which initially should be performed during Pre-Milestone 0 activities, assesses the strengths and weaknesses of a military force when confronting a postulated threat in specific scenarios vital to defense. It is imperative that this analysis be performed by the operational experts of the military commands responsible for taking the systems into combat. These analyses should address the basic mission tasks and assess the responsible military organizations' ability to perform each task in terms of the current and projected capability, the present and projected threat, the operational environment, and any other constraints which may limit solutions to identified needs. It is important to clearly identify the needs, to establish the deficiencies of existing systems in meeting those needs, and to identify opportunities for satisfying needs and alleviating deficiencies. It is important to link the need to strategic national security planning through the Defense Planning Guidance. The scenarios used should conform to those in the Defense Planning Guidance in terms of assumptions regarding the threat and U.S. and allied involvement.

2.3.3 Environment

The analysis of the environment should address the potential contribution of allied forces, the terrain, weather, ocean, and other pertinent operational parameters affecting environment. It should describe the allied concepts of operational projected force structures and associated capabilities. If allied forces are to operate in close proximity to the new U.S. system, the analysis should address how their role in the battle would be affected by the system's introduction and how U.S.

performance would be affected by allied contributions. The analysis should be supported by meteorological data describing both normal and abnormal weather conditions under which the new U.S. system would be expected to operate. The data should include temperature, altitude, winds, precipitation, cloud cover, visibility limits, ambient acoustic noise, sand, and dust, as well as other conditions that could affect operational employment. In most environments, alternative methods to survive will exist. For example, in the nuclear threat environment, hardening, avoidance, deception, proliferation, reconstitution, redundancy, or a combination thereof can be employed. For each threat environment, there should also be expected mission capabilities as affected by that environment. Each of these affords the opportunity to formulate multiple alternatives to be considered in estimating costs and determining the most operationally effective solutions.

2.3.4 Constraints

The constraints and assumptions that limit the set of viable alternatives should be addressed. They should be carefully defined and explicitly stated in terms of how each alternative is affected. Often, presumed constraints can be alleviated once they are identified and addressed. Constraints and assumptions can change over time. Understanding when and how the changes take place, and the consequences of these changes, is important to the analysis.

2.3.5 Operational Concept

The operational concept provides guidance for posturing and supporting forces and describes the intended employment of the forces in combat. It addresses both doctrine and tactics in explaining how the competing systems would be employed to accomplish national objectives. A preliminary systems operational concept is formulated for each alternative solution considered between Milestone 0 and Milestone I. Sometimes, a single plan (or modified version thereof) can accommodate the entire group of alternatives. As the acquisition process continues, the preliminary concepts are refined and updated for the associated solutions. Once the final solution/system is chosen, the operational concept is updated at each key decision point. Sometimes, field experimentation may be necessary to update or refine a concept.

2.4 Alternatives

2.4.1 Performance Objectives

Performance objectives should be based on the essential activities required to fight and win in combat. They should quantitatively describe the minimum acceptable operational requirements for performance of the proposed concept/system. They should be structured to show the impact of changes at the margin in performance and mission accomplishment as related to the Operational Requirements Document (ORD). The ORD is developed by the operational user and serves as a bridge connecting the Mission Need Statement to the acquisition program baseline and the specifications for the concept or system. Performance objectives depend upon the type of concept or system at issue. For example, when analyzing transportation systems, the objectives are stated in terms of the movement of forces and equipment; for firepower systems, they reflect the types of targets to be engaged and destroyed. The effectiveness of each system alternative is then measured in terms of the degree to which the objectives are achieved and its probable impact on the conflict. It is important to understand this relationship—how meeting basic operational objectives depends upon the performance of the alternative system. In the end, differences in system performance must be assessed against differences in system costs. The analysis should result in a clear understanding of the objectives established for the system in relation to the overall situation. Without this understanding, the MOEs and criteria for success used to compare the alternatives are unlikely to be relevant.

2.4.2 Description of Alternatives

The alternative concepts or systems represent the methods by which the performance objectives can be achieved. All reasonable options should be represented. It is important to identify the most promising and viable proposed solutions for comparison. Identifying the advocated solution along with several "weak alternatives" is not an acceptable approach. When structuring the set of alternatives, consider both current systems and improved versions, along with systems in development by the other Services or Allies and conceptual systems not yet on the drawing board.

Normally, the uncertainty associated with the capability and availability of a system will depend on its state of development, with the risks and uncertainties greater for systems in the early development stages and very great if not yet on the drawing board. A frequent weakness in COEAs results from devoting inadequate attention to the potential that can be provided by existing systems after appropriate modifications. For example, when feasible, the COEA should always contain at least one alternative that capitalizes on the capabilities and technologies that already exist rather than only those cosmic solutions that represent costly development and high risk.

When generating the set of alternatives, a reference alternative (or *base case*) funded in the current defense program should be included. This alternative represents the currently programmed capability (i.e., the status quo) against which other alternatives can be compared. The baseline case helps in understanding the absolute value for all of the alternatives (i.e., they can be compared against the base case). The proposed alternatives should represent a range of potential solutions to the extent possible. It should include alternatives where doctrine and tactics, rather than just hardware, are varied, since organizational and operational plans can change. The analysis should consider including alternatives with the potential to mitigate significant, environmentally driven performance limitations. Each alternative should be fully defined in terms of specification of material, organization, personnel, and tactics. The organizational and operational concept for each alternative system should be described, along with the units within which it would be embedded. It is important to explain how the system or unit would operate in conjunction with other systems or units in accomplishing the objectives. New alternatives may emerge as a result of insights gained from the ongoing analysis and from resources outside the analysis team. The analysis should allow for new alternatives to be considered as it progresses.

2.5 Analysis of Alternatives

Once the performance objectives and the alternatives to achieve the objectives are clearly defined, the alternatives should be compared in terms of effectiveness, costs, and risks. Analytical modeling and simulation is generally an essential tool in these comparisons. MOEs should be established to measure how well each objective can be achieved by each alternative. The costs associated with providing the capability of each alternative must be estimated. Trade-off analyses should be conducted to help understand the implications of various performance levels and costs. Finally, decision criteria should be established to help determine whether the alternatives can achieve the mission and to compare them in a meaningful manner.

2.5.1 Models

Models are simplified, stylized representations of the real world that abstract the cause-and-effect relationships essential to the issues studied. They may range from a purely verbal description of the situation to highly complex mathematical equations and relationships that employ computer solutions. The models used in COEAs can take a variety of forms, from simple "back of the envelope" calculations to large-scale force-on-force computer simulations. The output of these models is used to help quantify the MOEs and performance, as well as costs for the various alternatives. The type of model most useful for the analysis depends upon such things as the objective being addressed, the measure being applied, and the degree of validity and accuracy required, and thus is highly situation dependent. Models used in the analysis should be screened and tested through a formal Verification, Validation, and Accreditation (VV&A) process. Verification is the process of determining that a model implementation accurately represents the intended conceptual description and design specifications. Validation is the process of determining the extent to which a model is an accurate representation of the real world from the perspective of the intended uses of the model. Accreditation is the official determination by the decision maker that a computer model is acceptable for a specific purpose. VV&A is a continuing process and is applied throughout the model life cycle. VV&A also applies to the data sets required for model solutions. Many of today's problems with models and data stem from the nonexistent or sporadic application of VV&A. VV&A is best accomplished through an independent, qualified reviewer. The extent and results of the VV&A effort for the models applied should be of paramount interest to decision makers and thus an integral part of the COEA report. VV&A will be addressed additionally in Chapter 4. The DoD has established guidelines (listed below) to consider when selecting models for use in COEAs.

1) Like weapon systems, models are rarely entirely "good" or "bad." They are suitable or unsuitable for particular purposes.

2) Models should help eliminate personal bias and preference. So, be cautious when using models that include a "man-in-the-loop."

3) A great number of models already are available in almost every mission area. Consider them before attempting to build new ones.

4) Keep the model simple. Often a simple mathematical equation can project the performance you are seeking to display.

5) Be sure to test the model to see if it describes the base case well. Generally, we know more about the base case, the existing system, than we do about the alternatives. If the model does not "predict" what we know the existing system can do, it is not likely that its other predictions will be sound.

6) Use several models. If different models yield similar results, one might gain confidence that the estimates are reasonable.

7) Run a "common sense" test. Are the results plausible? Are they within reasonable bounds?

8) Evaluate the quality of the environmental simulation and the environmental limitation evaluation. For systems using sensors with a known vulnerability to adverse environmental conditions, for instance, does the model adequately incorporate the adverse effects of those during the simulation?

2.5.2 Measures of Effectiveness

In military comparisons, MOEs are approximations for indicating the attainment of the performance objectives. MOEs are tools that assist in discriminating among the competing alternatives. Ideally, they are relevant, quantitative, and measurable. Quantitative measures are preferred for the analysis because they help minimize the contamination of personal bias. MOEs should show how the alternatives compare in the performance of the objectives and missions. MOEs should be developed to a level of specificity such that the system's effectiveness during developmental and operational testing can be assessed with the same effectiveness criteria as used in the COEA. This will permit further refinement of the analysis to reassess cost effectiveness compared to alternatives in the event that performance, as determined during testing, indicates a significant drop in effectiveness (i.e., to or below a threshold) compared to the levels determined in the initial analysis. Example MOEs include enemy targets destroyed per day, friendly aircraft surviving per day, and tons delivered per day. The MOEs selected should relate directly to a system's performance objectives and to mission accomplishment. As stated by DoD, decision makers need to know the contribution of the system to the outcome of the battle, not just how far it can shoot or how fast it can fly.

2.5.3 Costs

The purpose of the cost analysis in the COEA is to establish the costs of the output of each alternative. While MOEs gauge the military utility of specified outputs, cost analyses assess the resource implications of associated inputs. In this regard, the concept of life-cycle cost is important. Life-cycle costs reflect the cumulative costs of developing, procuring, operating, and supporting a system. They often are estimated separately by budget accounts, such as research, development, test, and evaluation (RDT&E); procurement; and operations and maintenance (O&M). It is imperative to identify life-cycle costs, nonmonetary as well as monetary, associated with each alternative being considered in the COEA. To conduct the analysis, separate estimates of operations and maintenance costs must be made, particularly manpower, personnel, and training costs. This includes the base case alternative, which provides for continuation of the status quo. Like effectiveness measurements, cost measurements are also estimates subject to approximation and error. The cost uncertainty should be described and bounded in some manner to aid in the decision process. Cost estimates are sensitive to quantities. The cost estimate should reflect the program quantity required. Cost data can be presented in a variety of ways. Presentation is important because decision makers must combine cost considerations with assessments of operational effectiveness and potential constraints in weighing the alternatives (e.g., timeline, political situations).

2.5.3.1 Types of costs.
Costs are often presented in terms of annual costs, cost streams as a function of time, or values representative of a given point in time. Cost methods should be sensitive to unequal lifetimes, build-up costs, and the time value of money. Build-up costs are operating costs incurred during the phase-in period before the new system reaches its full-force size.

One common method of cost comparison is based on life-cycle cost (sometimes referred to as "cradle-to-the-grave" costs). Life-cycle cost is the total cost of an item or system over its full life (acquisition cost + operation and support cost + disposal cost). It includes the cost of development, production, ownership (operation, maintenance, support, etc.) and, where applicable, environmental cleanup and disposal. Acquisition cost is the sum of development and production costs. Military construction costs should be included if incurred as a result of the introduction of the system. Initial spares should also be included.

Operating and support cost is the added or variable cost of personnel, materials, facilities, and other items needed for the peacetime operation, maintenance, and support of a system during activation, steady state operation, and disposal. Disposal cost is the cost of demilitarization and getting rid of excess, surplus, scrap, inventories, or salvage property under proper authority. Sometimes, systems can be sold at a profit during salvage. Under these conditions, disposal costs become negative instead of positive. Sunk cost is the total of all past expenditures or irrevocably committed funds related to a program or project. Sunk costs are generally not relevant to decision making as they reflect prior choices rather than current or future choices. They are sometimes referred to as prior-year costs. Sunk costs should be eliminated from all COEA analysis to show only the true future or incremental cost of each alternative. Costs can also be described in terms of constant dollars and then year dollars. Constant dollars present program costs normalized for inflation to a selected base year. An estimate is in constant dollars when prior-year costs are adjusted to reflect the level of prices of the base year and future costs are estimated without inflation. Costs can also be adjusted to account for inflation. Then year dollars are budget dollars that reflect the amount of funding required to pay costs at the time they are incurred.

2.5.3.2 Cost estimating.
Cost estimating uses a variety of methods to include parametric methods, estimating by analogy, engineering estimates, or some combination thereof. Parametric methods relate cost to parameters that specify a system within a class of systems, such as weight and maximum speed for fighter aircraft. Often cost estimating relationships are derived based on historical data using the mathematical technique of regression analysis. Parametric estimates are "most probable numbers" whose accuracy is subject to statistical errors resulting from small data bases, technology changes, and poor definition of data. Consequently, parametric estimates depend upon the ability of the analyst to establish

relationships (i.e., correlation) among the attributes or elements that make up the alternative. This requires a proper choice and description of the cost influencing factors of the alternative. These descriptions are called cost estimating relationships (CERs).

When estimating by analogy, one adjusts the known costs of existing systems similar to the one in question to arrive at cost projections. These estimates are usually adjusted for differences attributed to value versus time as well as differences in complexity and technical or physical characteristics.

Engineering or bottom-up cost estimates are made by pricing each component of a system. This requires an extensive knowledge of the system and its characteristics. The system or item of hardware is broken down into components, and cost estimates are made of each component. Parametric methods and analogy methods frequently are necessary to estimate the costs of the components. These results are combined with estimates of the costs of integrating the components to arrive at an estimate of the total system cost. Often, several methods can be used in combination to estimate a given cost. The analysts must determine which is most appropriate on a case-by-case basis.

2.5.3.3 Other considerations. It is important that the cost analysis include an examination and presentation regarding the uncertainty, sensitivity, and relation to the baseline cost estimate. Cost uncertainty is inherent in the analyses and stems from many factors to include the potential for unplanned system changes, technical problems, schedule shifts, and estimating errors. In the early stages of development, it can arise from the ranges in a key cost/performance relationship for a system. Cost uncertainty analysis should be conducted to determine the effect of uncertainty on the answers and to "bound the estimate." This can be done objectively, by statistical analysis, or subjectively, through the use of expert opinion. Cost sensitivity analysis is conducted to examine the degree to which changes in certain parameters cause changes in the cost of a system (this is also sometimes referred to as parametric analysis). Each potential change should be tested independently. Operating parameters that affect costs (such as activity rates and performance characteristics) should be examined for sensitivity to change. The results of each sensitivity analysis should be documented in the COEA. It is also important to link the analysis to a valid cost estimate for the baseline case (i.e., the status quo alternative). When this is done properly, it adds credibility to the absolute costs associated with the alternatives.

2.5.4 Trade-Off Analyses

Cost, schedules, and performance trade-offs are an important part of how the COEA aids the decision process. Trade-off analyses can be depicted by equal-cost or equal-capability packages; that is, they display the implications of "trading" one set of controlled variables (such as schedule or performance) for another (such as cost). The analyses should play an important role in both Milestone I and II decisions. A trade-off analysis should identify the areas of uncertainty, the sensitivities of the proposed alternatives to cost, and address how the proposed alternatives compare to established threshold values. Trade-off analyses identify areas of uncertainty and estimate their extent. The implications of the uncertainties can be examined using both cost models and effectiveness models. This serves to highlight for decision makers the areas in which the uncertainties most affect the analysis and, therefore, its results. Sensitivity analyses should show explicitly how military utility is affected by changes in system capability. They also should show how system characteristics (size, weight, etc.) drive performance, and how performance affects military utility or effectiveness. Parameters should be varied individually where it is reasonable to do so to determine their specific effect on output. The uncertainty inherent in estimating parameters and in determining their impact should be depicted as explicitly as possible. As a result of these efforts, the analysis should reveal the system performance "at the margin" and "where we are on the output curve"; that is, whether performance is stretching a system to the point where increases in costs add little to system performance, and whether increases in system performance add little to overall mission benefit. Certain characteristics, capabilities, and levels of effectiveness should not be considered absolutely essential in the acquisition of systems "regardless of costs." Sensitivity analysis can illuminate how important it is to incorporate these features into a system and how other options or mixes might achieve the same goals at less costs. It is also important to determine and understand thresholds. Thresholds are the maximum and minimum values of important measures that can be tolerated in a system (e.g., maximum cost or minimum acceptable performance). In order to approach thresholds and acceptability bands realistically, senior decision makers and users must be directly involved in reviewing the combinations of cost and performance that would be acceptable. They answer such questions as: How valuable is a given capability to the Service? How much would the Service be willing to give up in order to obtain that capability? At what point would it be preferable to drop the idea in favor of some other course of action? Performance thresholds may be more difficult to determine but are at least as important as cost thresholds. They show at what point degradations in performance yield outcomes that no longer satisfy the mission need. Together, cost and performance thresholds help in determining which alternatives are worthwhile and what combinations or intervals of performance and cost are acceptable.

2.5.5 Decision Criteria

Decisions made by responsible acquisition officials always involve fact and judgment. The COEA should attempt to clearly define the issues and alternatives and to provide decision makers with a full, accurate, and meaningful summary of as many as possible of the relevant facts so that they can make well-informed judgments. The COEA is not a substitute for judgment by responsible defense authorities.

The COEA should suggest criteria for selecting among the alternatives. In the cost effectiveness context there are two generally accepted standard criteria. First, a system is the preferred choice when it provides more units of effectiveness for a fixed level of cost. Second, it is the preferred choice when it costs less than the other alternatives to achieve a fixed level of effectiveness. More specific criteria dealing with the absolute values of effectiveness and costs are situation and decision maker dependent. These criteria are generally in terms of the threshold values for the MOEs and other critical performance parameters described earlier. The actual decisions regarding these choices must take into account many factors to include the mission need, the specific operational requirement of the combat user, as well as the overall priority and affordability of the system in question.

It is easy to overstate the degree of assistance that the COEA can provide to decision makers under these complex conditions. At best, it can help them understand the relevant alternatives and the key interactions by providing estimates of effectiveness, costs, risks, and the associated timeframe for each course of action. Often, it can also lead to new and/or improved alternatives that end up in the acquired system.

Ideally, the COEA will focus the decision makers' attention on the important features, help increase their intuition, broaden their basis for judgment, and thus aid them in making better decisions. Value judgments, imprecise knowledge and estimates, intuition, and uncertainties about the nature and the actions of others result in the COEA doing little more than identifying the likely consequences of choice. The decision makers should never expect the COEA to conclusively demonstrate that a particular alternative is the best choice under all conditions and beyond all reasonable doubt. Ideally, the COEA will help the decision makers see what the important issues are and what judgments they must make in order to resolve them.

2.6 Scope by Milestone

The scope of a COEA is influenced by the acquisition stage to which the system has advanced, the milestone decision to be made, and the system's costs. An analysis conducted at Milestone 0 and to aid Milestone I decisions occurs when knowledge of the program under consideration is limited. The analysis must address a range of alternative concepts to satisfy the identified mission need. Performance expectations and costs are normally expressed as intervals (i.e., between this low value and that high value), with high reliance on parametric estimating techniques. Cost estimates must take into account advanced development and engineering development. In addition, gross estimates of investment (procurement) costs are required. It is generally difficult to obtain accurate organizational and operational cost projections for a Milestone I analysis, and rough estimates may be the best that can be provided. These early estimates or cost intervals should be qualified to highlight the weaknesses inherent in them and any possibility for gross error. To the extent known, the characteristics of each concept that drives the cost intervals or uncertainties should be highlighted.

A Milestone II analysis is accomplished toward the end of Phase I, Demonstration and Validation, after the most promising system concept has been demonstrated and validated. There is generally sufficient knowledge of the system to use point estimates of costs from the bottom-up (engineering) estimating techniques. A Milestone II cost assessment should include total life-cycle costs, expressed in both constant and current dollars. Point estimates should be bounded by an uncertainty range—"possible low" to "possible high" costs. Life-cycle estimates should be provided for all alternative design approaches.

At Milestone III, a decision must be made to produce, cancel, or continue development of a system. The design approach generally has been chosen. A cost and operational effectiveness assessment is required only when conditions have changed sufficiently to render previous cost-effectiveness determinations invalid. Because costs are more likely to have changes, Milestone III analyses often provide only updated estimates of life-cycle costs. If a change is of sufficient magnitude to cause the Defense Acquisition Board to revisit its Milestone II decision, the full Milestone II COEA is updated.

At Milestone IV, a decision must be made whether to upgrade or modify a system currently in production or one that has completed production. The analysis prepared for this decision should address the costs and effectiveness of all alternatives to include maintaining the status quo.

2.7 Cost and Effectiveness Comparisons

Cost and effectiveness comparisons must communicate essential facts as well as the program uncertainty to the decision makers. Throughout the acquisition process, scientific and engineering methods are applied to transform the operational need into a description of system performance parameters and a system configuration through an iterative approach that applies definition, synthesis, analysis, design, and test and evaluation. This effort must integrate related technical parameters and ensure the compatibility of all physical, functional, and program interfaces in a manner that optimizes the total system definition and design. Reliability, maintainability, safety, survivability, human factors, and other important attributes must be taken into account as a total management and engineering effort to meet cost, schedule, and technical performance objectives. Cost and effectiveness comparisons are an important tool and can contribute significantly to decision making during this process. They should be made at a variety of analytical levels, and employ a wide range of techniques. For illustrative purposes, we will examine several comparative techniques at four important levels of analysis: 1) the system concept level, 2) the system design level, 3) the system comparison level, and 4) the force structure level. It is important that the COEA appropriately address the issues at each level during the acquisition of a system.

2.7.1 System Concept Level Comparisons

These analyses are directed at providing information at the decision point for Milestone I, Concept Demonstration Approval During the conceptual phase, it is important to ensure that the up-front analyses lead to product and process decisions that can meet the mission requirements and be acquired within an affordable cost and schedule. It is important to evaluate cost to at least the same level, if not to a higher level of management concern as performance and schedule. The primary goal of the analysis at this point in time should be on defining and evaluating the feasibility of the alternative concepts, and providing the basis for assessing the relative merits of the concepts. The most promising concepts should be defined in terms of the initial objectives for cost, schedule, and performance and overall acquisition strategy. The acquisition strategy should provide for the validation of the technologies and capabilities required to achieve critical characteristics and meet operational constraints. It should also address the need for concurrency and for prototyping, considering the results of technology development and demonstration. Figure 2-3 depicts one example of the results of concept comparison. For this example, one of the performance goals is to destroy X_1 targets per sortie as depicted on the chart. Concept I clearly provides this capability at a least cost of C_1.

Additionally, it can be seen that concept I provides this capability at a decreas-

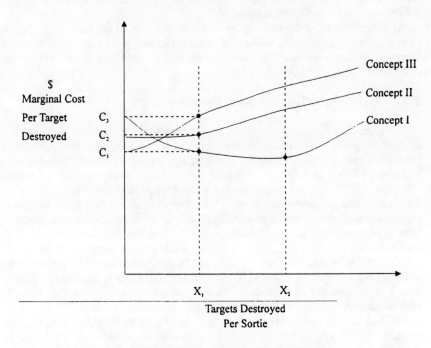

Figure 2-3. System Concept Comparisons

ing marginal cost, which indicates that if the requirement was X_2 instead of X_1, the cost of each target kill would decrease.

2.7.2 System Design Level Comparisons

These analyses are directed at the selection of the configuration, or characteristics, of single systems and have primary application during preparation for the decisions affecting Milestone II, Development Approval. These efforts attempt to better define the critical design characteristics and expected capabilities, demonstrate that the technologies can be incorporated, prove that essential processes are attainable, and establish a proposed development baseline for cost, schedule, and performance. During these analyses, costs are used to indicate the achievable characteristics which maximize performance for various possible funding levels. For example, Figure 2-4 depicts the derivation of a line of minimum cost for system performance in terms of survival based on the reduction of radar cross section. There are several important features regarding these curves. First, each of the curves (A, B, C, and D) depicts a given configuration and performance level versus cost, with curve A representing the highest level. Initially, the overall systems' cost decreases as radar cross section is reduced for each of the performance levels (i.e., A, B, C, and D). However, at some point, the cost of each performance level starts to increase because of diminishing returns. When minimum points on the cost curves occur at about the same value of the characteristic being examined, it is referred to as a dominant solution. That is, there is a value of the characteristic

COST AND OPERATIONAL EFFECTIVENESS

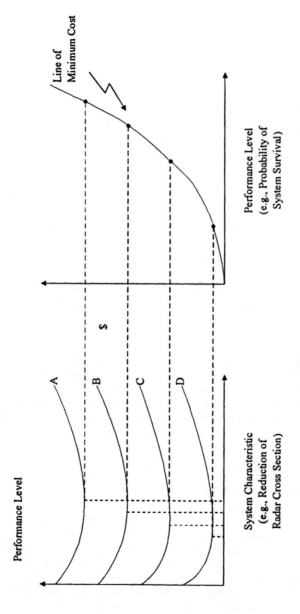

Figure 2-4. System Design Trade-Off of Radar Cross Section and Probability of System Survival

preferred under all the conditions examined. This is not the case for the situation depicted. The least cost solutions occur at different values of radar cross section reduction. Even when the solutions are not dominant, these types of trade-off analyses provide considerable insight into the level of performance that can be achieved, and at what cost.

The system design level analysis is probably most frequently conducted by indusstrial organizations. It attempts to manipulate performance equations (or other methods) with cost estimating relationships in a manner that identifies "optimized" solutions for the important design characteristics of the system.

2.7.3 System Comparison Level

The system comparison analyses are directed at identifying the best system from among competing systems. These analyses are appropriate throughout the acquisition process whenever essential data are available. During these comparisons, the alternative systems being addressed should have been "optimized" individually for the mission to the degree practical, and it is desired to differentiate among competing system concepts and/or designs at the system level or among competing final systems. As discussed earlier, these comparisons can be made by holding costs at some fixed budget value and seeking the alternative system that provides the greatest effectiveness (i.e., the maximum effectiveness solution) or, conversely, by holding effectiveness constant at some desired performance level and seeking the alternative system that costs least (i.e., the least cost solution). If it is not possible to achieve approximately equal effectiveness solutions for the comparisons, the effectiveness achieved by each alternative should be identified and the differences must be taken into account by the decision makers. For illustrative purposes, Figure 2-5 uses two system cost comparison methods and addresses the approach that provides the solution (i.e., desired level of target destruction) at least cost. The top chart is a simple comparison of life-cycle cost for three systems (A, B, and C) that all meet the desired level of effectiveness. It can be seen that the system C provides the desired capability at least cost. The lower graph also depicts the cost comparison of the three systems, with X_1 representing the desired level of target destruction. From these curves it can be seen that system C is the system of least cost to provide the X_1 level of effectiveness. However, if the level of effectiveness required was decreased low enough, system A would become the system of least cost.

2.7.4 Force Level Comparisons

Most decisions in acquisition deal with components that are a part of a larger system. Good decisions are based on the goal of having each component work together to serve a larger purpose. The ground troops, ships, and aircraft are all part of a larger "system" that will not work well if missing any of its parts. During force level analyses, one cannot understand how many of what components are needed without understanding what is to be done in the aggregate and what each component is expected to contribute. Likewise, it is important to look for options that can achieve the desired force effectiveness at less cost or provide more capability for the funds expended. During force level comparisons, the goal is to devise alternative force levels that can achieve the intended purpose as a "system"

COST AND OPERATIONAL EFFECTIVENESS

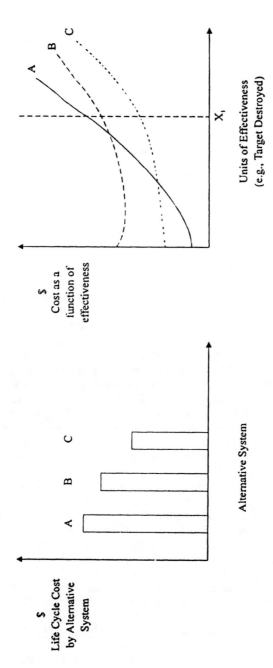

Figure 2-5. System Comparisons

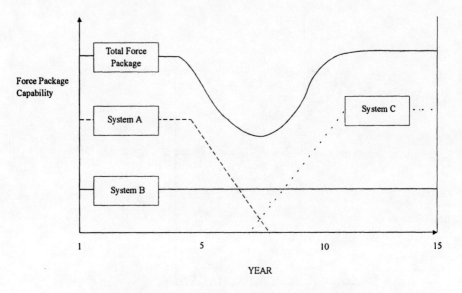

Figure 2-6. Force Package Effectiveness over Time

and to identify the best choices. These choices are typically arranged in terms of defense mission packages such as the general purpose forces. During force level analyses, both cost and effectiveness are each time-phased to the extent possible to allow estimates of the annual funds required and the military effectiveness achievable. The thrust of the analysis is to seek a balanced force posture over the entire planning period. Figure 2-6 illustrates the problem of maintaining a given level of effectiveness over time as weapon systems are phased in and out of the force structure. The chart represents one force mix maintained at constant cost, and its effectiveness stream is as indicated at the top of the chart. If the apparent dip in effectiveness at years 7–8 is not acceptable, it might be eliminated by either keeping system A longer or bringing on system C sooner (or some other option). This type of analysis is further complicated by the uncertainty of the future with regard to the enemy threat, as well as what actually could be achieved with systems A and C. Either solution would have to be feasible and would likely increase the cost of the package.

The decision to keep system A longer or to bring in system C sooner (if both were programmatic and technically feasible) could be examined in terms of the cost of two separate force packages: 1) keep system A longer and 2) bring in system C sooner. These costs (assumed to be for equal effectiveness force packages) are depicted in terms of annual cost and cumulative cost in Figure 2-7. From this analysis, it can be seen that, initially, more funds would be required for force package 2. However, as system A ages, its cost is increasing and the cumulative cost of the two packages is approximately the same near the year 10 point.

These analyses are designed to provide insight into both costs and force postures over a selected planning period. Force level decisions can affect the phasing in and phasing out of individual weapon systems at appropriate points in the plan-

COST AND OPERATIONAL EFFECTIVENESS

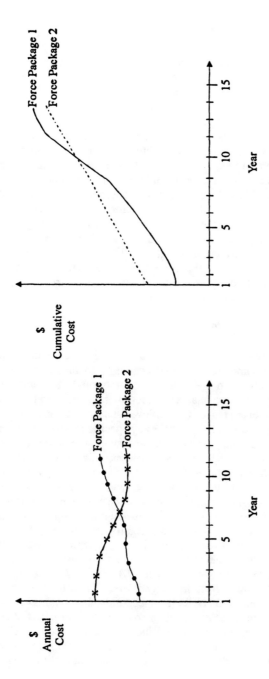

Figure 2-7. Cost of Force Packages over Time (Annual Cost and Cumulative Cost)

ning cycle, the options of extending the service life of various systems, as well as the alternative of bringing in new systems. They can help identify system mixes that achieve and maintain the planned level of total force capability throughout the period of consideration.

2.8 Addressing Risks (See References 17 and 18)

Risks in weapon system acquisition include the potential failure to meet program goals for system performance, time, costs, and schedule. These potential outcomes should be addressed both quantitatively and qualitatively. It is important that the COEA identify where the risks are and, when known, how they might affect the overall decisions being made.

A degree of uncertainty exists in the estimates of both costs and effectiveness presented in all COEAs. In most cases, additional information can be presented on the nature and extent of these uncertainties to help decision makers cope with these difficult circumstances. As a general approach, it is important that the COEA provide an objective analysis of the full situation and describe what and where the significant program risks are in terms of each alternative. Even basic information acknowledging that a risk exists and some qualitative assessment and discussion of the degree of its potential impact are useful. It is also important to acknowledge and discuss all known factors that can mitigate or eliminate the known risks. These analytical excursions that address the mitigation or elimination of risks often result in associated increased program costs that need to be clearly identified to the decision maker.

2.8.1 Uncertainty in Estimates

Effectiveness and cost data normally can be depicted in a manner that provides the decision makers with knowledge of the uncertainty contained in the estimates. Figure 2-8 depicts two ways in which the uncertainty might be addressed in the estimates. For example, the top chart compares alternatives A and B in terms of estimates of the effectiveness that can be achieved at a given fixed cost. Appropriate estimates of this type could be made for any specific subsystem performance objective or for objectives relating to an overall system. It can be noted that three estimates are presented for each alternative—low, most likely, and high. System B has the highest most likely value and smallest range between the low and high values. System A has a slightly lower most likely value and a smaller low value but, if all goes well, it has the potential to achieve the overall highest value. If the overall cost constraint were raised to a higher value and appropriate measures taken to reduce risks, this chart might change significantly, with system A becoming a clear preference. The lower chart in Figure 2-8 depicts effectiveness as a function of costs. The effectiveness presented could be for two single performance parameters (A and B), or for two entire systems (A and B). An interval (or band) is used to illustrate the uncertainty in the estimate. The effectiveness estimate represented within the band could be for the uncertainty in such things as the threat, future system performance capability, and/or costs. For lower levels of effectiveness, alternative B is clearly least costly. When the effectiveness level is raised high enough, alternative A becomes less costly. The intersection of the two curves represents an area where there is uncertainty as to which alternative is least costly.

COST AND OPERATIONAL EFFECTIVENESS

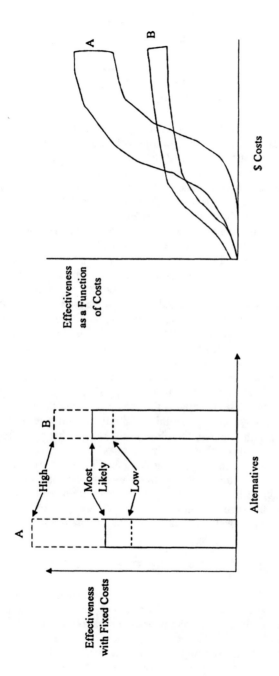

Figure 2-8. Effects of Uncertainty

2.8.2 Risks in Terms of Probability

Under certain conditions, it is possible to quantify the risks in terms of an associated statistical probability. For example, the band placed on the cost estimate in the top chart of Figure 2-9 illustrates a 90 percent statistical confidence interval on the cost estimate. One-sided estimates could also be established where the statistical confidence is 90 percent that the cost will not exceed the limit value. The lower chart in Figure 2-9 illustrates a function relating probability to costs. For example, there is a probability of .9 that the system (or performance objective) cost being examined will reach a value of X_1.

Risk comparisons for alternatives can also be approached from a risk-cost modeling perspective. Risk-cost modeling is an approach to quantify and capitalize on the probability estimates of the risks and their associated costs. For example, suppose a system program has 30 key performance objectives identified with the potential failure to meet program goals. Suppose that performance objective #1 (reliability) has a risk corrective action cost of $10M if the risk event were to occur (i.e., failure to meet the reliability goal). Further, suppose the risk of failing to meet this reliability goal is assessed at .2. Then the cost of this risk at the present point in time would be assessed at $2M (i.e., $10M × .2 = $2M). The cost of the total program risks at this point in time can then be estimated as the sum of all risk costs for the alternative. This general approach can be applied to compare risk costs among alternative systems. However, when making these comparisons, it is recommended that all risk factors associated with all of the alternatives be put into a single risk-cost comparison model for the evaluation. This allows all of the risks to be evaluated against all of the alternatives and should provide for a more objective comparison, given that all key risk factors have been identified and are considered in the model. The preferred system in terms of risks then becomes the system with the least total risk costs, given all else as being equal.

Risk-cost modeling is approached from another perspective by industry in the development of a system. Here, the goal of the producer is to continually monitor and manage the risk costs by working to minimize them for each specific program. Risks are managed by addressing and implementing actions that either lessen or eliminate the specific conditions that result in the risks.

Despite the difficulties and lack of standardization, it is extremely important to address the risks and uncertainties involved in the estimates of both effectiveness and costs. Considerable effort is now being expended within the DoD to better address such things as the verification, validation, and accreditation of analytical modeling and simulation, as well as other important uncertainties in the weapon system acquisition process.

2.9 Other Cost and Effectiveness Considerations

Cost analysis in the Department of Defense is normally performed in support of weapon system analysis, mission force structure analysis, and total force structure analysis. The relevancy of costs and their classification into various categories depend upon the kind of effectiveness comparisons to be made. In general, for the purpose of peacetime planning, the relevant costs are those incurred in providing the capabilities necessary to achieve and maintain a given level of potential wartime effectiveness. The efficiency of a system should be estimated as a

COST AND OPERATIONAL EFFECTIVENESS

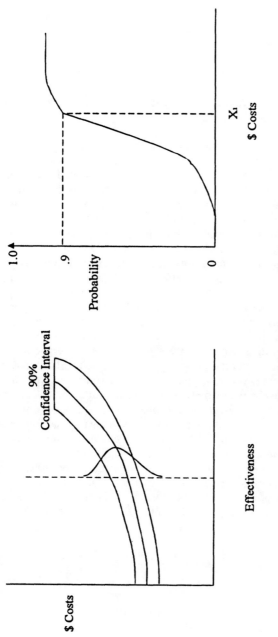

Figure 2-9. Risks in Terms of Probability

function of its peacetime costs and its potential wartime effectiveness (i.e., the level of effectiveness that is on hand during peacetime and ready for application in war).

Both cost and effectiveness can be viewed as streams of values time-phased into the future. The nature and relationship of these streams (or some other form of consideration and summarization of the streams) are the influencing data that can help the decision maker make choices. The point in time at which the decision must be made, the time frame or length of the cost and effectiveness streams to be considered, and the methods of accounting for differences in the value of both effectiveness and costs over time are all important factors that could appropriately influence the decision maker's choice. The COEA effort should be directed at putting both the effectiveness and cost streams into appropriate categories and formats that aid in the decision process.

2.9.1 Costs and Effectiveness over Time

One of the most difficult problems to address in a cost and effectiveness analysis is whether one time-phased effectiveness profile of competing systems is equivalent to another. This occurs because we do not have an acceptable methodology to place a value on the relative worth of effectiveness that can be achieved now versus that which can be achieved in the future. This difference in value is only measurable in view of the preference of the decision maker. Consequently, the practice often followed is to ensure that the effectiveness streams provided by competing systems meet the needed effectiveness level over time, and to summarize the decision data in terms of the system with least cost.

Cost streams can be summarized because of the preference of the economy for money today versus the future. This time preference for money is captured by the rate of interest that must be paid on borrowed funds. For any level of the output of a military force in different years, there generally are several alternatives with different cost streams over the planning period. Unless costs for one alternative force are lower in every year of the planning period, some weighting of the costs in different years must be made to provide a meaningful basis for decision making among the alternative forces. The technique used to weigh, sum, and express the cost streams into a single value is known as discounting (or establishing the present value of the cost of a system). This technique evaluates future expenditures in terms of their present value; that is, future expenditures are reduced to a single point estimate and are evaluated as if they occurred today rather than at some future date. Present value represents the amount C that would have to be set aside at an assumed rate of return of r to be worth C_t at the time the future expenditure must be made. It can be expressed in terms of the standard compound interest formula where

$$C_t = C(1 + r)^t$$

and

$$C = C_t / (1 + r)^t$$

where

C = single point estimate of present value of cost stream
C_t = value of investment at time t
r = interest rate
t = time period of investment

This method of aggregating annual costs links the price of a dollar in some future year, t, to the present-value cost through the interest rate to borrow money. This technique will be illustrated in a simplified example and discussed in additional detail at a later point.

2.9.1.1 System life-cycle costs.
The life-cycle cost of a system can be described in terms of its acquisition cost, operation and support cost, and disposal cost. Acquisition cost is expressed as the sum of research and development cost and production cost. Figure 2-10 depicts an example of the relationship of these cost streams over time. It can be seen that early in a program, the primary cost incurred is research and development. At a later point in time after the system has been designed, a production decision is made and the dominant cost becomes production costs. As the systems are fielded, the operations and support become the major costs. Engineering changes made very early in a program during development and design normally cost much less than those made after systems have been produced and placed in operation. Generally, a $1/$10/$100 rule applies. That is, for every $1 spent to make a change during design, it would cost $10 to make the same change during production and $100 after the system has been placed in field operation. Some systems wear out over time and their operations and support costs increase after extended time in the inventory. Also, modifications to some systems are needed to keep them capable of performing their mission and such costs are additive to those depicted. Finally, when a system is phased out, costs can be incurred to dispose of the system (sometimes referred to as salvage costs). In some cases, the systems might actually have positive dollar value at the end of their life and these costs could actually decrease the overall life-cycle cost of the system.

2.9.1.2 Example cost comparison.
Consider the problem of comparing the costs of two alternative systems (A and B) over their life cycles (see Figure 2-11). Assume that system A is currently in the inventory and providing a given level of effectiveness for the annual cost indicated. Also assume that we are at the program decision point indicated (i.e., year 6). System B can be developed and could provide effectiveness equal to system A at year 11 for the cost stream indicated in the chart. From Figure 2-11, it should be noted that the program decision point is year 6 and all costs prior to this point are considered sunk costs; that is, they are already expended and are irrelevant to the decision. A disposal cost for each system in the example is included during the last 2 years of the program. Table 2-1 presents an example format for aggregating the costs of a system.

Table 2-2 provides a summary of the costs for the systems (A and B) depicted in Figure 2-11. The present-value cost for each system was estimated for this simplified example by use of the following relationship:

$$C_n^* = \sum_{t=n}^{k} \frac{C_t}{(1+r)^{t-n}}$$

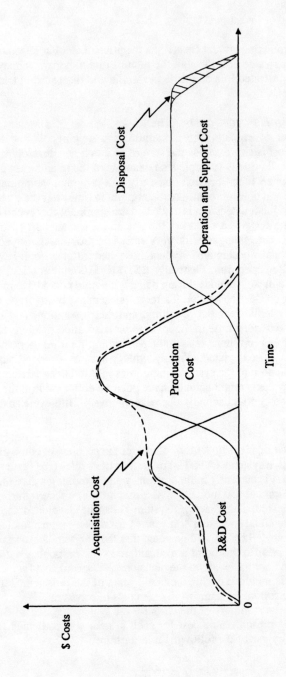

Figure 2-10. Life Cycle Costs over Time

COST AND OPERATIONAL EFFECTIVENESS 43

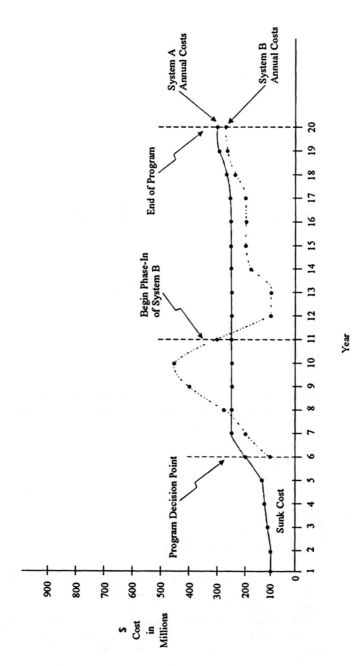

Figure 2-11. Life Cycle Cost Comparison of System A Versus System B

Table 2-1. Example Format for Aggregating the Cost of a System

Years	Acquisition	Operations and Support	Disposal	Total Annual
1	A_1	O_1	–	C_1
2	A_2	O_2	–	C_2
.	.	.	–	.
.	.	.	–	.
.	.	.	–	.
t	A_t	O_t	D_t	C_t

Table 2-2. Summary of Costs for Example Systems A and B

	System A	System B
Total Acquisition	$0.2B	$1.5B
Total Operations and Support	$3.3B	$2.0B
Total Cumulative Cost of Program	$3.5B	$3.5B
Present Value of Total Costs	$2.2B	$2.1B

1. Decision point was set at year 6 with operation to year 20.
2. System B was assumed to start phase-in at year 11.
3. Present-value computation used an 8 percent discount rate of interest.
4. Disposal cost of $100M and $60M was assumed for systems A and B, respectively.

where

C_n^* = present value cost of program at year n
n = year at which program decision is being made (i.e., year 6 for this example)
C_t = total annual cost to be incurred at year t
t = program year at which cost C_t is incurred
k = end year of program (i.e., year 20 for this example)
r = interest rate (i.e., 8% assumed for this example)

The summary in Table 2-2 indicates that there is not an overpowering preference between the systems based on costs. System B is a new system costing considerably more for acquisition with a lower operations and support cost. System A has minor cost expenditure for acquisition (i.e., modification costs) but has considerably higher operations and support costs. Both systems appear to be about equal in terms of the overall total cumulative program costs (i.e., nondiscounted costs). System B has a slight edge in terms of the overall present-value cost of the systems (i.e., costs after discounting).

In a case like this where neither the effectiveness nor cost can significantly differentiate between the alternatives, ideally, other alternatives should be identified and pursued. However, it might be possible for the decision maker to make a reasonable choice if the two programs under consideration are significantly different in terms of risks, past performance of competing contractors, or other acceptable contract performance discrimination factors.

2.10 Optimization and Partial Analysis (See References 19–31)

Because winning in warfare is paramount, one can argue that seeking an optimal mix and application of warfare assets may be more academic than practical. However, the scarcity of resources available for defense systems dictates that the optimal mix of capabilities be provided to the war fighting forces and that these capabilities be applied in the most effective manner once combat begins. An insight into the task of optimization can be gained from an examination of the theoretical aspects of the problem. Although the systems mix problem is discussed here in terms of continuous functions with highly divisible units and Lagrange multipliers other analytical approaches can be taken to address cases where such special conditions do not exist. Assuming that one can construct mathematical functions of output (i.e., effectiveness) and associated costs, these functions can be solved to identify the optimal mix of capabilities in terms of a constrained maxima or minima problem. Other analytical approaches such as linear programming, nonlinear programming, and dynamic programming should result in solutions that provide a similar interpretation regarding optimization.

Most analysis problems addressing solutions over the long term deal with the trade-off of one competing performance objective (or a factor within a single performance objective) versus others on the basis of output and costs. This problem can be addressed in a general context by examining the maximization of output as a function of performance objectives (or factors) subject to a cost constraint.

An equally important approach in operations analysis treats the short-term allocation or application problem. For this problem, inventories and resources are considered fixed, and the objective is to optimally allocate the available fixed resources toward winning. This problem is further complicated by the interactions and dynamics of force application (the timing of the application by each side of the various assets as well as the other components of the force). The quantitative techniques of operations research (such as mathematical modeling, computer simulation, and force integration testing) play an important role in the quantification and understanding of this process. Because constructing a win function may be too theoretical and abstract at this point in the sophistication of our analysis capability, one may choose to apply a lower level measure of success (for example, probability of survival). This more practical solution would be addressed by maximizing the probability of survival during mission accomplishment, given available resources. Although the ability to obtain an optimal solution for any specific system requires that we be able to mathematically establish the specific output and cost functions for the system, much can be learned from the general examination of how factors affecting output trade off in terms of their associated costs. The generalized constrained maximum problem is to maximize the output function $O(X_1, X_2, \cdots, X_n)$ subject to the constraint that only those values of $(X_1, X_2, \cdots, X_n)$ that satisfy the equation $C_T = C_1X_1 + C_2X_2 + \cdots + C_nX_n$ are permitted. For example, X_1 represents a unit of input that costs C_1 and can raise the level of output O. The budget is constrained at a total cost of C_T which is not to be exceeded. We are seeking the largest value of output (O) that can be achieved by various combinations of the X_i values without exceeding the C_T budget. This problem can be solved by forming a Lagrange multiplier function and using partial differential calculus to obtain the maximum for the function. The Lagrange multiplier used here is denoted by λ. Maximizing the Lagrangian function (L) shown below is equivalent to maximizing $O(X_1, X_2, \cdots, X_n)$ subject to $(C_T - C_1X_1 - C_2X_2 - \cdots - C_nX_n) = O$.

2.10.1 Lagrangian Function (L)

$$L = O(X_1, X_2, \cdots, X_n) + T(C_T - C_1X_1 - C_2X_2 - \cdots - C_nX_n)$$

2.10.1.1 First order conditions for maximization.
The first order conditions are obtained by taking the first partial derivatives of L and setting them equal to zero. Also, since they are all equal to λ, they are equal to each other. O_1 represents the first partial derivative of $O(X_1, X_2, \cdots, X_n)$ with respect to X_1, O_2 represents the first partial derivative with respect to X_2, etc. O_{11} represents the second partial derivative with respect to X_1, etc.

$$\frac{\partial L}{\partial X_1} = O_1 - \lambda C_1 = 0, \quad \lambda = \frac{O_1}{C_1}$$

$$\frac{\partial L}{\partial X_2} = O_2 - \lambda C_2 = 0, \quad \lambda = \frac{O_2}{C_2}$$

$$\vdots$$

$$\frac{\partial L}{\partial X_n} = O_n - \lambda C_n = 0, \quad \lambda = \frac{O_n}{C_n}$$

$$\frac{\partial L}{\partial \lambda} = C_T - C_1X_1 - C_2X_2 - \cdots - C_nX_n = 0,$$

$$C_T = C_1X_1 + C_2X_2 + \cdots + C_nX_n$$

2.10.1.2 Equating the first order conditions.

$$\frac{O_1}{C_1} = \frac{O_2}{C_2} = \cdots = \frac{O_n}{C_n}$$

2.10.1.3 Second order conditions.
These conditions are obtained by taking the second partial derivatives and forming determinants by bordering the principal minors of the Hessian determinant with a row and column containing the first

partial derivatives of the constraint. The element in the lower right corner is set to 0.

$$\begin{vmatrix} O_{11} & O_{12} & -C_1 \\ O_{21} & O_{22} & -C_{21} \\ -C_1 & -C_2 & 0 \end{vmatrix}, \begin{vmatrix} O_{11} & O_{12} & O_{13} & -C_1 \\ O_{21} & O_{22} & O_{23} & -C_2 \\ O_{31} & O_{32} & O_{23} & -C_3 \\ -C_1 & -C_2 & -C_3 & 0 \end{vmatrix}, \dots, \begin{vmatrix} O_{11} & O_{12} & \cdots & O_{1n} & -C_1 \\ O_{21} & O_{22} & \cdots & O_{2n} & -C_2 \\ \cdot & \cdot & \cdots & \cdot & \cdot \\ \cdot & \cdot & \cdots & \cdot & \cdot \\ O_{n1} & O_{n2} & \cdots & O_{nn} & -C_n \\ -C_1 & -C_2 & \cdots & -C_n & 0 \end{vmatrix}$$

For a maximum, the signs of these determinants from left to right must be +, −, +, etc. For a minimum, all the determinants must be negative.

2.10.2 Partial Analysis

The first order conditions resulting from the constrained maximum problem give rise to a very powerful tool in economics and in partial analysis. When the maximum output is achieved, the ratio of marginal output to marginal cost for each of the competing factors is equal. Prior to achieving this equilibrium, an increase in any given input factor likely would have resulted in an increase in overall output. However, according to the law of diminishing returns, continued increase in a given factor would result at some point in fewer units of output being achieved per unit of input added. Additionally, before the maximum is achieved, the factor with the highest ratio of marginal output to marginal cost is preferred, and units of it should be added in the maximization problem until its ratio is equal or starts to become lower than one of the others. This basic notion has wide application and can be used in a large number of cost effectiveness comparisons.

The optimization problem can also be examined graphically in two dimensions in terms of economic output curves called isoquants, and lines representing equal costs called isocost lines (see Figure 2-12). The graph in Figure 2-12 could represent the output of an overall defense system containing two parts, X_1 and X_2. All points on the isoquant output curve represent a single level of system output that can be achieved by various mixes in the quantities of X_1 and X_2. Higher levels of output (not shown) could be represented by additional curves further from the 0, 0 point of the graph. For any given isoquant curve, the two factors X_1 and X_2 are considered substitutes (i.e., alternative choices) for achieving that level of output. The more easily the two factors may substitute for one another, the less curvature of the isoquant. At the other extreme (no substitution), the curve will be represented by two straight lines intersecting to form a right angle. The economic concept that addresses the number of inputs of one factor (X_1) necessary to substitute for another factor (X_2) and maintain the same level of output is commonly referred to as the marginal rate of substitution. The isocost line is drawn by solving for the number of units of X_1 that could be purchased if the entire budget were used on X_1 (i.e., $X_1^* = C_T/C_1$). After obtaining a similar solution for X_2 (i.e., $X_2^* = C_T/C_2$), a straight line is drawn to connect the two points (X_1^* and X_2^*). The isocost line represents all possible cost mixes of units of X_1 and X_2 that can be purchased for C_T. The slope of the isocost line will be determined by the values of C_1 and C_2 (i.e., ratio). The point of tangency of the isocost line to the isoquant output curve de-

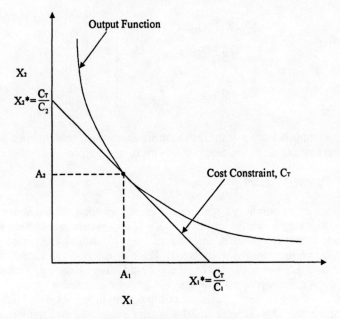

Figure 2-12. Graphic Depiction of Optimization of Output Subject to a Cost Constraint for Two Factors

picts the optimum mix of X_1 and X_2 to achieve the highest level of output subject to the budget constraint, C_T.

The values of X_1 and X_2 achieved in the solution are noted on the graph as A_1 and A_2. The point of tangency of the isocost line with the isoquant output curve represents the condition of maximization previously addressed where the ratios of marginal output to marginal cost for each factor are equal. At this point, the marginal rate of substitution of X_1 and X_2 is equal to the ratio of their costs (i.e., C_1/C_2). Also, the equilibrium condition exists for the ratios of marginal output to marginal costs as previously demonstrated in the solution for the general case.

$$\frac{\frac{\partial O}{\partial X_1}}{\frac{\partial C}{\partial X_1}} = \frac{\frac{\partial O}{\partial X_2}}{\frac{\partial C}{\partial X_2}} = \ldots = \frac{\frac{\partial O}{\partial X_n}}{\frac{\partial C}{\partial X_n}}$$

In defense comparisons, the goal should be to achieve the minimum cost solution for a given level of required effectiveness. The ability to address the discussed optimization issues for future systems centers around the problem of being able to better estimate these important functions quantitatively and to validate the complex models used in analysis. Future measurements made by instrumentation in test and evaluation, training, tactics development, and laboratory environments must play a significant role in such quantification.

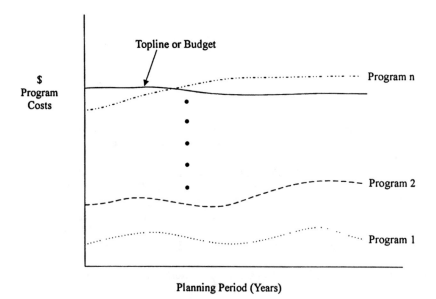

Figure 2-13. Example Graph Depicting Affordability

2.11 Affordability Assessments

Affordability is the characteristic of a product (or system) selling at a cost that is equal to or less than its functional worth and that is within the customer's ability to pay. Affordability analysis is conducted to estimate the functional worth of the product and its relative priority within the budget from the customer's perspective. Affordability, to a large degree, is a judgmental issue from both a commercial and governmental viewpoint. Decision makers allocate resources based on their perception of the relative urgency of needs and the resources available to meet those needs within budgetary constraints. Good affordability assessments can prevent situations from occurring where the funding streams available in a program are insufficient to result in the actual fruition of the program.

Once the costs of programs have been estimated and the programs have been prioritized relative to each other, their affordability can be determined. The affordability of a given system can be examined by plotting a graph of the topline program funds expected to be available versus time (i.e., total investment budget). The costs of individual programs can then be plotted on the chart to examine their specific affordability in accordance with their priority (see Figure 2-13). As additional programs are added, their summed value should not exceed the topline budget. From this examination, it should be clear how far down the priority list can be funded within the expected budget. In the example depicted in Figure 2-13, the nth program has exceeded the budget line and cannot be afforded unless it is reprioritized or some other funding arrangement can be made.

Individual program plans for new acquisition efforts must be consistent with the overall DoD planning and funding priorities. Affordability constraints should be established at the beginning of a program and assessed at each milestone decision point throughout the program. Decision makers are responsible for ensuring that a program is not approved to enter the next acquisition phase unless sufficient resources, including manpower, are or will be available to support the projected system development, testing, production, fielding, and support requirements.

References

[1] DoD Directive 5000.1, "Defense Acquisition," Office of the Secretary of Defense, Under Secretary of Defense for Acquisition, 23 February 1991.

[2] DoD Instruction 5000.2, "Defense Acquisition Management Policies and Procedures," Office of the Secretary of Defense, Under Secretary of Defense for Acquisition, 23 February 1991.

[3] DoD Manual 5000.2-M, "Defense Acquisition Management Documentation and Reports," Office of the Secretary of Defense, Under Secretary of Defense for Acquisition, 23 February 1991.

[4] DoDD 5000.3, Test and Evaluation. Establishes policy and guidance for the conduct of test and evaluation in the Department of Defense, provides guidance for the preparation and submission of a Test and Evaluation Master Plan and test reports, and outlines the respective responsibilities of the Director, Operational Test and Evaluation (DOT&E), and Deputy Director, Defense Research and Engineering [Test and Evaluation (DDDR&E (T&E)].

[5] ANSI Z39.18-198X, Scientific and Technical Reports: Organization, Preparation, and Production (American National Standards Institute, 1430 Broadway, New York, NY 10018, December, 1986) The accepted reports standard for DoD as well as a part of academia for the private sector.

[6] Goldman, T. A. (ed.), *Cost Effectiveness Analysis; New Approaches in Decision-Making.* New York: Praeger, 1967.

[7] Sugden, R., and Williams, A., T*he Principles of Practical Cost-Benefit Analysis.* Oxford, England: Oxford University Press, 1978.

[8] Alchian, A. A., "The Meaning of Utility Measurement." *American Economic Review,* Vol. 43, March 1953, pp. 26-50.

[9] Samuelson, P. A., *Foundations of Economic Analysis.* Cambridge, Massachusetts, Harvard University Press, 1948.

[10] Baumoul, W. J., *Economic Theory and Operations Analysis* (2nd ed.). Englewood Cliffs, New Jersey: Prentice-Hall, Inc., 1965.

[11] Fasal, J. H., *Practical Value Analysis Methods.* New York: Hayden Book Company, Inc., 1978.

[12] Allen, R. G. D., *Mathematical Analysis for Economists.* London, England: Macmillan, 1938.

[13] Clark, F. D., *Cost Engineering.* New York: Marcel Dekker, 1978.

[14] Stewart, R. D., *Cost Estimating.* New York: John Wiley and Sons, Inc., 1982.

[15] Michaels, J. V., and Wood, W. P., *Design to Cost.* New York: John Wiley and Sons, Inc., 1989.

[16] Friedman, M., and Savage, L. J., "The Utility Analysis of Choices Involving Risk." *Journal of Political Economy,* Vol. 56, August 1948, pp. 279-304.

[17] Knight, F. H., *Risk, Uncertainty, and Profit.* Boston, Massachusetts: Houghton Mifflin, 1921.

[18] Perlis, S., *Theory of Matrices*. Cambridge, Massachusetts: Addison-Wesley, 1952.

[19] Baumoul, W. J., *Economic Dynamics*. New York: Macmillan, 1951.

[20] Henderson, J. M., and Quandt, R. E., *Microeconomic Theory, A Mathematical Approach*. New York: McGraw-Hill Book Company, 1958.

[21] Bierman, H., and Smidt, S., *The Capital Budgeting Decision*. New York: Macmillan, 1975.

[22] Broadway, R.W., "The Welfare Foundations of Cost-Benefit Analysis." *Economic Journal*, Vol. 84, 1974.

[23] Samuelson, P., and Solow, R., *Linear Programming and Economic Analysis*. New York: McGraw-Hill, 1958.

[24] Millward, R., *Public Expenditure Economics*. New York: McGraw-Hill, 1971.

[25] Musgrave, R.A., *The Theory of Public Finance*. New York: McGraw-Hill, 1959.

[26] Raiffa, H., *Decision Analysis*. Reading, Massachusetts: Addison-Wesley, 1968.

[27] Schelling, T.C., *The Strategy of Conflict*. Cambridge, Massachusetts: Harvard University Press, 1960.

[28] Von Neumann, J., and Morgenstern, O., *Theory of Games and Economic Behavior*. New York: John Wiley and Sons, Inc., 1947.

[29] Morgenstern, O. (ed.), *Economic Activity Analysis*. New York: John Wiley and Sons, Inc., 1954.

[30] Dantzig, G.B., *Linear Programming and Extensions*. Princeton, New Jersey: Princeton University Press, 1963.

[31] Hicks, J.R., *Value and Capital*. Oxford, England: Clarendon Press, 1946.

3
Basic Principles

3.1 General

Important parts of the acquisition and fielding of new defense systems are the test and evaluation (T&E) of effectiveness and suitability, the development and evaluation of tactics and procedures, and the training of personnel who will employ the systems in combat (i.e., the end users). The first of these T&Es is critical from the standpoint of measuring the operational capabilities of new systems and determining whether these capabilities can be expected to satisfy combat requirements. Follow-on T&Es help provide for the continuous improvement of the products over time. The latter T&Es are also important from the standpoint of developing tactics and procedures and training users to effectively employ the systems in combat.

3.1.1 Effectiveness

It is important to establish the effectiveness of any product. Effectiveness determines whether the product can do what it is expected to do. If effectiveness is not known, then it is not possible to ascertain whether the product is acceptable and whether it can and should be employed and/or improved. In military T&E, effectiveness is a measure of the degree to which a component, system, concept, or approach is able to accomplish its mission when used by qualified and representative personnel in the environment planned or expected for combat employment. Employment considers a wide variety of factors to include organization, doctrine, tactics, survivability, vulnerability, and threat.

3.1.2 Suitability

The suitability of a product is also critical to its being acceptable and successful in use. Suitability establishes whether the product can be kept serviceable for its intended use. Suitability is a broad measure of the degree to which a component, system, concept, or approach can be placed satisfactorily in field/flight use, with consideration given to availability, compatibility, transportability, interoperability, reliability, wartime usage rates, maintainability, safety, human factors, manpower supportability, logistics supportability, software supportability, documentation, and training requirements.

3.1.3 Tactics, Procedures, and Training (See References 1 and 2)

Tactics, procedures, and training are important to the successful application of any military system. Tactics and procedures are the specific way in which a weapon system is employed to accomplish its assigned mission in combat. This includes the deployment of forces, tasking of units, mission planning, and operational employment of the system. Training is the act, process, or method by which one becomes fit, qualified, or proficient in the application of a weapon system and the carrying out of a military mission. Successful training cannot be achieved without adequate measurement and feedback to those being trained regarding success in mission accomplishment.

3.2 Defining the Test and Evaluation Problem

In order to conduct a successful T&E, it is first necessary to sufficiently define the T&E problem. In military T&E, it is necessary to fully understand the weapon system's potential applications and its operational mission and environment, and to define the critical issues and objectives that must be addressed. The systems, subsystems, equipment, facilities, support elements, tactics, techniques, procedures, and doctrines which are developed, tested, and evaluated during T&E programs are interrelated components or elements of a total military operational capability. This overall capability can be formulated and described in terms of a general structure of which all lesser operations and related support elements are an integral part. The overall military structure should, in turn, be used to assist in the planning process. The structuring procedure recommended is basically a scheme that first identifies the appropriate spectrum of military operations and then, for the specific equipment undergoing T&E, aids the T&E project team in relating tactics, personnel, and resources to those operations. Under this procedure, the planner should partition those operations considered significant to the T&E into tasks and subtasks. This effort should lead to the formulation of critical operational issues and specific objectives, appropriate measures of effectiveness for those objectives, and a systematic examination of resources in relation to the tasks. Within each military operation, the operational flow sequences for premission, mission, and postmission events should be constructed to aid in identifying how pertinent classes of factors affect tasks within the operation and their relationship to the overall operation. The approach to examining systems in terms of their overall military structure is important to achieving a total system evaluation that accommodates both compatibility and interoperability issues. This approach is essential to effective T&E and provides for a systematic and careful examination of the operational equipment, mission factors, mission tasks, support equipment, and operational procedures in terms of their capability to effectively perform the total military mission.

The clear specification of the correct issues and objectives for the overall T&E problem is essential. If the incorrect issues and objectives are chosen, the wrong T&E will be conducted and wrong answers will be rendered, no matter how good the testing might be. There are a number of approaches that can be taken and planning tools that can be applied to help properly define the T&E problem.

It is essential to thoroughly understand the mission that must be performed, the environment in which it will be performed, interrelationships of the systems involved in the mission, and interfaces and measures of success for the item undergoing T&E. This effort must be accomplished with an open mind. The traditional, conventional, or plausible way of carrying out the mission may not be the best. It is important to look for new ideas and concepts, as well as to look for facts and relationships that explain those already known. It is important not to exclude

BASIC PRINCIPLES

methods, tactics, and concepts merely because they run contrary to past practice or superficially seem impractical. This examination can be aided by the construction of a set of flowcharts for the T&E item that describe the significant premission, mission, and postmission sequences of events believed to be necessary for mission success. Careful study and examination of these flowcharts by a team of highly qualified operational, technical, and analytical experts should allow for the identification of critical issues and objectives related to actions, communications, organization of components, and decisions essential to the success of the mission. Figure 3-1 shows an example of a simplified flowchart for an Air Force air-to-air mission.

Once broad issues and objectives are specified, it is generally necessary to decompose them into lesser parts and specific data that can be addressed by a given test activity. One method that has gained wide acceptance for this purpose is the "dendritic" or "tree" method. This approach provides an underlying structure to the program level analysis. It involves a methodical breakdown of the issues and objectives into smaller elements or actions necessary for objective resolution. It links program level issues and objectives to the lowest level information and data that must be collected to answer the issues and objectives. The dendritic process can be viewed as the root structure of a tree with no root segment disconnected from the trunk. The structure should be exhaustive to the extent that lower level data requirements fully resolve the primary issues and objectives. The decomposition normally begins with the question, "What must be known to answer this objective?" Objectives derived at each stage lead to lower level attributes at the next stage. When attributes can be quantified or qualified, they are referred to as either a measure of effectiveness (MOE) or measure of performance (MOP). In some cases, it may be necessary to calculate or estimate the MOE and MOP values based on lesser attributes quantified in the test. MOEs depict how well the test item performs its operational mission, whereas MOPs are measurements of engineering technical parameters. When the data element level of decomposition is reached, "what must be known" becomes a measurable quantity that can be obtained in one of the planned T&E activities.

3.3 Establishing an Overall Test and Evaluation Model

It is essential to establish the overall structure or model by which the total T&E problem will be addressed. For complex weapons systems, it is generally necessary to use a wide variety of testing activities in a complementary fashion to achieve cost-effective T&E (see Figure 3-2).

The modeling and simulation (M&S) approach to T&E will be addressed in greater detail in Chapter 4. It includes the specification of what issues and objectives can best be addressed in measurement facilities, integration laboratories, hardware-in-the-loop test facilities, installed test facilities, field/flight tests, and analytical M&S efforts. Measurement facilities provide a capability to quantitatively characterize such things as antenna patterns, dynamic and static radar cross section, and other important measurements. An example system would be the precision antenna measurement system located at Rome Air Development Center, Griffiss AFB, New York. Integration laboratories are located at numerous government and contractor locations and provide an environment in which several hardware and software components can be integrated and tested together. Hardware-in-the-

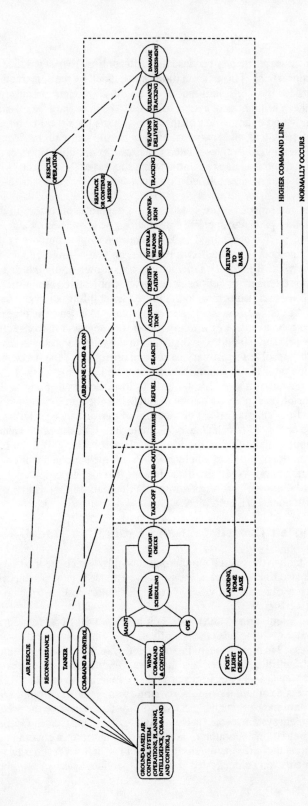

Figure 3-1. Simplified Flow Sequence Diagram of the Counter Air (Air to Air) Operation

BASIC PRINCIPLES

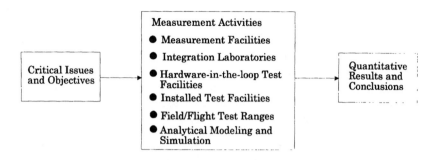

Figure 3-2. Example Overall T&E Model

loop test facilities provide an environment that allows a hardware component such as an electronic jammer or missile to be tested as if in operation or flight. An example hardware-in-the-loop test facility is the Air Force Electronic Environment Simulator located at Carswell AFB, Texas. An installed test facility is one where an entire weapons system such as an aircraft with an electronic jammer and weapons installed is subjected to tests inside of an anechoic chamber. The Preflight Integration of Munitions and Electronic Systems facility located at Eglin AFB, Florida, is an example; it enables ground intra-aircraft electromagnetic interference checkout in a simulated flight environment. Field/flight tests involve actual physical testing of the equipment and systems under simulated combat conditions. These types of tests are conducted by the Services at various locations throughout the United States. Example field/flight test ranges are discussed in Appendix 2. Finally, analytical modeling and simulation involve an examination of the equipment and systems through some form of models and simulations.

Obtaining the right balance of these activities is important. Too much instrumentation can drive the cost and time required for T&E up to impractical levels, while too little instrumentation does not support adequate quantification and understanding of the results. Additionally, it is necessary to properly execute and analyze each T&E activity identified in the overall T&E model in an effective and cost-efficient manner.

3.4 Measurement (See References 3–5)

"Measurement" means many things. Measurements treated in physics, psychology, economics, accounting, philosophy, mathematics, military science, and operations research, to name a few, are not clearly connected by a unified theory of measurement. A complicating factor is that the broader the scope of a system evaluation, the more likely it is to require multidisciplinary measures. In physics, chemistry, and engineering, measurements are usually made on the more directly perceptible physical phenomena—for example, thermal phenomena, mechanical stress, or electromagnetic phenomena. In most cases, physical measurements are more understandable to the human senses. In some sciences they are more abstract. It is easier to visualize or assimilate the measure of length as opposed to a probability measure. As a result, physical scientists have developed, over the generations, ways of naming, "scoring," and, in general, expressing the characteristics of these phenomena. In engineering, these characteristics are expressed in dimensional units. The technique of combining dimensional units to form a more composite description of physical relationships of some phenomena is dimensional analysis. In mathematics, the study of how any units or characteristics connect is measure theory. In the natural or behavioral sciences, measurements or descriptions of phenomena are handled in similar ways, but do not enjoy the status of names like dimensional analysis or measure theory. However, regardless of the discipline involved, once the physical units or characteristics to be observed during a test are established, the problem of counting or assigning values to them must be solved. Test variables that can be measured in numerical units are termed quantitative, whereas those that must be measured in purely judgmental units are termed qualitative. For the purposes of military T&E, the assignment of a quantitative or qualitative value is considered a measurement. To be more precise, a quantitative measurement is the assignment of a particular set of numbers to a set of conceptual entities in such a way as to 1) permit an unambiguous number to be assigned to each entity, and 2) allow the arrangement of the numbers in some order. In addition to these requirements, it is important to the tester to know, within reason, the error involved in each measurement. In reality, engineering and scientific measurements allow the translation of observations of physical objects into mathematical concepts. This is the application of measure theory.

3.4.1 Physical Measurements (See Reference 6)

The physical measurements of system components are relatively easy to perceive; however, as components or subsystems are synthesized, measurements become somewhat more obtuse. For example, consider a physical system, where a "system" is a broad enough term to cover all processes. Assume this system can be described at any time by means of a finite set of quantities, $x_1(t)$, $x_2(t)$,...$x_n(t)$. These quantities, called "state variables," constitute the components of a vector $x(t)$, the state vector. Generally, $x(t)$ will be multidimensional, as in the study of mechanical systems where the components of x are points in phase space (positions and velocities), as in electrical circuits where components may represent currents and voltages, and as in logistic systems where the components may be productive capacities and stockpiles of interdependent maintenance and depot facilities. In the study and testing of systems, many of these components may be described in terms of probability distributions. A fundamental question that arises in the construction of a descriptive model of the behavior of a physical system is how to relate the time change in the system to the state of the system. When characterized and understood, an operational flow sequence describing military systems can be developed and used as an aid in performing the integration function. By studying an operations sequence, it can be observed that many of the operational tasks represent a change in the state of the system. For example, an aircraft is in a different operational state when it is undergoing a preflight inspection as compared to when it is attacking a target. In practice, when the system or a component of the system makes a transition from one state to another, a change in the physical effects may have an implicit effect on the measurements that are most appropriate to describe the success of the operation within that state and at that point in time.

3.4.2 Measurement Process (See References 7–9)

The process of measuring consists of transferring information about a system from the system itself—the source—to another system—the measuring system (see Figure 3-3). This involves the notion of transforming physical observations into test data. These observations can be thought of as having either a transparent or nontransparent relationship to the measurement system. When no energy is transferred from the system under test, information transfer can be considered totally transparent. This might occur in an analytical M&S examination of certain system characteristics and their relationship to the military operations structure. On the other hand, under field/flight test and evaluation conditions, energy is transferred through the measurement system to provide some analog of the system, and nontransparent observations result. For example, the measurement system might consist of a test observer whose presence has adverse effects on the outcome of the test trial. Therefore, in the field/flight test, information normally would not be transferred without energy also being transferred from the source system to the measurement system. Further, energy cannot be drawn from the source system without changing the behavior, at least to a minor degree, of the source system.

The device used in engineering and physical science measurements to convert energy from the source system to produce measurement values is called a transducer.[9] The function of the transducer is to transform the physical quantity to be measured (as it exists in one form of energy) into some other physical quantity more easily measured and interpreted and to provide minimum disturbance of the source system's normal environment and operation. Our definition of a transducer is broad enough to include physical observers, as well as any other means used to obtain test data. Selecting, applying, and displaying data from transducers in a measurement scheme is the process of instrumentation.

The measurement systems required for the components of many tests and evaluations have already been identified and their associated dimensional units rigorously established. When new devices for a weapon system are introduced, and new measurements are developed, the dimensional units for these measurements must be established before the devices are evaluated in T&E. The T&E measures must capture the effects of personnel, procedures, and organization. Statistical measures, developmental measures, operational measures, and optimizing measures necessary to enumerate test MOEs and MOPs must allow this as well as provide a measurement of total system effectiveness that documents the success of the overall solution.

3.4.3 Validity of Measurements

The concept of validity in test measurement techniques involves a very large number of terms in common use. In many cases, these terms are used without general agreement in reference to exact definitions. As a result, different experts often act independently in assigning definitions which best suit their purposes. The meanings selected for use herein have been chosen from among those which have gained some scientific recognition. Unfortunately, they are not necessarily universally accepted.

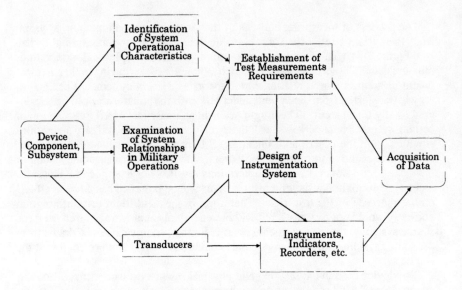

Figure 3-3. Measurement Process

3.4.3.1 Static considerations. Static considerations involve the characteristics of the measurement system under a set of conditions not varying with time. These characteristics can be described in terms of accuracy, precision, sensitivity, and discrimination or resolution.

Accuracy may be regarded as the degree of correctness with which a measured value agrees with the true value. Even though this definition emphasizes agreement, it is common to express accuracy in terms of the disagreement or difference between the measured value and the true value. In other words, the numerical expression of accuracy is most conveniently expressed in terms of error.

If we take a series of observations of exactly the same event with an instrument, we would likely obtain a different value on each observation. Variations such as these are usually attributed to such things as friction, hysteresis, and other minor causes. The capability of the instrument to repeat a reading of the same energy source signifies the degree to which successive measurements yield identical measured values. It represents the *precision* of the instrument and relates only to the performance of the instrument under essentially constant conditions. On the other hand, if we assume that a recent check of the instrument against a known source* revealed in the range of values being used that the instrument always was $+\Delta x$ units in error, then we would have present a systematic error of $+\Delta x$ units. Any value read using the instrument in this range would be $+\Delta x$ units in error.

A distinction is often made between instrument error and measurement error. Instrument error always includes random or accidental errors as well as systematic

---

*The process of checking an instrument against a known source and recording the systematic differences in measurement is referred to as calibration. In some cases, this may involve adjusting the instrument to remove systematic error.

BASIC PRINCIPLES

error, and it expresses the *accuracy* of readings taken with the instrument. However, the observer who uses this instrument usually applies a correction factor for the systematic error to any reading taken. This leaves the measurement error, which expresses the *precision* of the instrument. In scientific measurements, it is accepted practice to apply corrections to data before considering the measurement to be complete. The systematic error is therefore usually not of great importance, as long as it is known and it is not excessively large.

The choice of instrumentation to perform a required measurement is, of necessity, based to a considerable degree on the precision with which the value to be measured must be known. In military T&E, the precision required is driven by the operational significance* of the parameter or parameters being measured, as well as certain other characteristics of the instrumentation. One important characteristic is *sensitivity*. Fundamentally, sensitivity is the ratio of the magnitude of the instrument response to the quantity being measured (e.g., inches per degree). Another characteristic is *discrimination* or *resolution*. The smallest increment in the quantity being measured that can be detected with certainty by an instrument represents its discrimination or resolution.

The accuracy of a measurement may be expressed in several ways. A common method is to specify "accurate within x percent." This is an abbreviation, which means "accurate to within $\pm x$ percent of instrument span at all points of the scale, unless otherwise stated." Often, however, accuracy is based on instrument range and instrument reading instead of instrument span. Obviously, these and other methods of specification differ greatly. Therefore, when test measurements are being conducted, it is extremely important for the test team to consider and understand the method of specification and the overall accuracy of the measurements involved.

In the specification of system measurements, we are normally concerned with overall accuracy. For example, suppose we are dealing with a measurement system composed of several elements (A, B, C,..., etc.). Let the accuracy of each element be within $\pm a$, within $\pm b$, within $\pm c$, and so on. The *least accuracy* involved in the system measurement would be within $\pm (a + b + c +, \text{etc.})$

Because it is not probable that all elements of the system will have the greatest static error at the same point at the same time, another method of expressing accuracy called root-square accuracy is commonly used. The root square accuracy would be determined as follows:

$$\text{Within } \pm \sqrt{a^2 + b^2 + c^2}, \text{ etc.}$$

3.4.3.2 Dynamic considerations.
Measurement devices rarely respond instantaneously to changes in the measured variable. Instead, they exhibit a characteristic slowness or sluggishness caused by such things as mass, thermal capacitance, fluid capacitance, or electric inductance and capacitance. Moreover, pure delay in time is often encountered where the measurement system "waits" for some reactions to take place. Military T&Es are conducted in most cases where the

---

*The term "operational significance" reflects the determination of how much of a change in the parameter being measured is required to significantly affect the operation. This principle is discussed further in Chapter 4.

measured quantities fluctuate or change with time. Dynamic and transient behavior of the measurement system is therefore as important as (and often more important than) static behavior.

The dynamic behavior of a measurement system is determined primarily by subjecting its primary element to some known and predetermined variation in the measured quantity. The three most common variations are either a step change, a linear change, or a sinusoidal change. The dynamic characteristics of the device are speed of response, which is indicative of measurement system lag, and fidelity, which is indicative of dynamic error.

Speed of response is the rapidity with which the measurement system responds to changes in the quantity being measured. Measuring lag is a retardation or delay in the response of the measurement system to changes in the measured quantity. *Fidelity* is the degree to which the measurement system indicates changes in the measured variable without dynamic error. *Dynamic error* is the difference between the true value of a quantity changing with time and the value indicated by the measurement system when static error is assumed not to be present.

In general, testing can be instrumented to measure and record almost any quantity to almost any desired precision. For economic reasons, instrumentation requirements should be determined chiefly on the basis of the precision required in the result. To go beyond this precision may require additional or more expensive instruments or a change in measurement technique to one involving more test resources, time, care, and skill. For these reasons, it is extremely important that the tester be aware of the precision required and available on each measured variable. Generally this determination can be made by having some insight into how much difference in the parameter being measured makes an operationally significant difference. For example, if a 1-mile greater detection range in a weapon system has a significant impact on mission success, then the measurement system should be able to accurately discern detection range in at least 1-mile increments. In cases where the required precision cannot be established, estimating it, or documenting the specific conditions under which the data are obtained during T&E, will provide an invaluable input to the analysis of the results.

3.4.3.3 Classification of errors.

Since no measurement system can operate with complete accuracy, an introduction to some of the possible error sources is essential in the discussion of T&E. It is normally desirable to separate these disturbing effects from the main elements that are being investigated. The fact that we are concerned with the subject of errors does not mean that we wish or expect to make all observation and measurements to an extreme degree of accuracy. T&E measurements cannot be called good or bad merely on the basis of the degree of accuracy, but must be judged on their adequacy to support an analysis that answers the critical issues and satisfies the specific objectives under investigation. In general, greater accuracy (i.e., smaller errors) can be attained by additional time, care, and expense; whether the improvement is justified in a given case is, of course, another question. In any event, we must document what the accuracy of the results are—whether high or low; otherwise we will produce data which we do not know how to evaluate.

A discussion of errors is also important as a step in finding ways for the T&E design to reduce them, and as a means of estimating the reliability of the final data. A T&E quantity is normally measured in terms of a standard—perhaps

BASIC PRINCIPLES

in some cases established by an individual using his own expert judgment, which in itself cannot be perfect. In addition, errors can and do occur in the processes of description and comparison. T&E errors may originate in a large variety of ways, but they may be grouped under three main headings; gross, systematic, and random.

3.4.3.3.1 Gross errors. Gross errors cover, in the main, basic mistakes in the reading and recording of test data. These errors are normally made by the data collector and are the result of such mistakes as gross misinterpretation of test instrumentation output or of an observation, the incorrect transposition of data in recording the result, or a combination of these mistakes. We all make mistakes of this sort at times, and only by extreme care can they be prevented from appearing in the data records. Errors of this type may be of any size and, unfortunately, are not subject to mathematical treatment.

Two precautions can be taken to avoid such difficulties. First, as mentioned earlier, one can exercise extreme care in reading, interpreting, and recording the data. The other precaution against the possibility of gross errors is the making of two, three, or more determinations of the desired quantity, preferably at slightly different measurement points to avoid rereading with the same error. Then if the measurements at a given point show disagreement by an unreasonably large amount, the situation can be investigated and hopefully the bad measurement eliminated. When dealing with critical measurements, complete dependence should never be placed upon a single observer or determination. Indeed the advantage of taking at least three readings at critical points in the data lies not so much in the use of an average as in the confidence given by the general agreement of values that no gross error has been committed.

3.4.3.3.2 Systematic errors. Instrumentation: All instruments and standards possess inaccuracies of some amount. As supplied by the manufacturer, there is always a tolerance allowance in calibration, and additional inaccuracies may of course develop with use and age. It is important to recognize the possibility of instrumentation errors when making test measurements, for it is often possible to eliminate these, or at least to reduce them greatly, by methods such as the following:

1) Careful planning of instrumentation procedures (interchange of equipment, use of multiple test ranges and facilities, etc.).

2) Determination of instrumentation errors and application of correction factors.

3) Careful recalibration.

Misuse of instrumentation can also result in significant measurement errors. Often, shortcomings in measured data can be traced to the operator or tester rather than to the equipment. Good test instrumentation used in an unintelligent way will most likely yield poor results (e.g., measuring the wrong thing, incorrect calibration, and circuit loading).

Environmental: Another source of systematic error in test data can be caused by conditions external to the measuring equipment. This includes any condition in the region surrounding the test area that has an effect on the instrumented or observed measurements. For example, some measurement devices may be affected by humidity, barometric pressure, the earth's magnetic field, gravity, stray electric and magnetic fields, and others. Several kinds of actions can be taken to eliminate, or at least lessen, the effects of these undesirable disturbances:

1) Experimental arrangements to keep instrument conditions as nearly constant as possible (e.g., environmental control for instrumentation).

2) Selection of instrumenting equipment largely immune to such effects.

3) Use of computed corrections (computed correction should be avoided when possible, but sometimes such corrections must be made).

It should be understood that the best we can hope for from any of these methods is that they neutralize the major part of the error to a "first order" of approximation; "second order" or "residual" errors remain.

Observational: The term "observational error" recognizes that an observer possesses individual personal characteristics so that several people using the same equipment for duplicate sets of observations do not necessarily produce duplicate results. One observer may tend characteristically to rate performance lower than others do, possibly because of his own personal qualifications, standards, or experience. Tests have been conducted in which a number of persons observed the same event under carefully controlled conditions, and consistent individual differences in the recorded outcome were found. In measurements involving the timing of an event, one observer may tend to anticipate the signal and read too soon. Considerable differences are likely to appear in determinations of the difficulty of accomplishing certain tasks or the adequacy of a given procedure. Important assessments of this type should be shared by two or more observers to minimize the possibility of a constant bias.

3.4.3.3.3 Random errors. It has been found repeatedly that data from T&E show variations from observation to observation, even after close attention to all sources of error previously discussed. There is no doubt a reason, or rather a set of reasons, for these variations, but we are unable to explicitly identify them. The physical events that we are measuring in a test are affected by many factors occurring throughout the military structure or universe, and we are aware of only the more obvious. To deal analytically with the remainder, we usually must lump them together and term them "random" or "residual" errors.

The errors considered in this classification may be regarded as the residue of errors when all known systematic effects have been taken into account; hence the term "residual." Test conditions are subject to variations because of a multiplicity of small causes that cannot be traced separately. Also, when corrections are made for known effects, the correction is likely to be approximate and thus leave a small residue of error. The unknown errors are probably caused by a large number of small effects, each one variable, so that they are additive in some cases and subtractive in others in their effect. In many observations, the positive and negative effects may be nearly equal, so that the resultant error is small. In other cases, the positive (or negative) errors may dominate and thus result in a comparatively large error. If we assume the presence of a large number of small causes, each of which may give either a plus or a minus effect in a completely random manner, we obtain the condition of "scatter" around a central value. This condition is frequently encountered in T&E data; thus we are justified in using this concept as a basis for our studies of those discrepancies to which we cannot assign a known cause. We also see why the term "random error" is an appropriate description of the situation. The supposition of randomness is used in permitting correlation with the mathematical properties of probability and statistics, and thus allows analytical treatment of this type of error.

3.5 Statistical Nature of Test and Evaluation Data (See References 10–12)

The results obtained from any T&E sample, or group of samples, are relatively unimportant. What are important are the conclusions that can be drawn regarding the untested remainder of the population and the assurance that can be associated with those conclusions. The most desirable basis for establishing those conclusions consists of quantitative measurements taken under conditions that are operationally realistic. Planning for the proper use of statistics in the analysis of both quantitative and qualitative data is an important part of the T&E. Fortunately, whenever problems involve the use of data that are subject to appreciable experimental errors,* statistical methods offer a sound and logical means of treatment; there is no equally satisfactory alternative. When testers rely solely on their own judgment to assess the results of their work, they apply, or should apply, criteria similar to those which might have formed the basis of a statistical test; however, such judgments do not impose the numerical discipline of a statistical test. As long as the tester is dealing with effects that are large compared to the random variations present, such a procedure may be satisfactory. When the random variations become considerable compared to the effects being measured, such mental judgments usually become misleading and cannot be considered satisfactory. For these reasons, it is highly important that statistical methods be an integral part of the technique used in the interpretation of T&E data. Statistical methods allow the analysis and interpretation of data obtained by a repetitive operation. In many cases, the operation that gave rise to the data is clearly repetitive. This would be true, for example, in the case of the various times involved in an aircraft repair process, or the errors involved in the delivery of weapons. In other sets of data, the actual operation may not be repetitive, but it may be possible to think of it as being so. This would be true for such things as the ages of certain equipment at wear-out, or for the number of mistakes an experimental set of aircrews make on the first pass as they attempt to attack a given type of target. Experience dictates that many repetitive operations behave as though they occurred under essentially stable circumstances. Games of chance, such as dice and roulette, usually exhibit this property. Many experiments and operations with complex military systems do likewise.

For a variety of reasons, it is often impossible to obtain quantitative measurements which totally fulfill certain test objectives. Consequently, the informed assessment of the expert operator (although qualitative) is always a highly valuable test data and judgmental source. At times, quantitative measurements may be inadequate, too expensive, or insufficiently accurate—or all three—to serve as the sole basis of evaluation. Hence, planning during T&E design must include the proper use of qualitative data. Qualitative data, in some cases, might have to stand alone on their own merits and/or serve as a guide for further quantitative measurement and analysis. We should not overlook the fact that in certain instances a

---

*The term "experimental error" describes the failure of two identically treated experimental units to yield identical results.

qualitative description of what happened may be more valuable than large amounts of quantitative data. For example, an aircrew's report that an air-to-air IR missile "appeared to guide at a cloud that was approximately 25 degrees right of the target at launch" might very well be more valuable than an extensive amount of quantitative measurement of trajectory parameters. The information that the missile appeared not to discriminate against the cloud background falls into a different category—one that possibly requires further engineering investigation of the missile's discrimination capability and/or possibly a corresponding refinement of weapons delivery tactics. Thus, there are instances when unquestioning acceptance of quantitative measurements would mask or even misrepresent the actual performance of an item or system undergoing testing. While qualitative data have an important place in testing, properly measured quantitative data usually have less variability, are highly credible with most decision makers, and should be obtained whenever feasible.

The amount of statistical information that can be extracted from a test depends upon the variability of the data σ^2* and the number of observations in the testing sample n.* Variance might be thought of as a measure of the quantity of information per measurement; that is, the greater the variance, the less information per test measurement. Sample size can be thought of as the bits of information in the test. Thus, the amount of statistical information derived from any given test may be increased by increasing n or by decreasing the σ^2.

This basic notion gives rise to two principles involved in generating a greater amount of T&E information for a given cost. First, the T&E can be designed with an attempt to reduce the variability of measurements, and, second, it can be designed to increase the number of measurements of the parameters of specific interest. Because the tester pays a cost for the information obtained from the test, he should always be aware of and attempt to follow one, or a combination, of these principles in the design and execution of the test.

3.6 Typical Range System Measurement Capabilities (See References 13–33)

3.6.1 Time-Space-Position Information

Time-Space-Position Information (TSPI) is the backbone of T&E data and is normally measured and recorded on instrumented military and industrial ranges for test items of interest to allow detailed analysis of mission results. TSPI measurements provide basic position, velocity, and acceleration data along with the measurement of the occurrence of significant mission events. These data are often processed and displayed in some fashion in real time (or near real time) as the mission occurs, and are also processed into analytical formats to support postmission

---

*σ^2 = variance: the average of the square of the deviations of the observations (measurements) about their mean or average μ. $\bar{x}$ = the unbiased estimator of μ. S^2 = the unbiased estimator of σ^2, computed from a sample of obersvations, where $x_1, x_2, ..., x_n$ represent the n observations in the sample, and,

$$\bar{x} = \sum_{i=1}^{n} \frac{x_i}{n}, \quad S^2 = \sum_{i=1}^{n} \frac{(x_i - \bar{x})^2}{n-1}$$

BASIC PRINCIPLES 67

analysis. TSPI systems are generally capable of acquiring and accurately tracking items, such as missiles (with or without beacons), rockets, aircraft, nose cones, boosters, tankage assemblies, instrument packages, and debris, and of providing associated trajectory and event data. Several types of sensors are commonly used in combination with range timing and processing systems to allow the measurement of TSPI. Some of the more common sensor methods are radar, laser, time and phase difference systems, optical systems, and global position systems. All these systems have strengths and weaknesses, and can make measurements to varying degrees of accuracy.

3.6.1.1 Radar TSPI. One of the more common primary tracking radars integrated into many range tracking complexes is the AN/FPS-16 monopulse radar. The accuracy specified for this system is 0.1 mil rms in azimuth and elevation and 5 yd in range, with a signal to noise ratio of 20 db or greater. Some of these radars have been modified to provide a 1,000 nm unambiguous tracking range and may be operated at 1 MW or 5 MW peak power output.

3.6.1.2 Laser TSPI. One example laser tracking system is the precision automated tracking system (PATS). The PATS, a self-contained, mobile laser tracking system, measures the range, azimuth, and elevation of a cooperative target relative to the PATS site location. The exact PATS site location on the ground would normally be established by a first order survey. The major system components are the optical transmitter, receiver and mount, the data processing subsystem, the target acquisition aid, and the target-mounted optical retroreflector units. The transmitter illuminates the target, an optical retroreflector array mounted to the tracked object. The receiver measures the time interval between transmitted pulse and received pulse to determine range data. Angular data are obtained from the shaft angle encoders on the optical mount. The data subsystem processes, time-correlates, and records the data.

3.6.1.3 Time and phase difference systems for TSPI. Several types of TSPI systems are based on time and/or phase difference, and other measurements among an array of tracking stations to provide for a multilateral solution. One example system is the Air Combat Maneuvering Instrumentation System (ACMI) located at numerous testing and training facilities throughout the world. This system was designed to provide aid in reconstructing air combat engagements for training purposes. In addition, because of specified accuracy requirements, it can also support selected test activities. The ACMI provides all-attitude, interaircraft data within a specified airspace. These data are processed to provide a pseudo three-dimensional ground display of aircraft maneuvers, as well as aircraft and interaircraft parameters. In addition to a real time display, a complete, permanent record of the air-to-air engagement is provided so that simultaneous training and evaluation can be achieved during postmission review. A later version of this type of system, called the Red Flag Measurement and Debriefing System, which combines measurements for both air and ground players, is located at Nellis AFB, Nevada.

3.6.1.4 Optical TSPI. One basic instrument used to obtain precise optical solutions to the TSPI problem is the cinetheodolite. This cinetheodolite was developed by Contraves AG, Switzerland, to meet the demand for an instrument of high dynamic accuracy. The pointing accuracy of Contraves cinetheodolites has been demonstrated to fall between 5 to 15 seconds of arc. A complex of six cinetheodolites has demonstrated target position to +1.5 feet, velocity +1.5 feet per second, and acceleration of +2.5 feet per second per second in the test of a specific aircraft. The cinetheodolites are generally employed in an array of three or more to allow a multilateral solution to the TSPI problem.

3.6.1.5 Global Positioning System for TSPI.

The Global Positioning System (GPS) has worldwide application and consists of three segments: 1) space segment, 2) ground control segment, and 3) user segment. The operational space segment consists of 24 satellites with synchronized atomic clocks in six orbit planes. This results in from 6–11 satellites visible above 5 degrees at any time from any point on the Earth's surface. At least four satellites must be in view of the user segment for a good solution (three if the user knows his altitude). The ground control segment monitors the satellite navigation signals, provides commands to the satellites, and operates the overall system through a master control station. The user segment measures the time of arrival of specially coded signals from the four satellites and solves four simultaneous equations for four unknowns, three for his position and one for the bias of his quartz clock. The GPS user segment can be internally installed on weapons systems or carried in an external pod. The standard GPS system is advertised to achieve a position accuracy of approximately 10–15 meters. A more accurate variant of the system is called differential GPS. It employs a surveyed reference ground station to measure and transmit principal errors to the user segments which, in turn, correct their observed positions and obtain an accuracy of 1 to 5 meters. The obvious advantage of GPS as a TSPI system is its immediate coverage of test areas and environments throughout the world.

3.6.2 Electromagnetic Environment Testing

A number of electromagnetic test environments available in the U.S. Department of Defense support developmental and operational agencies in the evaluation of electromagnetic warfare (EMW) devices, systems, tactics, and techniques. These test environments provide a sophisticated range complex of highly instrumented systems operating in different frequency bands and modes supporting a flexible test facility for the evaluation of electromagnetic warfare devices and techniques. The complex of ground systems is integrated into an operable facility with a complex instrumentation system, a ground communications system (intersite and intrasite), and an air-to-ground communications system. Normally, a centralized computer laboratory is interconnected with the system and has a real time capability to receive, record, process, and transmit data required in support of test missions. The overall environment consists of range tracking radars and a simulated air defense system containing early warning, threat acquisition, and threat tracking radar systems. The environment also contains a capability for end game calculations and telemetry. All telemetry sites have the capability of receiving, recording, and demultiplexing telemetry data as completely independent stations. The range signal and control system ties the stations together as one integrated system and can present the telemetry data in real time. Range telemetry operations in remote areas are accommodated by mobile receiving stations. The system also contains a Frequency Control and Analysis (FCA) Facility which can monitor and record signals in the radio frequency bands. Desired radio frequency (RF) signals of guidance, control, destruct, and others are monitored and recorded for postmission analysis, as applicable. Interfering or undesirable RF signals and/or radiation are detected, located, and reported to enable the suppression of undesirable radiation interference. In addition to the fixed facilities, a mobile FCA van supplements this capability and assists in locating undesirable signals.

3.6.3 Engineering Sequential Photography

This is generally defined as the recovery of information on photographic film which also records a time base to provide event versus time at recording rates higher than can be obtained with video or cinetheodolites and which can be correlated with other instrumentation. For this purpose a wide variety of camera lens combinations is generally available for fixed installations or to be used on tracking mounts as determined by the mission support requirement. Typical cameras are the Mitchell, Fastax, and Nova. A specialized camera used for terminal effects of aircraft-released munitions is the CZR Ribbon Frame Camera. These cameras are usually mounted in trailers and can be arranged at specified intervals along a prescribed flight path or even installed aboard aircraft. The shutters are activated in proper sequence for full coverage of the event.

3.6.4 Range-Timing Systems

Range-timing systems are used to provide a common time among the various instrumentation systems. Timing system accuracy depends upon the specific systems. The basic timing system for each ground-based system usually contains a timing receiver, a dual-trace oscilloscope, and an interrange instrumentation group (IRIG) B time code generator. National Bureau of Standard Time broadcasts (WWV) are normally used for time synchronization. Additionally, when time insertion is required into certain airborne instrumentation, a portable, battery-powered time code synchronization unit can be employed. This unit contains a time code generator, an oscillator, and a remote control readout unit. The time code generator installed in the airborne unit is normally configured to derive time synchronization from the oscillator unit.

3.6.5 Aerospace Environmental Support

Most range complexes provide the means of obtaining needed meteorological and other geophysical measurements during tests. These data are generally gathered throughout the land and sea range complex and span the atmosphere from the Earth's surface to heights above 100,000 feet. Measurements are made to some extent by local range personnel, but are furnished primarily by the Air Weather Service (AWS).

3.6.6 Calibration and Alignment

The processes of calibration and alignment are necessary support functions for many electronic instruments in a range system test complex. The calibration process attempts to remove any tracking or timing biases present in the TSPI system. For a system attempting to collect TSPI on airborne objects, an example calibration process might consist of flying an aircraft with a beacon and a synchronized strobe light over the range area of interest and collecting synchronized data from all the available range systems (radars, lasers, time/phase difference systems, and the cinetheodolites). A comparative calibration process using the most accurate system as the standard might then be applied to identify and remove any biases revealed in the tracking system. A geometric advantage is often gained by flying the test source closer to the calibrating instrumentation than the instrumentation being calibrated. Other aligning and calibrating equipment is also usually available for radar boresighting and precision leveling. Boresighting can be performed with cameras, small telescopes, and closed circuit television systems which have been aligned with the principal axis of the radar antenna. Star calibration is also common for both radar systems equipped with optics and pure optical trackers. All these methods, including data processing schemes, should be applied as appropriate to help reduce any systematic errors present in the range measurement systems and to achieve the required range measurement accuracy.

References

[1] Bilodeau, E.A., and Bilodeau, I.M., "Motor Skills Learning." *Annual Review of Psychology*, 1961.

[2] Killion, T.H., "Electronic Combat Range Training Effectiveness." *AFHRL/TSR*, 1986.

[3] Churchman, C.W., and Ratoosh, P., *Measurement Definitions and Theories*. New York: John Wiley and Sons, Inc., 1959.

[4] Malvino, A.P., *Electronic Instrumentation Fundamentals*. New York: McGraw-Hill Book Company, 1967.

[5] Herrick, C.N., *Instruments and Measurements for Electronics*. New York: McGraw-Hill Book Company, 1972.

[6] Cox, D.R., and Miller, H.D., *The Theory of Stochastic Processes*. New York: John Wiley and Sons, Inc., 1965.

[7] Liddicoat, R.T., and Potts, P.O., *Laboratory Manual of Materials Testing*. New York: The Macmillan Company, 1957.

[8] McMaster (ed.) for the Society for Nondestructive Testing. *Nondestructive Testing Handbook*. New York: The Ronald Press Company, 1963.

[9] Perry, C.C., and Lissner, H.R., *The Strain Gage Primer*. New York: McGraw-Hill Book Company, 1962.

[10] Brownlee, K.A., *Statistical Theory and Methodology in Science and Engineering*. New York: John Wiley and Sons, Inc., 1960.

[11] Hodges, J.L., Jr., and Lehmann, E.L., *Basic Concepts of Probability and Statistics*. San Francisco, California: Holden-Day, Inc., 1964.

[12] Hoel, P.G., *Introduction to Mathematical Statistics* (4th ed.). New York: John Wiley and Sons, Inc., 1971.

[13] Ehling, E.H., *Range Instrumentation*. Englewood Cliffs, New Jersey: Prentice-Hall, 1967.

[14] Herrick, C.N., *Instruments and Measurements for Electronics*. New York: McGraw-Hill Book Company, 1971/72.

[15]Seminar Proceedings, *Laser Range Instrumentation*. U.S. Army White Sands Missile Range, El Paso, Texas: 1967.

[16]Society of Photo-Optical Instrumentation Engineers, *Photo and Electro-optics in Range Instrumentation*. Bellingham, Washington: 1978.

[17]Strobel, H.A., *Chemical Instrumentation; A Systematic Approach* (2nd ed.). Reading, Massachusetts: Addison-Wesley Publishing Co., 1973.

[18]Harvey, G.F. (ed.), *Instrument Society of America, Transducer Compendium* (2nd ed.). New York: 1969.

[19]Instrument Society of America, *National Telemetry Conference*. New York: American Institute of Electrical Engineers, 1959.

[20]Jones, E.B., *Instrument Technology*. London, England: Butterworths Scientific Publications, 1953.

[21]Gouw, T. (ed.), *Guide to Modern Methods of Instrumental Analysis*. New York: Wiley-Interscience, 1972.

[22]Skoog, D.A., *Instrumental Analysis* (2nd ed.). Philadelphia, Pennsylvania: Saunders College, 1980.

[23]Society of Motion Picture and Television Engineers, *Instrumentation and High-Speed Photography*. New York: The Society, 1960.

[24]Lion, K.S., *Instrumentation in Scientific Research*. New York: McGraw-Hill Book Company, 1959.

[25]Jaques, C.N., *Instrumentation in Testing Aircraft*. London, England: Chapman and Hall, 1957.

[26]Instrument Society of America, *Instrumentation in the Aerospace Industry*. Research Triangle Park, North Carolina: ISA, 1978 (80).

[27]Washburn, B. (ed.), *Proceedings of the 22nd International Instrumentation Symposium*. San Diego, California: ISA, 1976.

[28]U.S. National Bureau of Standards, *Guide to Instrumentation Literature*. Washington, D.C.: U.S. Government Printing Office, 1955.

[29]Baumol, W.J., *Economic Theory and Operations Research* (2nd ed.). Englewood Cliffs, New Jersey: Prentice-Hall, 1965.

[30]Dorfman, R., "The Nature and Significance of Input-Output." *Review of the Economics and Statistics*, Vol. 36, May 1954, pp. 121-133.

[31]Brunk, H.D., *An Introduction to Mathematical Statistics* (2nd ed.). Waltham, Massachusetts: Blaisdell Publishing Company, 1965.

[32]Charnes, A., Cooper, W.W., and Henderson, A., *An Introduction to Linear Programming*. New York: John Wiley and Sons, Inc., 1953.

[33]Bellman, R., *Dynamic Programming*. Princeton, New Jersey: Princeton University Press, 1957.

4
Modeling and Simulation Approach

4.1 General

The primary purpose of modeling and simulation (M&S) in military test and evaluation (T&E) is to provide an analytical representation (or framework) that can be used to investigate the basic functioning of a system in its combat environment. Often the system that is being investigated is extremely complex, the environment within which the system must operate is extremely complex, and the mission that must be performed is extremely complex.

Modeling, in its simplest terms, can be defined as a collection (or set) of assumptions that describe how something works. The model provides a representation of the system and its environment that allows study to predict how the real system would work under similar circumstances in the real world. Iconic models may be in terms of scaled physical objects, whereas abstract models are often represented by mathematical equations and logical relationships, and visual models are pictorial or graphical representations. When these models represent extremely complex situations, a computer is often used to numerically evaluate the models and to estimate characteristics and solutions for the problems being addressed. This method of model solution is referred to as "simulation" and the models used in this process are called "simulation models," "simulations," or "simulators." The computer approach to the study of these highly complex models allows a vast number of input and output situations to be economically addressed in a relatively short period of time.

M&S provides a scientific approach to address these highly complex problems by allowing the problem to be defined and viewed in its entirety as well as decomposed into its vital parts and addressed at various levels in terms of what is important to the specifics of the analysis at those levels. Modeling enhances decomposition of the overall T&E problem and aids in the decision process to optimize the best use of the available T&E facilities and environments (i.e., laboratory tests, hardware-in-the-loop tests, installed test facilities, system integration tests, open air test ranges, etc.). When all of these analytical methods are combined in an optimum manner, they provide a powerful tool in the acquisition, T&E, and application of modern weapons systems.

4.2 Model Concept (See References 1–37)

The primary purpose of a model or simulation is to provide a representation of a system or relationships which can then be used to investigate the basic functions of the system or relationships. A model, in a broad sense, allows the abstraction of a study of a problem in such a way that 1) fundamental processes of the problem and their influences and relationships can be better understood, 2) predictions or extrapolations from the outcomes of current problem conditions to potential outcomes of future problem conditions can be made, and 3) relative comparisons of alternative systems or solutions in meeting stated goals and objectives can be made.

Modern modeling has been extended to the examination of extremely complicated problems ranging from military war games to vastly complex systems like those proposed for ballistic missile defense. Modeling has been defined as encompassing "...the development of axiomatic systems, the formulation of social theories, the derivation of physical first principles, and the drafting of laws. It is thus an art natural to mankind, and focusing this art on the domain of military science conceptually encompasses the principles of war, strategy, tactics, the laws of warfare, and the structure of military forces" (Reference 1).

Scientists and engineers employ models as a means of mathematically or logically expressing the relationships between variables. Figure 4-1 is a simplistic representation of this process. The simple model depicted in Figure 4-1 can be thought of as somewhat analogous to a scientific hypothesis based on a priori knowledge which is accepted as correct and from which inferences can be drawn. The model may contain axiomatic systems, social theories, and physical first principles and laws, and always requires certain assumptions. When certain input data and conditions (e.g., scenarios and data bases) are provided to the model, it operates on the inputs to produce certain outputs that can be described in terms of desired measures of effectiveness (MOEs), measures of performance (MOPs), etc. An MOE is an operational term defined as the measure of a system's operational capability or a characteristic that indicates the degree to which it performs a task or meets a requirement under specified conditions. An MOP is a developmental or engineering term defined as a measure of a system's engineering performance or a characteristic that indicates the degree to which it performs a function or meets a contractual specification. The form of the model used to address the MOEs and MOPs may be analog, digital, hybrid, man-in-the-loop, hardware-in-the-loop, or some other variant; and it may be deterministic, probabilistic, or a combination of both (see References 2–24).

A model or simulation often takes on a hierarchical structure for application to very complex problems. This structure may take the form of a very detailed model which addresses each of the fundamental processes of a problem. The outputs from these models are used to provide input to the next level of the hierarchy which may treat several of these fundamental processes and their influences and relationships to each other. The outputs from this level of the hierarchy feed the next level, etc. Each level of the hierarchy addresses a larger problem but usually in more general and less specific terms. This facilitates treating extremely complex problems in a more structured, rigorous, and transparent manner than would otherwise be possible.

Modelers and simulators sometimes describe their applications of this hierarchical approach to modeling in terms of levels of analysis. Over the past decade, some analytical communities have adopted four levels of model analysis which are defined in "A Methodology for the T&E of Command, Control, Communications, and Intelligence (C3I) Systems" (draft), published by the Deputy Director, Defense Research and Engineering (Test and Evaluation) as part of an Implementation Program Plan (IPP) on "The Use of Modeling and Simulation to Support the T&E of Command, Control, Communications, and Intelligence (C3I) Systems,"

MODELING AND SIMULATION APPROACH

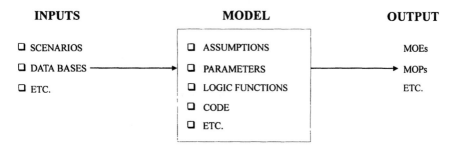

Figure 4-1. Simplistic Representation of the Modeling Process

11 April 1990. Table 4-1 illustrates these levels for the investigation of the contribution of electronic combat in air warfare.

A Level I model, for example, by definition "examines the performance of an individual engineering subsystem or technique in the presence of a single threat." The Level I model requires as input engineering parameters and characteristics of the subsystem in sufficient detail to ascertain the actual end game effects of the electronic countermeasures (ECM), chaff, flares, and/or maneuvers employed by a single aircraft on the performance of the system or missile. These effects can be determined internal to the model and are not required inputs. By definition, Level I outputs can be combined and fed into Level II analyses to evaluate the installed, aggregate performance of a number of specific engineering subsystems or techniques against a specific threat. This definition implies that the Level II model has

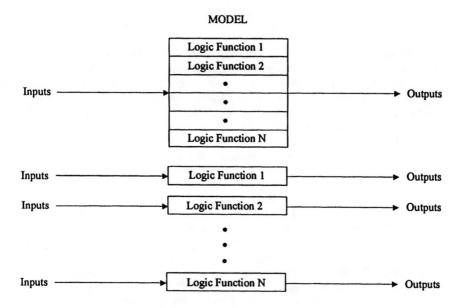

Figure 4-2. Decomposition of the Model

Table 4-1. Example Levels of Model Analysis

Level	Short Name	Key Process	Modeling Output
I	Engineering	Electromagnetic Signal Flow	Electronic Combat (EC) Performance
II	Platform	Weapon System Engagement Performance	Weapon System Performance
III	Mission	Multiweapon System Operations	Weapon System Effectiveness
IV	Theater/Campaign	Force-on-Force Targeting	Force Effectiveness

the appropriate logic to treat the effects of ECM, chaff, flares, and/or maneuvers of the individual systems and how they combine and behave within specific threats. Levels III and IV are defined to aggregate at higher orders of complexity and conflict.

4.2.1 Decomposition of a Model

Models and simulations, regardless of the level of complexity, can be thought of in terms of a number of interrelated component parts which function together to take input information and operate on it according to specific model functions. Model functions encompass all things internal to the model (e.g., assumptions, logic, algorithms, parameters, and coding). During design, an attempt is made to structure model functions in an optimum manner to produce the desired outputs. Decomposition of a model can be thought of as a reversal of the initial design process.

The understanding of a system is greatly facilitated by decomposing its model into component parts and subparts. Decomposition of a model is analogous to the application of engineering systems theory, whereby a total system is broken down into a number of subsystems described by transfer functions. In fact, engineering systems theory can often be employed in this process. Each subsystem is then characterized by its own model functions and associated inputs and resulting outputs. Full system examination is facilitated because a large intractable problem is broken up into a number of smaller manageable problems.

The decomposed model can be examined in terms of model functions as depicted in Figure 4-2.

Decomposition allows the modeler to view and better understand the operation of the internal workings of the system. For example, our model might be the flyout model of a specific missile. Our first level of decomposition could be examining the model function for the pitch steering channel of the missile. Decomposition might be achieved to lower levels as depicted in Figure 4-3. Here, we take the specific model functions and break them down into components. The model function for missile pitch steering might be further decomposed into seeker unit, processor, torque converters, and so forth. Neither of these two graphic depictions is intended to imply that certain parallel and/or serial relationships within the model not be accounted for (e.g., the adequacy/completeness of defining subsystem interdependency). In fact, the model functions themselves (and/or compo-

MODELING AND SIMULATION APPROACH

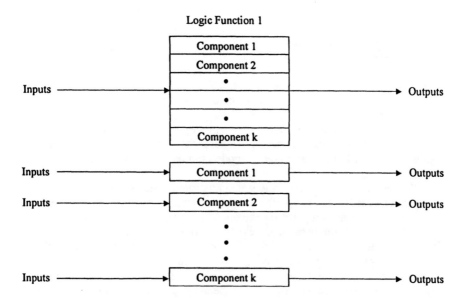

Figure 4-3. Decomposition of Model Functions

nents) and their interactions within the model must account for subsystem interdependency to achieve the appropriate output.

Model outputs are generally selected based on how well they represent the military performance and utility of a system (i.e., MOEs and MOPs as defined earlier). MOEs and MOPs represent different sets of system measures of interest from the perspective of operators and developers, respectively. As depicted in Figure 4-4, these two sets are not necessarily mutually exclusive. For example, it is highly probable and desirable that both operators and developers have a keen interest in some of the same measures for certain systems and situations. Also, some form of functional relationships normally exist between the MOEs and MOPs of interest, even though they may not be well defined or explicit.

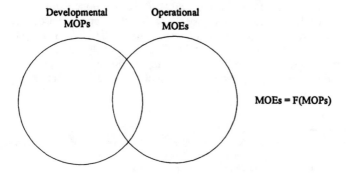

Figure 4-4. Output MOEs and MOPs

4.2.2 Applications

Models and simulations can play an important role throughout the life cycle of a weapons system. The simplified relationship in Figure 4-5 depicts how a model and simulation can be used to study a system. Modeling and simulation can be used to help identify the most promising options to achieve a given level of effectiveness and to aid in identifying the characteristics that systems and components should possess. Once the components are designed and built, they can be tested and compared against model results. Both the components and models can be improved through an iterative comparative process. Once the system is designed, built, and tested, it too can be compared against the system model results and both can be improved through an iterative model, build, test, feedback, and improvement process.

When employed during the early stages of system definition, models and simulation can be used in T&E to perform early operational assessments, and to examine various concepts and performance options for satisfying an operational need without having to actually build and evaluate prototypes for each option. Later on in the acquisition process, M&S may allow the developer to better define those areas needing more extensive investigation or refinement.

M&S provides a convenient vehicle for planning and designing physical testing. It can be used to better define the scope of physical testing. Once test results are available, it can be used to extend those results to more complex situations.

Models have several distinct advantages which make them attractive for use in evaluating systems. Once the model has been developed and refined to the point at which users of its output have sufficient confidence in its performance, it can be used to conduct extensive parametric analysis in an expeditious and efficient manner.* That is not to say that the developing of the model in the first place is not necessarily time consuming and costly. However, once available, the model provides a convenient vehicle to do extensive "what if" type analysis. Provided appropriate models with the required credibility are available, they can also be used to perform higher level survivability analyses and force structure analyses.

4.3 Verification, Validation, and Accreditation

When applying models and simulations, it is always important to know how much credence can be placed in their results. Verification, Validation, and Accreditation (VV&A) is the formal process within the Department of Defense by which one attempts to establish this creditability. This chapter provides broad guidance for the VV&A of models and simulations. It also includes information for developing a detailed model VV&A plan, conducting an appropriate model VV&A, and communicating the results of such activities to officials responsible for model accreditation.

---

*Parametric analysis is a process whereby the output of a system is examined when a single test parameter is varied over a range of values while all other test parameters are held constant. It helps the modeler to understand how the problem solution changes as a function of the individual input parameters. The overall process is referred to as sensitivity analysis because it reveals the sensitivity of the output of the model to changes in input.

MODELING AND SIMULATION APPROACH

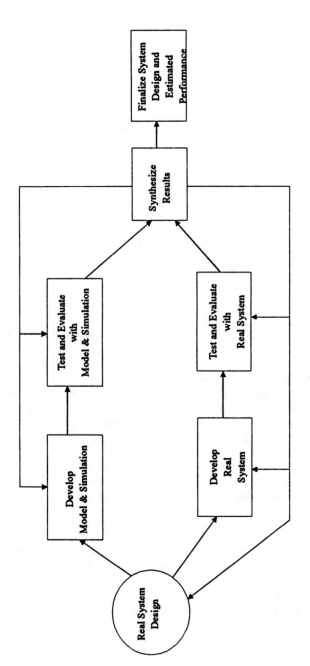

Figure 4-5. Simplified Example of How a Model and Simulation Can Be Used To Study a System

Models vary in the degree of accuracy with which they depict the true state of nature. All are subject to error, in some cases large and in others small. However, since accuracy is a relative concept, a model can be considered accurate—that is, good enough—if it is able to predict the variable of interest (e.g., MOEs or MOPs) and do so with an error that will be acceptable for practical purposes. Consequently, one may rightfully ask the question "How do I know when I should believe the output of a given model?" The process whereby one answers this question is called "VV&A." Because the uses of models and simulations cover a wide range of purposes, complexities, and activities, the degree to which VV&A can be achieved in a practical manner for each model use will vary. Moreover, since many models are improved as they are used, VV&A of a model should be conducted throughout its life cycle to accumulate information regarding its accuracy and potential uses.

Model verification can be distinguished from model validation in that *verification compares the model against its design specifications*, whereas *validation compares the model against the real world*. The term "real world" is used herein to characterize actual objects or situations, or our best representation of them. Model validation can be distinguished from model accreditation in that valuation is a comparison process whereas *accreditation is a decision to use a model based on some level of verification and validation* (V&V) (Reference 2). The effort put forth in VV&A should quantify and qualify the model accuracy and precision associated with each MOE and MOP of interest. Once this information is available and documented, the accreditor can examine it in terms of the needs of his/her analysis problem to establish whether the M&S should be accredited for that specific use.

The VV&A of a model is an extremely important process. It should be systematic, tractable, repeatable, and describable. It should establish the degree to which a model is an accurate representation of the real world. If the specifications to which a model has been developed accurately reflect the real world, the processes of V&V should yield essentially the same results. However, there may indeed be differences between the model and the real world—some intentional and some not intentional. The VV&A process formally identifies and establishes the degree of the important differences. Finally, VV&A should be accomplished from the perspective of the intended uses of the model.

Embedded in the VV&A process is the implied responsibility to identify and document both the proper use and the potential misuse of a model. Ultimately, for each model usage, VV&A is accomplished and documented for the specific classes of objects [e.g., scenario(s), mission(s), and weapon systems], specific levels of investigation (e.g., end game, platform performance, and campaign), specific inputs and conditions (e.g., parameters and data bases), and the specific outputs of interest. As stated earlier, in military M&S, the outputs of interest derived from the models are generally described in terms of MOEs and MOPs.

Because of the broad, all-encompassing definition of models and simulation, the specifics of each VV&A effort must be tailored to the given problem the model is being used to solve, technical situation, or operational application. Consequently, model VV&A can be limited in scope. Although tailoring for the specifics of the problem is required, the basic VV&A framework put forth in this chapter is generally applicable to all types of models and simulations.

Model inputs, outputs, and internal functions all vary in the degree of accuracy

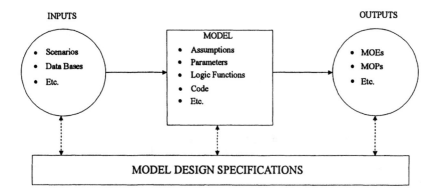

Figure 4-6. Potential Comparisons for Verification

with which they represent or describe the known or agreed upon state of nature. The model V&V process systematically identifies and documents this degree of accuracy. The accreditor must then decide whether the documented accuracy is sufficient to address the specifics of his or her problem.

4.3.1 Verification

Verification is the process of determining that a model implementation accurately represents the developer's conceptual description and specifications. As described earlier, decomposition of the model greatly enhances the V&V process. The verification process compares both the logic (i.e., basic equations, algorithms, etc.) and the code/object implementation of the model against the design specifications. This process is depicted in Figure 4-6. The model design is a structural description of the model. It specifies the elements to be modeled, their functioning, and the appropriate interrelationships. The model code is written by a computer software programmer and is the instructional language that is assembled or compiled by the computer and used to instruct the computer as it executes the program to arrive at a model solution. During verification, it is important that the basic model logic functions, data bases, code, and documentation all be compared and verified against the design specifications for the model.

4.3.2 Validation

Validation has often been used in the broadest sense as a measure of how much credence one should place in the output of a given model. The term is defined here more narrowly to refer to the process of checking out the model against real world information. Supposedly, if the model has been validated, one can believe its output. But what is the process of validation? Who does it? When does it take place? From our perspective, validation is part of determining the credibility of the model for a particular set of uses, not the totality of all possible uses. The validation process is designed to increase knowledge about how well the model represents reality and to aid users and decision makers in determining whether the results obtained from the model sufficiently represent what they would observe if the situation or entity were actually played out in the real world.

There are at least two basic schools of thought on validation Most approaches to validation apply one or the other, or some variation or combination, of these methods. The first deals with an a priori examination of the input data, the basic principles of the model, and its assumptions. Are they exhaustive enough, and are they reasonable for the types predictions being made? The second school deals with the collecting of real world data and compares it with the output of the model to measure how well it predicts. In complex and difficult modeling situations, the requirements for the second school of thought can almost never be met. This is because of the inability to actually conduct a real world test of the item itself (i.e., lack of specific threat knowledge, lack of threats, insufficient environment, safety, etc.). Strangely enough, it is this inability to ascertain real world knowledge that often drives us to use models and simulation in the first place.

There are numerous dimensions by which one can partition the comparisons that can be made in validating a model. The spectrum of potential comparisons includes elements of the inputs, the outputs, and the model itself. As illustrated in Figure 4-7, we have partitioned the outputs of the model into a domain called "output validation" and the model inputs along with the model internal functions into a domain called "structural validation." Validation of complex models should include some combination of both structural and output validation.

Output validation is the most credible form of validation and should be conducted at the full model level to the extent possible. When it is not possible to conduct output validation for the full model, the model can be decomposed and output validation accomplished for parts of the model to the extent practical.

Structural validation should also be accomplished for those aspects of the model critical to the model's use. The planned application of the model should always be a key driver in establishing the details of its specific validation.

4.3.2.1 Output validation. In general, output validation usually contributes the most convincing evidence for establishing the credibility of a model. It is the process of determining the extent to which the output (outcomes or outcome distributions for the model and/or submodels) represents significant and salient features of real world systems, events, and scenarios it is supposed to represent. Output validation involves collecting real world data and comparing them with the output of the model (i.e., MOEs and MOPs of interest) to assess how well the model results reflect those of the "real world" (i.e., the actual system or process being modeled).

In extremely complex and difficult modeling situations, the requirements for comparing real world results and model results may be difficult, if not impossible, to meet. This difficulty usually arises from the inability to actually conduct a realistic exercise of the system being modeled because of certain constraints (e.g., insufficient environmental conditions, resources constraints, safety considerations, insufficient threat representation, and inability to replicate real world conditions). Even though comparison at the output level of the full model may not be possible, it is often possible to make comparisons at lesser levels or with a scaled-down version of the system. Ironically enough, it is this inability to replicate (or even understand) the real world that usually drives one to the use of a model in the first place. Quantitative approaches to the comparisons usually provide the most con-

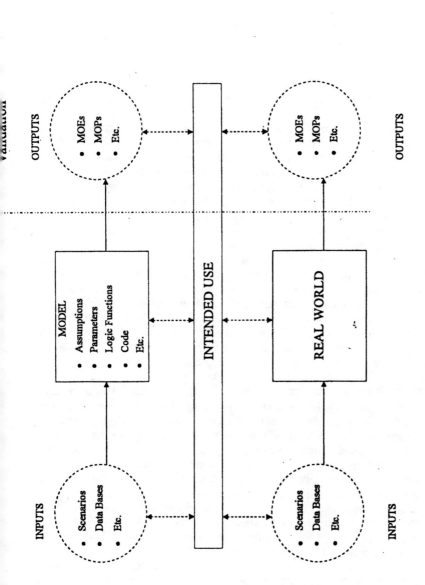

Figure 4-7. Potential Comparisons for Validation

vincing evidence about model validation. Quantitative output validation, however, requires that the outputs of the model be observable and measurable in the real world.

Qualitative assessments made by operational and technical subject matter experts are important inputs to the validation process. As discussed earlier, under certain conditions, quantitative comparisons may be prohibitive or limited. "Face validation" by experts and other qualitative methods (e.g., the use of focused-group interviews or a modified Delphi technique) for obtaining expert opinions on critical model issues should be applied. Findings from the qualitative methods should be used to supplement and reinforce the available quantitative comparisons.

Model outputs are generally selected based on how well they represent the military performance and utility of a system (i.e., MOEs and MOPs as discussed earlier). MOEs and MOPs represent different sets of system measures of interest from the perspective of operators and developers, respectively. As discussed earlier, these two sets are not necessarily mutually exclusive. For example, it is highly probable and desirable that both operators and developers have a keen interest in some of these same measures for certain systems and situations. Also, some form of functional relationships normally exist between the MOEs and MOPs of interest, even though they may not be well-defined or explicit.

It is extremely important that observations and measurements made in the real world be executed in such a way that they accurately represent the outputs of interest. This implies the application of the scientific approach to testing and experimenting and the inclusion of quantitative and qualitative statistical comparisons where appropriate. Such comparisons may be made based on data points, intervals, and distributions, and may involve both absolute and relative values. Data from testing should be model-compatible to the extent possible so the model-test-model approach to development can be implemented.

4.3.2.2 Structural validation.

Structural validation deals with an a priori examination of the model input data, the basic principles of the model, and its assumptions to determine the degree to which they are complete, logically consistent, and reasonable for the types of uses envisioned. To a large extent, structural validation is designed to increase the knowledge and confidence of developers, users, customers, decision makers, and independent reviewers in the model results by demonstrating that the model has internal integrity. Structural validation is the process of determining the extent to which the input data (e.g., scenario(s), mission(s), and data bases) and the model (e.g., assumptions, logic, algorithms, parameters, and coding) represent the significant and salient features of real world systems, events, and scenarios. Model decomposition greatly aids in this process.

Structural validation may be the primary form of validation that can be accomplished with extremely complex models, especially when observations and measurements of the outputs of interest are not possible or practical in the real world. Structural validation can occur through empirical measurement and comparison, as well as theoretical examination based on physical first principles, logic and axiomatic systems, and other scientific laws. Such comparisons can include qualitative approaches, such as expert opinions, or statistical tests based on both quantitative and qualitative data. As one proceeds to greater refinement and depth in structural validation, several dimensions on which to base these comparisons may

MODELING AND SIMULATION APPROACH

Table 4-2. Examples of Information Sources

- Functional Experts
- Operational Experts
- Scientific Theory (physics, engineering, behavioral, etc.)
- System Design Specifications and Information
- Laboratory Measurements (components, response functions, etc.)
- Special Test & Training Facility Measurements (hardware-in-the-loop, antenna patterns, radar cross section, infrared, millimeter wave, human factors, decision making, etc.)
- System Level Measurements
  - Developmental Test and Evaluation
  - Operational Test and Evaluation
- Live-Fire Test Measurements
- Combat History and Measurements
- Other Accredited Models

be examined. For example, one might pursue validating critical parts of the model structure in terms of theory, performance against other accredited models, scaled-down or limited component testing, and any available combat history relating to the specific portion of the model being addressed.

4.3.3 Sources of Information

Model VV&A requires the use of a broad range of sources of information (see Table 4-2). This information is both qualitative and quantitative and can vary from the opinion of functional experts, operational experts, and scientists to comparisons based on precise deterministic and probabilistic measurements. Scientific theory, system design information, and laboratory measurements can provide essential structural information. Special test and training facilities measurements, as well as system level measurements during developmental and operational test and evaluation, are excellent sources. Finally, live-fire test measurements, historical information and combat measurements, and data (which include both pertinent information from VV&A and relevant output data) from other accredited models used for similar applications are potential sources of validation information.

4.3.4 Sensitivity Analysis

The sensitivity analysis is a critical part of VV&A. Sensitivity analysis is a formal examination of how output variables of the model respond to changes in inputs, assumptions, parameters, and critical logic functions. Sensitivity analysis should be conducted on the total model and on all its decomposed parts. Sensitivity analysis can be used to check for prior responses to input variables and to identify marginal break points and special limiting values. It can be used to understand better how the model works and to help identify errors in the model structure and/or code. However, sensitivity analysis is limited to telling you only what the model is sensitive to; one has to go further with the comparative validation process to ensure that the model sensitivities are indeed representative of the real world.

A vital part of sensitivity analysis is to help one understand where the model results are extremely sensitive to changes in model algorithms, input data, parameters, and/or assumptions. These sensitivities should be of paramount interest to those who have to validate and accredit a model as well as those who must rely heavily on the model output for important decisions. This discussion addresses the situation where a change in a given part of the model is found not to make a significant or pronounced change in the major output of interest. For example, it might be that doubling the reliability of a given subsystem will change the overall total mission reliability only slightly because that particular subsystem does not really make a difference in being able to successfully complete the mission.

4.3.5 Tasks for Model Verification, Validation, and Accreditation

Example tasks that should be accomplished and the results documented during model VV&A are discussed below in terms of preparing for and conducting the VV&A for the model structure and output. These specific tasks are listed in Table 4-3.

During preparation for model VV&A, it is critical to specifically define the problem to be modeled and addressed (that is, the problem or class of problems to which the model is being applied). Definition of the problem will, to a large extent, establish the intended application of the model. It is important to remember that model VV&A is accomplished for a particular problem (application) or class of problems. One must develop and/or select the appropriate scenario(s) and mission(s) to address the problem and assess their realism in terms of the real world. The MOEs and MOPs required to address the specifics of the problem will be the primary model outputs examined. It is important at this stage to take into account whether or not this model will be used in a stand-alone role or as one level in a model hierarchy. This will impact the inputs and outputs and how model realism needs to be addressed.

Once the above tasks are completed, output and structural examination should take place. Generally, the amount of output examination that practically can be accomplished will influence the amount of structural examination necessary. Quantitative output examination normally is performed as a comparative test or experiment which provides a quantitative assessment of the agreement of the model with the real world. Sensitivities of the model output MOEs and MOPs to inputs, critical model logic, and assumptions should be identified and quantified to the extent practical. When output VV&A cannot be performed at the full model level, the model should be decomposed and structural examination conducted. As discussed earlier, this involves a comparison of input parameters, data bases, assumptions, and model functions with the design specifications for verification and with the real world for validation. Sensitivity analysis should always be a key tool during both structural and output VV&A.

4.3.6 Stakeholders in Model Verification, Validation, and Accreditation

The model developer, user, independent reviewer, customer, and decision maker all have an interest and responsibility in the VV&A of a model. If a new model is being developed for a given use or class of uses, the model developer should verify and validate the model to the extent practical, and document those results. Verification should be performed routinely as part of the programming and checkout

MODELING AND SIMULATION APPROACH

Table 4-3. Example Tasks for Conducting Model VV&A

Category	Tasks
Preparation	• Define specific function/system to be modeled and addressed. • Develop/select level of model required (end game, one-on-one, campaign, etc.). • Develop/select scenario(s), mission(s), etc. (address reasonableness versus real world). • Determine whether model will be used as stand-alone or as one level in a hierarchy. • Identify specific model output MOEs and MOPs required to address the problem. • Identify input from and output to other models. • Select and implement the appropriate category or categories of VV&A.
Output	• Quantify agreement of output MOEs and MOPs of interest versus specifications for verification and real world for validation. • Quantify sensitivities of model output MOEs and MOPs of interest to inputs, critical model logic, assumptions, and parameters. • Conduct face validation and other appropriate forms of expert qualitative assessments. • Compare input scenarios, parameters, and data bases versus design specifications and real world. • Address adequacy of inputs from other models and outputs to other models. • Address assumptions versus design specifications and real world.
Structural	• Address total and decomposed model functions versus design specifications and real world. • Address sensitivities of model output MOEs and MOPs of interest to inputs, critical model logic, assumptions, and parameters. • Address interdependency of logic functions. • Address adequacy/completeness of model logic. • Address adequacy of model in context of model hierarchy.

Based on V&V results, assess whether model and simulation is sufficient for the problem being addressed and should be accredited for the specific applications.

phases of a model's development. In the early stages, the validation effort likely will be more structural- than output-oriented. When a user other than the developer selects and applies a model, the user inherits a responsibility for properly applying the model as well as conducting any additional V&V necessary for the problem at hand. (The user will also want to ensure that the model or simulation has been accredited for his/her application.)

When critical issues are to be addressed by M&S, it is beneficial to have someone other than the developer and user (i.e., an independent reviewer) conduct additional model V&V. Independent V&V is designed to provide a separate and objective look and, hopefully, offsets any biases that the developer and user may have. The customers and decision makers also have a high stake in model V&V. For example, the decision to procure a major weapon system may be based largely on model results; and the validity of the model. Furthermore, those responsible for accreditation of the model will rely on V&V reports in arriving at their decision.

It is essential to maintain configuration control of models. When there are multiple users of a model and the various users are modifying the model to accommodate their particular needs or usage requirements, then each version should have VV&A accomplished for its particular application.

4.3.7 Model Verification, Validation, and Accreditation Plan

The VV&A of any given model should be a continuing process with appropriate documentation of the results at various key application points throughout the life cycle of the model. A formal plan should be developed to conduct V&V. The V&V plan could be developed sequentially over the lifetime use of the model, with a basic plan covering the initial V&V and supplements as needed to address each unique application and/or configuration update. The information set forth in the V&V plan should be sufficient to supplement other program and decision making documentation, as well as to serve as the road map for V&V of the model at specific points in time. The V&V plan and report should be the key documentation that supports the decision to accredit the use of a model. A sample format illustrating the types of information that should be included in the VV&A plan is provided below.

I. EXECUTIVE SUMMARY

II. BACKGROUND
   Purpose of the VV&A Effort
   General Description of the Model
   Previous and Planned Usage
   Program and Decision Making Structure

III. PROBLEM
   Specific Problem(s) Being Addressed
      MOEs and MOPs
      Critical Evaluation Issues
   Critical Validation Issues (Related to Critical Evaluation Issues)

IV. APPROACH
   V&V Task(s) To Be Addressed
   General Approach to V&V
      Scope
      Limitations
   Specific Approach to V&V for Each Task

V. DESCRIPTION OF V&V EXPERIMENT
   Output
   Structural

VI. ADMINISTRATION
V&V Management and Schedule
Tasking and Responsibilities
Safety
Security
Environmental Impact

VII. REPORTING

VIII. ATTACHMENTS (As Required)

The background information in the VV&A plan should sufficiently describe the model, its present configuration, and previous and planned applications. It should also relate how the model fits into the overall program and decision making structure. The specific problem that the model is to address should be clearly delineated along with the MOE(s) and MOP(s) of interest. Critical evaluation issues related to the problem, along with the specific tasks planned for the VV&A process, should be addressed. Critical VV&A issues should be identified and addressed in terms of how they relate to the critical evaluation issues that must be addressed by the model. The general and specific approaches to be used for V&V tasks should be addressed. Planned V&V experiments for both output and structural examination should be described. Finally, such tasks as scheduling, planning for administration and management, tasking and responsibilities, and reporting on the VV&A efforts should be formally documented in the plan.

4.3.8 Documentation of the Model Verification, Validation, and Accreditation Efforts

Formal documentation of the model VV&A activities and reports for each model VV&A effort are essential for proper life cycle management of the model. The documentation activity and reports should be directed at assisting the customer and the decision maker in the model accreditation process (i.e., it should provide information that helps the decision maker decide whether the model is good enough for the specific application and problem being addressed). Documentation of prior VV&A efforts also assists those tasked to conduct subsequent VV&A efforts (i.e., it should provide the basis for accumulating VV&A information). The documentation efforts include collection of information and data during execution of the VV&A plan and analysis and reporting of the model V&V results.

A sample format illustrating the types of information that are of interest in the VV&A report is provided below. Ideally, the executive summary will concentrate on the model, critical issues regarding its application(s), and its strengths and weaknesses for addressing the specific problem(s) in terms of the real world comparisons made during VV&A. Sections II through V provide background on what was planned for model VV&A and, except as modified, could be extracted from the plan. Sections VI, VII, and VIII address both general and specific results of the VV&A effort along with a detailed accounting of the specific findings. Model trends and sensitivities, as they relate to the problem and the critical evaluation issues, should be described. Quantitative methods are highly desirable for the V&V comparisons and should be documented. Qualitative methods also are useful and, because they usually include more subjective judgments, may require even more documentation.

I. EXECUTIVE SUMMARY

II. BACKGROUND

III. PROBLEM

IV. APPROACH

V. DESCRIPTION OF V&V EXPERIMENTS
    Output
    Structural

VI. RESULTS OF V&V EFFORTS
    General
    V&V by Task
    Discussion of Critical Issues
    Discussion of Model Trends and Sensitivities

VII. SPECIFIC V&V FINDINGS

VIII. ATTACHMENTS (As Required)

4.3.9 Special Considerations When Verifying, Validating, and Accrediting Models

Model V&V will always require some level of judgment but, to the maximum extent possible, empirical comparisons should be made. As discussed earlier, the process of verifying and validating a model will never be exhaustive. There will always be some things not addressed during V&V, and others not addressed to the degree that some would like. Consequently, for model VV&A to be a productive process, it must concentrate on the specific uses of the model and the actual V&V issues addressed regarding those specific uses. The V&V comparisons should present a reasonable, systematic examination of the model and an objective picture of its true capabilities and limitations in that application. Both the strengths and weaknesses of the model for addressing the stated problem(s) must be communicated to the accreditor by the VV&A process.

The VV&A process should be extensive and robust enough to properly consider the findings and views of neutrals, advocates, adversaries, and other interested parties. The goal should be to communicate all important findings regarding model comparisons and critical V&V issues to the accreditor. When serious competing views emerge on critical VV&A issues, it may be necessary to pursue further validation efforts that can provide additional objective comparisons and information for consideration by the responsible accreditation authority.

There should be test data available for comparison on each critical issue to be addressed by the model. If feasible, it is desirable to collect two sets of real world data—one for structural comparisons and another for output comparisons. The VV&A process should be such that when data derived from realistic field and development testing raise questions about prior assumptions and/or propositions of the model, these questions are addressed. The process of VV&A must shed light on what is known and not known about the model's structural content, its internal functions and capabilities, and its output accuracy. The analysis of conflicting or discrepant information often can lead to the insights necessary for improving the models and obtaining better answers to difficult questions.

Independent technical and operational experts can examine the model processes, its assumptions, inputs, and outputs, to arrive at their opinions of the appropriateness and validity of the model and its associated results. In the academic world and in the field of operations research analysis, this independent review is often performed by a separate unbiased party (i.e., a referee) who is responsible for helping maintain the objectivity of the analysis. Unfortunately, when dealing with highly complex and often classified systems and techniques, this objectivity can be somewhat limited, especially if documentation is not adequate. Therefore, it is incumbent upon all interested parties (e.g., model developers, users, decision makers, and other responsible authorities) to ensure that the VV&A process is objective, comprehensive, and well documented.

4.3.10 Summary

The Military Operations Research Society advocates the formal determination and documentation of model credibility through a three-part investigation involving VV&A. Model verification is the process of determining that the implementation of the model or simulation accurately represents the developer's description and specifications. Model validation is the process of determining the degree to which a model is an accurate representation of the real world from the perspective of the intended uses of the model. Model accreditation is the official determination that a model is accepted for a specific purpose. The effort put forth in VV&A should quantify and qualify the model accuracy and precision associated with each MOE and MOP of interest. Once this information is available and documented, the accreditor can examine it in terms of the needs of his/her analysis problem to establish whether the M&S should be accredited for that specific use.

A model can be portrayed in terms of its inputs, the model itself, and its outputs. Thus, when V&V comparisons are made against the design specification and the real world, or physical theories and laws associated with the real world, they can be addressed in terms of model inputs, the model itself, and model outputs.

Model V&V has been partitioned into two parts: 1) output and 2) structural. Output V&V is the most credible form and consists of comparing the output of the model against real world observations. Structural V&V involves determining the extent to which the input data (e.g., scenarios, missions, and data bases) and the model (e.g., assumptions, logic, algorithms, parameters, and code execution) represent the significant and salient features of the design specifications and the real world systems, events, and scenarios. Decomposition of the model into fundamental model functions and components aids in the process of structural V&V.

VV&A of complex models requires an appropriate understanding of the problem being addressed and a combination of both structural and output examination. Maintaining model configuration control and essential documentation are important. A formal plan for model VV&A, along with adequate reporting and documentation of the results as described herein, are vital parts of the model VV&A process.

4.4 Establishing an Overall Test and Evaluation Model (See References 31–51)

Once the critical issues, objectives, measures, and data elements of a T&E are identified, a determination must be made of the best combination of activities to use to obtain the required data. Often this will involve collecting data and information from one or more testing activities to address each issue and objective. This test structure can be approached and viewed as an overall T&E model. Within the overall T&E model, analytical M&S is generally involved to some degree to help with planning and to help create the environments used in the various test activities. However, it is not appropriate (or wise) to use anything other than the real system as the test item during operational T&E. The final acceptance or rejection of any system in testing should require that the real system withstand the rigor of realistic physical T&E.

Analytical M&S can provide a highly valuable tool for evaluating early conceptual systems before hardware and software are available. These evaluations can demonstrate the potential operational utility of a system, given that the hardware and software design and associated production efforts can result in the system becoming a reality. Analytical models and simulation can also be used to conduct sensitivity analyses and to provide insight into areas where developmental work and operational testing are required. As the hardware and software become available, laboratory testing can be used to aid in system integration and to help verify the design and the associated capabilities of components. Hybrid simulations with appropriate man-in-the-loop simulation testing can also be used to provide insight into how the man and system can be expected to perform under highly advanced threats and dynamic environments. As discussed earlier, the final step in the proof of a system's capability is its performance in live physical field or flight testing. Thus, all T&E environments used to complement each other are extremely important to the overall T&E model and process.

Ideally, the T&E environment would be physical field or flight testing totally representative of combat. The exception to this condition would be the situation where total employment and exercising of the weapon system would result in nonacceptable consequences regarding environmental hazards, safety, and costs. The environment should be instrumented so that data collection is transparent to the military operations taking place and the instrumentation is capable of providing a description of the combat processes to a degree of accuracy sufficient to allow mission refinement and optimization of personnel and equipment. However, such ideal conditions can rarely be achieved due to limitations resulting from the lack of threat knowledge, resources and costs, timeliness, safety, and a host of other practical constraints.

4.4.1 System Test and Evaluation Environments

To address the diversity of testing requirements, T&E must generally be achieved in a combination of environments over a system's life cycle. This combination is necessary to generate the essential results in a cost-effective manner. This multienvironmental testing is required to evaluate whether the system meets total system requirements and to support predictions of the system's combat performance. Such complex T&E normally requires a combination of measurement

MODELING AND SIMULATION APPROACH

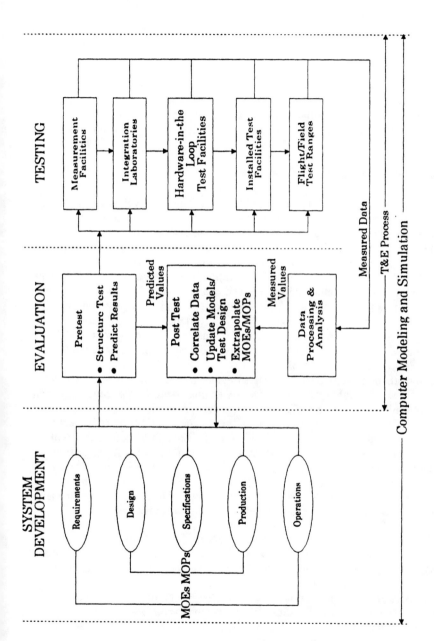

Figure 4-8. System T&E Environments

facilities, integration laboratories, hardware-in-the-loop facilities, installed test facilities, and field/flight test ranges. Analytical M&S can be applied in the overall T&E approach to help establish where field/flight testing should occur and to help address the sensitivities as well as the full range and scope of the issues being examined. Figure 4-8 depicts the allocation of the T&E problem to the various measurement activities in relation to the system development, evaluation, and testing process. This allocation should be achieved based on the consideration of data validity, timeliness, safety, and costs.

The MOEs and MOPs must be based on the system operational and developmental requirements, and the system testing and evaluation should be broad-based, quantitative to the degree practical, and objective and unbiased. When models and simulation are applied in T&E, they should be fully examined and supported by a formal VV&A process.

4.4.2 Other Modeling and Simulation

4.4.2.1 Mathematical and statistical models.
In T&E, it is also highly desirable to construct satisfactory mathematical models of repetitive operations which generate data that can be used to address the critical issues and objectives of the test. These models can then be used to aid in the study of how various factors and properties will likely affect the operation, and to assist in the T&E design. As the T&E progresses, data can be collected to verify the models and to substantiate their properties. Mathematical models and computer simulations are especially useful devices for studying problems when the models realistically represent the actual operations involved and when the results produced are highly correlated to the operations. It often happens that models may prove useful even though the models are somewhat abstract and the results produced are indicative of only trends and directions rather than absolutes.

The mathematical models that testers might select for repetitive operations are those that would enable them to make predictions about the frequency at which certain results can be expected to occur when the operations are repeated a number of times. For example, in the dropping of bombs under various sets of operational conditions, models can be used to predict the percentage of bombs that will hit within a specified distance from the target aim point or to assist in determining which variables contribute most significantly to bombs missing (or hitting) the target. In these cases, statistical models provide an invaluable tool and can be applied by the tester to extract the maximum amount of information from the testing processes. A variety of these models are discussed in Chapter 6.

4.4.2.2 Distributed interactive simulation (DIS).
Distributed interactive simulation (DIS) is a concept that has broad application and has significantly enhanced the capability to conduct T&E. A distributed system is one that has multiple processing elements that are interconnected but can also run independently. Each processing element (or node) contains at least a control processing unit and a memory. The interconnectivity allows the systems to communicate and synchronize while running processes in parallel. The processing elements are fault-tolerant in that they fail independently and can recover from failures by maintaining a shared state. DIS provides for the netting of models and simulations located at various points throughout the world into a common system whereby problems and

issues can be addressed that far exceed the capabilities of any single simulation or location. This type of connectivity allows for various test ranges and simulation facilities to be linked and interacted in a manner to create challenging and extensive testing environments. An important aspect of these simulations is the real time interaction with human participants. One of the major challenges in today's T&E design is to ascertain what can best be addressed where and to plan for use of the necessary distributed interactive capabilities.

4.5 Future Challenges

Analytical M&S (both analog and digital) has long been a tool used by government and industry in support of the development and design process. There has been no technical limit on how much and what kind of M&S the developer and implementor can use throughout the acquisition cycle. Only the competition for development dollars within a program constrains this effort. Analytical M&S traditionally has been used to support both objective and advocacy types of analyses in the development and design of systems. Rarely have major systems failed T&E based on poor results obtained from analytical M&S (i.e., results are usually fed back to help correct system problems prior to fielding). The results obtained from models and simulations can be extremely sensitive to changes in model algorithms, input data, and/or assumptions. These sensitivities must remain of paramount interest to those who would rely solely on the output of the models and simulations for critical program decisions without sufficient VV&A.

Historically, the development and acquisition of military systems have been complex and difficult endeavors that have drawn their share of criticism from Congress down to military operators. The T&E community (government and industry) is continually at work sorting out what sensibly can and should be achieved now and in the reasonable future by models and simulations. The virtues of M&S must be carefully exploited where practical and affordable. Models must be modified and/or developed and linked at sufficient levels to get the job done. A meaningful and formal process for VV&A is essential to all important T&E applications.

References

[1] Battilega, J.A., and Grange, J.K. (eds.), *The Military Applications of Modeling*. Wright-Patterson AFB, Ohio: Air Force Institute of Technology Press, 1978.

[2] Ritchie, A.E. (ed.), *Simulation Validation Workshop Proceedings (SIMVAL II)*, Military Operations Research Society. Alexandria, Virginia: Institute for Defense Analyses, 31 March-2 April 1992.

[3] *Catalog of War Gaming and Military Simulation Models* (7th ed.). Washington, D.C.: Joint Chiefs of Staff, Studies, Analysis and Gaming Agency (SAGA-180-77), 1977.

[4] *Hierarchy of Combat Analysis Models*. McLean, Virginia: General Research Corporation (AD 906 982), 1973.

[5] Koopman, B.O., *A Study of the Logical Basis of Combat Simulation*. Operations Research 18:855-882, 1970.

[6] Robinson, R.C., McDowell, A.C., and Rowan, J.W., *Simulation and Gaming Methods for Analysis of Logistics*, Part II (SIMALOG II): Capability Analysis

System. McLean, Virginia: Research Analysis Corporation (LD 16779R), 1971.

[7]*Models of the U.S. Army Worldwide Logistics System*, Volumes 2A and 3. Vienna, Virginia: BDM Corporation (LD 3829OMC and AD-A040-804), 1977.

[8]Hausrath, A.H., *Venture Simulation in War, Business, and Politics*. New York: McGraw-Hill Book Company, 1971.

[9]Taylor, J.G., "A Tutorial on Lanchester-Type Models of Warfare," in *Proceedings of the 35th Military Operations Research Symposium* (July 1975). Alexandria, Virginia: Military Operations Research Society (AD C005 592), 1975.

[10]*Techniques for Evaluating Strategic Bomber Penetration: An Aggregated Penetration Model—PENEX*. Washington, D.C.: United States Air Force Assistant Chief of Staff, 1972.

[11]Neu, R.M., *Attacking Hardened Air Bases (AHAB): A Decision Analysis Aid for the Tactical Commander*. Santa Monica, California: Rand Corporation (AD 786 879), 1974.

[12]*NMCSSC Strategic War Gaming Support With the Quick-Reacting War Gaming System (QUICK)*. Arlington, Virginia: Defense Communications Agency (TM 90-74), 1974.

[13]Emerson, D.E., *UNCLE-A New Force-Exchange Model for Analyzing Strategic Uncertainty Levels*. Santa Monica, California: Rand Corporation (R-430-PR), 1969.

[14]Bonder, S., and Farrell, R., *Development of Models for Defense Systems Planning*. Ann Arbor, Michigan: Systems Research Laboratory, University of Michigan (AD 715 646), 1970.

[15]Fish, J.R., *ATACM: ACDA Tactical Air Campaign Model*. Arlington, Virginia: Ketron (AD B008 645), 1975.

[16]Stockfisch, J.A., *Models, Data, and War: A Critique of the Study of Conventional Forces*. Santa Monica, California: Rand Corporation (R-1526-PR), 1975.

[17]*Tactical Warfare Simulation Evaluation and Analysis Systems (TWSEAS)*. Quantico, Virginia: Marine Corps Development and Education Command (AD B007 867L), 1976.

[18]Amdor, S.L., "TAC-WARRIOR—A Campaign Level Air Battle Model," in *Proceedings of the 37th Military Operations Symposium* (June 1976). Alexandria, Virginia: Military Operations Research Society (AD C009 755L), 1976.

[19]*The 37th Military Operations Symposium* (June 1976). Alexandria, Virginia: Military Operations Research Society (AD C009 755L), 1976.

[20]Anderson, H.C., "Command and Control Considerations Relevant to the Combined Arm Simulation Model (CASM) (U)," in *Proceedings of the 39th Military Operations Research Symposium* (June 1977). Alexandria, Virginia: Military Operations Research Society, 1978.

[21]Spaulding, S.L., Samuel, A.H., Judnick, W.E., et al., "Tactical Force Mix Versatility Planning with VETCOR-I and UNICORN (U)," in *Proceedings of the 38th Military Operations Research Symposium* (June 1976). Alexandria, Virginia: Military Operations Research Society (AD C012 958L), 1977.

[22]*Operation Test and Evaluation Simulation Guide*. Vienna, Virginia: Braddock, Dunn, and McDonald, 1974: Volume I: Summary (AD A001 692); Volume II: Hardware (AD A001 693); Volume III: Classified Simulator Devices (U) (AD C000 021); Volume IV: Software Simulators (AD A001 694).

[23]Davis, H., *Maintenance of Strategic Models for Arms Control Analysis - Description of Strategic International Relations Nuclear Exchange Model*. Chicago,

Illinois: Academy for Interscience Methodology (AD-780-490), 1974.

[24]DeSobrino, R., *A Comparison of Two Strategic Forces Evaluation Models*. Menlo Park, California: Stanford Research Institute (SED-RM-5205-53), 1969.

[25]Judnick, W.E., "Use of Experimental Design With Theater-Level Models (U)," in *Proceedings of the 39th Military Operations Research Symposium* (June 1977). Alexandria, Virginia: Military Operations Research Society, 1978.

[26]Pritsker, A.A.B., and Pegden, C.D., *Introduction to Simulation and SLAM*. New York: John Wiley and Sons, Inc., 1979.

[27]Law, A.M., and Kelton, W.D., *Simulation Modeling and Analysis*. New York: McGraw-Hill, Inc., 1991.

[28]Mullender, S. (ed.), *Distributed Systems*. New York: Addison-Wesley Publishing Company, 1989.

[29]Khuri, A.I., and Cornell, J.A., *Response Surfaces*. New York: Marcel Dekker, 1987.

[30]Kleijnen, J.P.C., *Statistical Techniques in Simulation, Part II*. New York: Marcel Dekker, 1975.

[31]Montgomery, D.C., *Design and Analysis of Experiments* (2nd ed.). New York: John Wiley and Sons, Inc., 1984.

[32]Rubinstein, R.Y., *Monte Carlo Optimization, Simulation, and Sensitivity of Queuing Networks*. New York: John Wiley and Sons, Inc., 1986.

[33]Biles, W.E., and Swain, J.J., *Optimization and Industrial Experimentation*. New York: John Wiley and Sons, Inc., 1980.

[34]Box, G.E.P., and Draper, N.R., *Empirical Model-Building and Response Surfaces*. New York: John Wiley and Sons, Inc., 1987.

[35]Box, G.E.P., Hunter, W.G., and Hunter, J.S., *Statistics for Experimenters: An Introduction to Design, Data Analysis, and Model Building*. New York: John Wiley and Sons, Inc., 1978.

[36]Morgan, B.J.T., *Elements of Simulation*. London, England: Chapman and Hall, 1984.

[37]Bratley, P., Fox, B.L., and Schrage, L., *A Guide to Simulation* (2nd ed.). New York: Springer-Verlag, 1987.

[38]Ross, S.M., *Introduction to Probability Models* (4th ed.). San Diego, California: Academic Press, 1989.

[39]Conover, W.J., *Practical Nonparametric Statistics* (2nd ed.). New York: John Wiley and Sons, Inc., 1980.

[40]Gibbons, J.D., Olkin, I., and Sobel, M., *Selecting and Ordering Populations: A New Statistical Methodology*. New York: John Wiley and Sons, Inc., 1977.

[41]Anderson, T.W., *The Statistical Analysis of Time Series*. New York: John Wiley and Sons, Inc., 1971.

[42]Banks, J., and Carson, J.S., *Discrete-Event System Simulation*. Englewood Cliffs, New Jersey: Prentice-Hall, 1984.

[43]Billingsley, P., *Convergence Probability Measures*. New York: John Wiley and Sons, Inc., 1968.

[44]Chung, K.L., *A Course in Probability Theory* (2nd ed.). New York: Academic Press, 1974.

[45]Fishman, G.S., *Concepts and Methods in Discrete Event Digital Simulation*. New York: John Wiley and Sons, Inc., 1973.

[46]Fishman, G.S., *Principles of Discrete Event Simulation*. New York: John Wiley and Sons, Inc., 1978.

[47]Hogg, R.V., and Craig, A.T., *Introduction to Mathematical Statistics* (3rd ed.). New York: Macmillan, 1970.

[48]Wagner, H.M., *Principles of Operations Research*. Englewood Cliffs, New Jersey: Prentice-Hall, 1969.

[49]Welch, P.D., "The Statistical Analysis of Simulation Results," in *The Computer Performance Modeling Handbook*, Lavenberg, S.S. (ed.), pp. 268-328. New York: Academic Press, 1983.

[50]Naylor, T.H., Balintfy, J.L., Burdick, D.S., and Chu, K., *Computer Simulation Techniques*. New York: John Wiley and Sons, Inc., 1966.

[51]Lindley, D. V., *Making Decisions*. New York: John Wiley and Sons, Inc., 1973.

5
Test and Evaluation Concept

5.1 General

The development of a good concept for a Test and Evaluation (T&E) requires a full understanding and appreciation of the intended use of the data and information to be derived from the T&E (i.e., its purpose). T&E is generally accomplished to gain data and information on which to make decisions (i.e., reduce risks in decisions), to gain knowledge and learn about what is being tested, or both. Sometimes a definitional distinction is made between the terms "test" and "experiment" based on whether it is a pass/fail activity (i.e., a test) or a measurement and learning activity (i.e., an experiment). Although these definitions are not widely accepted, they do emphasize the idea that the problem being addressed and the use to be made of the data or information derived from the T&E are paramount in the development of the concept.

There is always a trade-off that must be made regarding how much information is needed to meet the purpose of the T&E and the cost of resources required to conduct it. Generally, this trade-off needs to be made during the development of the T&E concept so that appropriate resources can be acquired for the T&E. Work on the T&E concept should begin as early as the development of the COEA described in Chapter 2. How much T&E and how best to perform it for a particular test item or system should be a part of the cost and effectiveness decision process. Figure 5-1 illustrates the principle that decisions based on little or no risks will cost significantly more than those where some risk is acceptable. As more data and information are demanded from a T&E, the cost of T&E increases significantly. If the decision maker is willing to take high risks, little or no testing is required. When the decision maker is not willing to take any risks, T&E costs increase substantially. For the example presented, the last 20 percent of the information costs more than the first 80 percent.

Good T&E begins early in a program with the development and explicit documentation of the correct T&E concept. Development of an effective concept is one of the most important and difficult T&E tasks to accomplish. The T&E concept is a formal definition and documentation of the innovative thoughts or notions that serve as the basis of the test. The T&E concept must scale down the broad testing problem to one which can be adequately addressed within the available T&E resources and activities. Only a limited amount of field/flight testing normally can be conducted and the concept should identify and prioritize the testing activity. This can be achieved by examining what is already known, what can be meaningfully determined through modeling and simulation (M&S) and other T&E activities, and what must be derived through field/flight testing. The T&E concept provides the guiding information from which operational and technical

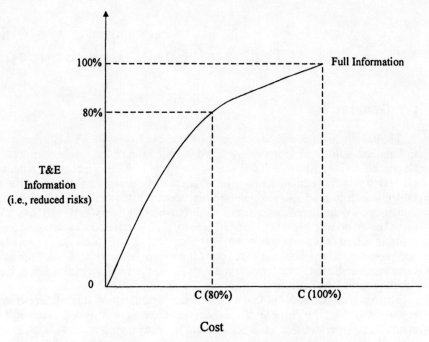

Figure 5-1. Cost of Decision Risk Reduction

experts can build the detailed T&E design and plan. The T&E concept provides the initial step in identifying the major critical issues and objectives that will be addressed; the overall measures of effectiveness (MOEs) and measures of performance (MOPs) that will be applied; the general criteria that must be met for success; the key data to be collected, analyzed, and evaluated; the primary test scenarios under which the system will be evaluated; and the scope and overall approach (i.e., overall T&E structure) by which the T&E will be conducted. This structure lays the groundwork for what will be tested and evaluated as well as how the overall T&E problem will be partitioned in terms of laboratory tests, hardware-in-the-loop tests, measurement facility tests, field/flight tests, analytical M&S, and other evaluation methodology. Other general information concerning participating organizations, responsibilities, and resources should be provided to the degree available at that point in time.

Throughout the T&E planning process, the T&E concept is refined, improved, expanded to greater detail, and used as a guide for the other more detailed test design and planning activities. A good T&E concept, by necessity, may require abstraction from the real world because of limitations in resources, environments, etc., but must be capable of producing valid and useful results that accomplish the purpose of the T&E. The T&E concept shows how each critical issue and objective will be addressed and clearly describes the linkage among the critical issues, areas of risk, the T&E objectives, the overall test approach (i.e., laboratory tests, field tests, analytical M&S, etc.), the data analysis, and the evaluation. Because a complex T&E normally requires a combination of multiple environments and activities for success, the concept must delineate how and where the various test activities supplement and complement each other in accomplishing the overall T&E project. All of this general information should be addressed in the concept to a level that is sufficient to guide future detailed test design and planning activities. Normally, the T&E concept can best be developed by a multidisciplinary team of war fighting and strategy experts consisting of operators, maintainers, logisticians, operations research analysts, designers, and testers.

5.2 Focusing on the Test and Evaluation Issues (See References 1–5)

An important part of the acquisition and fielding of new systems is the T&E of operational effectiveness and suitability for components, subsystems, and systems; the development and evaluation of tactics and doctrine; and the training of operational and maintenance personnel. The first of these evaluations is critical from the standpoint of measuring the operational capabilities of new test items and systems and determining whether these capabilities satisfy user requirements. The latter evaluations are important from the standpoint of developing tactics and training personnel to effectively employ the systems in combat. During the development of a T&E concept, it is important to obtain the right balance in applying the various instrumentation and test environments. Too much instrumentation can drive the cost and time of T&E up to impractical levels; too little instrumentation does not support adequate quantification and understanding of the results. In seeking this balance, developers, testers, trainers, and tacticians agree that if one is to learn from doing, one must receive adequate and realistic feedback on the successes and failures.

The T&E issues depend to a great extent on the type of testing being conducted and on where the system is within the research, design, development, production, and operations/support cycle. Figure 5-2 depicts the major types of testing that occur during the life cycle of a weapons system when research and development (R&D) is required in the acquisition of the system. Operational Demonstrations (ODs) are often conducted by the operational using command to examine new concepts and to aid in the definition of system requirements. When conducting operational testing, the focus of the testing is on quantifying the MOEs that describe how well the test item or system performs its operational mission. Development Test and Evaluation (DT&E) is conducted by the developing command to research technologies, examine components within a design, and examine subsystems and systems. DT&E is conducted to ensure that the test items and systems meet the engineering specifications stated for the systems. This is normally done by testing for MOPs related to the system engineering design specifications. Operational Assessments (OAs) are conducted early in a weapons system's life by operational testers who are normally independent of the system developer and system user. Initial Operational Test and Evaluation (IOT&E) is also conducted by the independent testers and is directed at arriving at an independent judgment of whether a system meets its operational requirements and should be produced. Follow-on Operational Test and Evaluation (FOT&E) is generally conducted by the operational using command to gain knowledge about how best to bring the test item or system into the inventory and to apply it in combat. Sometimes, when the test item or system has significant problems in IOT&E, it is necessary for the

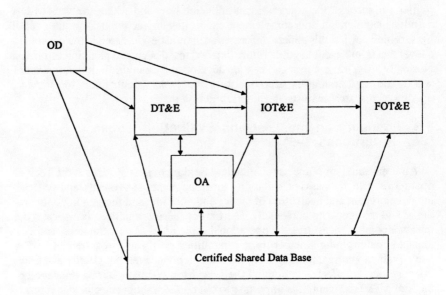

Figure 5-2. Spectrum of T&E When Research and Development Is Required

independent tester to conduct a limited FOT&E (called FOT&E-1) to verify that the system problems have been adequately corrected. Ideally, data from all types of testing should be properly labeled, certified, and placed into a common controlled data base for use by all qualified users. This shared data base approach has the potential to significantly reduce present duplication and costs in T&E.

When test items or systems do not have R&D costs associated with their acquisition, the T&E performed on them is called "qualification testing." This type of testing is depicted in Figure 5-3. Some example test items and systems are those developed outside of defense in the U.S. commercial market or by foreign governments. ODs performed on these products are similar to those performed on products that require R&D. Qualification Test and Evaluation (QT&E) is somewhat similar to DT&E with the exception that there are no associated R&D costs. QOT&E is conducted to determine if the test item or system is operationally effective and suitable to carry out the mission in the combat environment. FOT&E is similar to that described for systems that have R&D costs. Again, the certified shared data base concept could also be applied to T&E where no R&D cost is incurred and result in a similar reduction in duplication and costs in T&E.

5.3 Supporting Data and System Documentation (See References 6–10)

The T&E concept must be developed in light of what is known and not known about the system under examination. The mission needs statement (MNS), the operational requirements document (ORD), and the various types of system design information all contribute to the knowledge base required when the T&E concept is developed. The operations concept, the maintenance and support con-

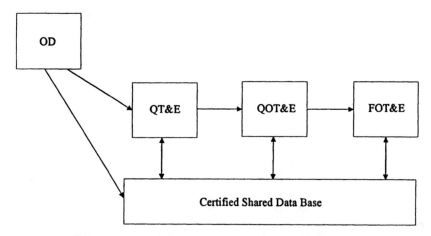

Figure 5-3. Spectrum of T&E When Research and Development Is Not Required

cept, and the design concept for the system are all important sources of information.

5.3.1 Design Concept

The design concept of a system can heavily influence how the T&E should be conducted. Knowledge about design and development difficulties should be used to focus the T&E activities on the critical areas of concern. Although the T&E must be broad-based to the extent that reasonable confidence can be achieved that the components and system will work under all required conditions, insight into the design problems can enable the T&E efforts to be focused on the most productive areas for T&E and corrective actions.

When models and simulation have been applied as an integral part of the design and development process, the T&E effort can place emphasis on the verification and validation of the models, with the longer-term goal of accrediting the models for a significant portion of the T&E activities. As such, the verification, validation, and accreditation effort is an integral part of the system design and development process. When models and simulation are not a part of the design and development process, a greater amount of physical testing of the components as well as full system testing will likely be required.

5.3.2 Operations Concept

The operations concept describes how the system fits into the overall operational structure and how it is planned to be employed in combat. A preliminary systems operational concept is formulated for each alternative solution considered early in the acquisition process. As the acquisition process continues, the system operational concept is refined and updated to address the alternative solutions. Once the preferred solution is identified, the operations concept for the resulting system is updated throughout acquisition and operational employment. Maintaining a current operational concept aids in all phases of planning for the T&E of components as well as the full system. The operations concept provides essential information for developing the overall T&E approach. More often than not, military weapon systems are deployed and operated as an integral part of a larger scheme directed at achieving certain combat goals. The tactics and procedures applied by each participating system are essential elements in achieving those goals. The operational concept describes when, where, and how the system will be applied in combat and provides other information essential to developing how T&E of the system should be structured.

5.3.3 Maintenance and Support Concept

The maintenance and support of any military system is vital to success in combat. The last place that military operators and supporters want to find out that systems don't function properly, or can't be supported, is in combat. The T&E concept must provide for a thorough examination of the maintenance and support concept. The maintenance and support concept addresses the planned methods and levels of maintenance and support. Technical orders, training, skill levels, logistics, and supply are all vital parts of the concept. The T&E concept must address the appropriate parts of these activities, as well as the collective performance of all support system activities.

5.4 Critical Issues and Test Objectives (See References 11-23)

Critical issues and associated test objectives are items of vital importance to successful operation and support of the system under test and should be explicitly stated in the T&E concept. Critical issues must be adequately addressed during T&E to allow the decision authority to confidently advance a system into the next acquisition or operational phase. They characterize system effectiveness and suitability in terms of what information is essential to render a judgment on whether the system is responsive and successful in the accomplishment of the mission, and whether it should proceed into the next phase of development, acquisition, and/or operation.

Objectives are more specific statements of what is to be tested than critical issues. Objectives should break the critical issues down into more clearly defined, manageable tasks or areas that can be examined during the test. Proper specification of the objectives and collection of the associated data are essential to allow statements to be made regarding the performance of a component or system within certain limits of error.

5.4.1 Relating Critical Issues and Objectives to the Military Mission

A careful examination of the military operations and tasks found to be pertinent to the purpose of the T&E should help the test team to construct a comprehensive list of applicable critical issues and test objectives for a given component, subsystem, or weapon system. Moreover, the critical issues and test objectives generated in this manner should be based on a broad examination of the system and should be specifically directed at the most important portions of the overall mission structure where the system is expected to contribute.

In some cases, the critical issues and test objectives may be formulated in terms of specific questions that, when answered, satisfy those aspects of the test. The flowchart method and the dendritic or tree method discussed earlier in Chapter 3 can greatly aid in this process. Additionally, it is sometimes desirable to specify the objectives in terms of the effect that a given factor might have on a specific military task being performed. The final T&E objectives must relate directly to the critical issues and be specific, yet comprehensive and relative enough to ensure the achievement of the purpose of the T&E.

During the formulation of critical issues and objectives, only those which can be investigated in the T&E should be selected. In other words, can measurements be made that are relevant to the hypothesis under test, and do they have a reasonable chance of yielding unambiguous results? It is during this formative period that fundamental questions, such as the following, should be raised and answered.

1) How do we conduct the right test?
2) What effects are to be described, compared, or estimated?
3) What hypotheses are to be tested?
4) What extrapolations of the test are desired?
5) What mathematical M&S might be applied and how?

5.4.2 Researching the Critical Issues and Objectives

All available program documentation should be thoroughly researched to help establish what the essential information is that must be derived from the T&E. As stated earlier, often critical issues can be identified from significant problems that are experienced in the design, development, and production processes. Generally, these issues involve what the decision maker must know to accept moving the system into the next phase of acquisition or deployment. A careful review by experts in operations, maintenance, and logistics is also important to provide insight into the potential critical issues that should be investigated during the T&E.

The list of critical issues and objectives initially prepared for consideration is likely to be larger than can possibly be accommodated by actual testing. The T&E team should make a thorough research of what is already known about the equipment from previous operations, studies, and tests. The T&E team should eliminate issues and objectives that have already been satisfactorily answered or that would require repeating tests for which acceptable results are already available. The certified common data base approach recommended in Section 5.2 could greatly assist in this process. In cases where little or no information on a particular subject exists, it may be necessary for the T&E organization to conduct a short pretest to help identify sensitive variables or to confirm opinions held by various technical and operations experts.

5.5 Measures of Effectiveness, Measures of Performance, and Criteria (See References 24–36)

At least one MOE or MOP must be identified for each critical issue and objective. It is important that the MOEs and MOPs be quantitative, relevant to the critical issues and objectives, and measurable in T&E. These measures are used during the T&E to determine the level of output the system is capable of providing with regard to the critical issues and objectives. The data collected for the MOEs and MOPs measured during the test will be expected to be within the accuracies required to satisfy both operational and statistical significance and to meet the criteria established for success if the system is to be considered effective. Operational significance is a determination that a quantified value (or a quantified difference in values) does indeed represent an appreciable gain or loss in military capability. Statistical significance is a determination that it is highly unlikely that the quantified value (or quantified difference in value) occurred due to chance. Both of these terms are addressed in additional detail in Chapter 6.

5.6 Data Analysis and Evaluation

Quantitative and qualitative data collected on each trial of the T&E must be analyzed to ascertain the degree to which each critical issue and test objective is being met. For each critical issue and objective being examined, one or more MOEs or MOPs are usually evaluated to accomplish this process. These measures provide a convenient summarization of the data in such a way that the evaluation can focus on the degree of achievement of the objectives and, in turn, the critical issues. When possible, it is highly desirable to develop a dominant measure for each critical issue and objective, that is, a single measure which, when evaluated, will consistently yield the same answer and preferred solution as multiple measures would yield.

The T&E concept should fully describe the evaluation process by which the overall significance or military worth of the test item or system will be established. For example, evaluation criteria should be defined in terms of quantitative and qualitative values that the MOEs and MOPs must reach if the system is to be judged successful. Usually, these values are called "test criteria" and are specified by the operational combat user in the system requirements. Ideally, they are established by the operational user very early in a program so that the developers, operators, logistical supporters, and professional testers all know what the system must achieve to be judged successful. Sometimes this may not be possible because certain information on "how much is enough" for a given system to be judged successful in a specific area of interest may not be known at the time of T&E initiation. Under such conditions, the operational user and professional tester should acknowledge these limitations in the T&E concept document and plan the T&E to yield insight into the needed information. As a minimum, when judgments of this nature are made after the T&E has been conducted, the criteria that were applied to reach the judgment and the logic leading to the judgment should be clearly evident in the test evaluation documented in the final report. Under no conditions are the developer or independent tester qualified to define the requirements and success criteria for the operational combat user, and they should not attempt to do so.

5.7 Scenarios

T&E scenarios are required to provide an outline or synopsis of the general structure and environment within which the T&E will be conducted. The realism of a T&E is often judged on the appropriateness of the scenarios for the weapons systems being applied and the related existing threat situation and world conditions. The T&E scenarios provide a planning aid that describes the military operations that are to take place, and the locations and timing of the weapons systems within those operations (i.e., layout of both friendly and enemy forces). A given test scenario may be a small slice of a larger overall scenario, with the size depending on how much must be included to be operationally representative of the situation being examined; that is, how much must be included to obtain operationally representative data. It is important that the enemy threat portions of the military scenarios be approved by the appropriate intelligence agencies and that the friendly forces portion be approved by operational experts from the operational service commands because of their extensive expertise in these areas. The test scenarios must be operationally representative and credible if the T&E is to be credible.

5.8 Requirements for Test and Evaluation Facilities

Once the critical issues, objectives, MOEs, MOPs, and scenarios are determined, requirements for the various T&E facilities can be established and identified in the T&E concept. This information can be used to assist in the development of more detailed requirements specified in other planning and programming forums. The test concept should be definitive enough to allow for long-range scheduling and sometimes even development of the required T&E facilities. It is important that sufficient lead time be planned to allow for the incorporation of new or improved facilities, weapons system instrumentation, and test range capabilities into the appropriate investment programs. Terrain, atmosphere, and other key environmental features are all important considerations in the T&E process. It is important in DT&E to be able to progressively test equipment and systems in a manner that allows understanding of the cause and effect relationships regarding failure (or success). The need for special instrumentation, data bases, and processing should be at least initially addressed in the T&E concept. The T&E concept should provide the initial estimate of the resources needed to effectively carry out the T&E. These estimates should continually be "fine-tuned" as the program progresses through the various stages of acquisition and operational employment.

5.9 Scope and Overall Test and Evaluation Approach

The scope and overall T&E approach delineates how critical issues and objectives will be addressed during the T&E. Often a general approach to the T&E can be devised that will allow multiple objectives to be addressed on a single T&E trial. In others, a single objective may require a multiple of different types of trials. As stated previously, complex T&E normally requires that testing be accomplished in a combination of measurement facilities, integration laboratories, hardware-in-the-loop facilities, installed test facilities, and field/flight test ranges. The allocation of T&E to the various types of test environments will be driven by what can best be achieved within the available constraints and resources. It is extremely important that the scope and overall approach provided in the test concept help establish these bounds and provide information to guide the specific test designs that will be applied in the various T&E facilities.

5.10 Non-Real Time Kill Removal Versus Real Time Kill Removal

Most military testing, other than combat or live-fire testing, does not actually kill the targets being engaged. Further, most analytical methods of assessing whether a kill could have been achieved do not occur in real time as the missions take place, but are achieved during complex postmission processing and analysis. Consequently, in the normal T&E environment, this lack of removal of a "killed" target can result in multiple simulated kills being achieved against the same target during a given operation or mission. The failure to be able to assess and remove a simulated killed target is believed by many experts to have a significant impact on both the accuracy of the analysis and the behavior of the T&E participants.

5.10.1 Non-Real Time Kill Removal Testing

Most often, the T&E and associated data processing are too complex to permit analytical assessment and removal of killed targets in real time. This problem can usually be mitigated to some degree by making certain assumptions in the postmission analysis of the data. For example, in certain scenarios, one might assume that the environment is so "target rich" that each killed target in reality is regenerated as a new target. Also, one might evoke certain "rules of engagement," such as that once a target has been engaged, a mandatory disengagement and physical separation of x miles or minutes must take place before that particular target can be reengaged. The impact of these assumptions and rules on the T&E results can best be examined in light of the specific purpose for which the T&E is being achieved. For example, if one is making a comparison of two weapons or jamming systems, the relative performance of the two can likely be established. On the other hand, if one is attempting to establish the absolute value of kill rate or attrition level for a given size conflict, the analysis problem becomes more difficult and subject to influence by the multiple kill process.

5.10.2 Real Time Kill Removal Testing

Some of the more modern military test ranges and associated instrumentation systems allow real time (in some cases only near real time) kill assessment and removal during testing. This process has at least two advantages. It does not allow the multiple counting of kills of the same target; and it instills the idea into participants that, if they are assessed as killed, they will be removed from at least that portion of the mission. There is a strong belief among portions of the test and training community that both of these features have a significant impact on test measurements and are highly desirable attributes of military T&E.

5.11 Use of Surrogates

In military T&E it is sometimes necessary to use substitute systems in place of real systems to create a representative test and threat environment. These systems are commonly referred to as "surrogates" or "surrogate systems" and may be selected for use because of their physical representation; radar, avionics, and electronics signal and processing representation; and/or some other form of emulation important to the characteristics and technologies of the test. Surrogate systems are normally used because the "real" threat systems cannot be obtained, are too expensive, or for some other reasons are not available. Surrogate systems may be in their standard configuration, or in some modified configuration to make them more functionally and operationally representative. Adequate surrogate systems are generally acceptable for most of the systems needed to create a realistic T&E environment, but they are rarely acceptable or used in place of the real system for which the test is being conducted.

Surrogate systems are sometimes used as friendly systems to examine new concepts, tactics, procedures, and doctrine. However, the most common use of surrogates is to represent potential enemy threat systems (e.g., enemy aircraft, radar, and missile systems). Surrogates are rarely totally representative of the real threat systems and must be carefully selected on the basis of adequately representing the specific characteristics and technologies that are the focus of the test.

Accurate surrogates must possess the specific characteristics and technologies of interest in the test. For example, if an electronic jammer that employs an electronic warfare technique designed to defeat the velocity tracking circuitry of a given threat radar is being tested, then it is important that the surrogate radar accurately reflect the threat signal characteristics and technologies to include the velocity tracking circuitry of the "real" threat system. If this cannot be accurately established, then the basis of the T&E of the jammer system becomes suspect.

The first step in the surrogate selection process is to thoroughly understand the scope of the T&E and the specific intended use of the surrogate. This requires an in-depth understanding of the characteristics, technologies, and operations actually present in the surrogate. An analysis should be made of the similarities and differences between the real system and the potential surrogate. This analysis should address and document the specific parameters which could impact the use of the candidate as a surrogate. The impacts should be carefully weighed and evaluated by a team of experts who make the determination of whether or not the characteristics, technologies, and operations of the surrogate are a valid enough representation to adequately address the critical issues and objectives of the test. When the decision is made to use surrogates, their use should be explained and thoroughly justified in the test concept.

5.12 Managing Change

By virtue of being developed early in the acquisition process, the T&E concept always has to be refined and updated to some degree as the system progresses through the various stages of development and operational employment. Although such changes will be essential, it is important that the T&E problem be identified early and addressed extensively to arrive at the best possible initial T&E concept. Once established, the initial T&E concept provides guiding information for numerous, more detailed planning activities and thus should remain stable to the degree practical and appropriate.

There are several lessons from past T&Es that should be important considerations in the development of the T&E concept. These involve assumptions, MOEs and MOPs, maturity of hardware and software, and the overall usability of the final product. It is extremely important to get an innovative and well-established "first cut" on all of these factors in the T&E concept. These factors are key "drivers" of the T&E and there will likely be considerable pressure to change them as the T&E progresses. Some of the reasons for change will be valid, and others will not. It is important that changes made to the original T&E concept be formal and that a careful audit trail be maintained.

Sometimes, a change in the assumptions regarding the T&E can cause a weapons system to meet or not meet the essential requirements of its mission. This is especially true for a complex system that is developed over an extended period of time. For example, if the system is designed to counter a threat at 3 miles, and the threat situation changes during the period of system development to one of 6 miles, the system is very likely to fail operational testing unless a considerable safety margin has been designed into the system, or the system is tested to an obsolete threat. Assumptions should reflect what is necessary as well as proper; yet it is essential that the resulting system be able to meet the demands of the combat mission.

Making sure that the concept MOEs, MOPs, and their associated criteria are established for each specific T&E situation is extremely important. Once the MOEs, MOPs, and their associated success criteria are established, only changes that are absolutely essential should be made. All changes that are made should be a part of the formal acquisition process and, as such, be reflected in the evolving system design as well as all related documentation. Changes to the MOEs, MOPs, and their associated criteria that are not reflected in the system design will likely lead to failure during T&E. Once testing and data collection for a given phase of T&E have begun, the configuration of the component or system, MOEs, MOPs, and associated criteria should remain fixed to allow for a meaningful evaluation.

Most complex systems require time to mature. Generally, software bugs and hardware reliability need some degree of real use experience before they are ready for the "pass/fail" testing associated with the intended operational combat environment. Progressive T&E, where system problems are allowed to be discovered and corrected, is essential. This type of nonretribution "test-fix-test" behavior is essential to orderly research and development. Attempts to avoid the necessary T&E steps and to proceed directly to examination of the full system in the intended operational environment can lead to failure. For these types of systems, it is important that the T&E concept reflect the appropriate approach to developing and fielding systems that have sufficient maturity to meet the needs of the operational mission.

During the acquisition process, many government officials are active in a wide range of endeavors to include stating requirements, influencing requirements, writing specifications, interpreting specifications, reviewing designs, conducting T&E, and independently conducting T&E. If some care is not taken, it is possible to end up with a system that the real customer (i.e., the soldier, sailor, pilot, etc.) cannot use. One way to avoid this catastrophe is to make sure the end user is heavily involved in the T&E of the system. It is extremely important to obtain and use the progressive feedback from operations, maintenance, and logistics experts during all phases of T&E. If the T&E concept does not provide for heavy combat user involvement, there is a good chance that the real customer (when he/she eventually gets a chance to see and operate the test item) will say, "This thing doesn't work right. I can't use it."

References

[1] Brewer, G., and Shubik, M., *The War Game, A Critique of Military Problem Solving.* Cambridge, Massachusetts: Harvard University Press, 1979.

[2] Dupuy, T., *Understanding War.* New York: Paragon House, 1987.

[3] Waterman, D.A., *A Guide to Expert Systems.* Reading, Massachusetts: Addison-

Wesley, 1986.

[4]Miser, H.J., and Quade, E.S., *Handbook of Systems Analysis: Craft Issues and Procedural Choices.* New York: Elsevier Science Publishing Co., 1988.

[5]Checkland, P.B., *Systems Thinking, Systems Practice.* New York: John Wiley and Sons, Inc., 1981.

[6]Bernstein, P.A., Hadzilacos, V., and Goodman, N., *Concurrency, Control, and Recovery in Database Systems.* Reading, Massachusetts: Addison-Wesley, 1987.

[7]Ceri, S., and Pelegatti, G., *Distributed Databases: Principles and Systems.* New York: McGraw-Hill, 1984.

[8]Ullman, J.D., *Databases and Knowledge—Base Systems,* Vol I. Computer Science Press, 1988.

[9]Satyanarayanan, M., *Integrating Security in a Large Distributed Environment.* CMV-CS-87-179, Department of Computer Science, Carnegie Mellon University, 1987.

[10]Burrows, M., *Efficient Data Sharing,* Ph.D. Thesis. Cambridge University Computer Laboratory, September 1988.

[11]Przemieniecki, J.S. (ed.), *Acquisition of Defense Systems.* AIAA Education Series, Washington, D.C.: American Institute of Aeronautics and Astronautics, 1993.

[12]DoD Instruction 5000.2, "Defense Acquisition Management Policies and Procedures," Office of the Secretary of Defense, Under Secretary of Defense for Acquisition, 23 February 1991.

[13]Roberts, G.W., *Quality Assurance in Research and Development.* New York: Marcel Dekker, Inc., 1983.

[14]Kivenson, G., *The Art and Science of Inventing* (2nd ed.). New York: Van Nostrand Reinhold Company, 1982.

[15]Stewart, R.D., and Stewart, A.L., *Managing Millions.* New York: John Wiley and Sons, Inc., 1988.

[16]Breiman, L., *Statistics with a View Toward Applications.* Boston, Massachusetts: Houghton Mifflin, 1973.

[17]Miles, L.D., *Techniques of Value Analysis and Engineering.* New York: McGraw-Hill Book Company, 1972.

[18]Defense Systems Management College, *Systems Engineering Management Guide,* 1982.

[19]U.S. Army Corps of Engineers Publication EPS1-3, *Value Engineering Programs,* April, 1987.

[20]Buffa, E.S., *Modern Production Management* (4th ed.). New York: John Wiley and Sons, Inc., 1973.

[21]Sohenberger, R.J., *Japanese Manufacturing Techniques.* New York: The Free Press, 1982.

[22]Loomba, N.P., *Linear Programming.* New York: McGraw-Hill Book Company, 1964.

[23]U.S. Army Procurement Research Office, APRO Report 709-3, 1978.

[24]Duckworth, W.E., Gear, A.E., and Lockett, A.E., *A Guide to Operational Research.* New York: John Wiley and Sons, Inc., 1977.

[25]Gribik, P.R., and Kortanek, K.O., *Extremal Methods of Operations Research.* New York: Marcel Dekker, Inc., 1985.

[26]Marlow, W.H., *Mathematics for Operations Research.* New York: John Wiley and Sons, Inc., 1978.

[27] Forrester, J.W., *Industrial Dynamics*. Cambridge, Massachusetts: M.I.T. Press, 1961.

[28] Hamburg, M., *Statistical Analysis for Decision Makers* (2nd ed.). New York: Harcourt, Brace, Jovanovich, Inc., 1977.

[29] Larson, H.J., *Introduction to Probability Theory and Statistical Inference*. New York: John Wiley and Sons, Inc., 1982.

[30] Augustine, N.R., *Augustine's Laws: Revised and Enlarged*. New York: American Institute of Aeronautics and Astronautics, 1983.

[31] Croskery, G.T., and Horton, C.F., "XM-1 Main Battle Tank." *Defense Management Journal*, September, 1974.

[32] DoD 4245.7M, *Transition from Development to Production*, 1985.

[33] MIL-STD-499A, Engineering Management.

[34] MIL-STD-881A, Work Breakdown Structure for Defense Material Items.

[35] MIL-STD-1388-1, -2, Logistics Support Analysis.

[36] MIL-STD-1528, Production Management.

6
Test and Evaluation Design

6.1 General

Test and Evaluation (T&E) design is the formal underlying structure specifying what must take place to adequately address the purpose of a test. It must be accomplished before the detailed T&E plan can be completed, and it establishes the basis for how the T&E will be conducted. It formally specifies the way in which measurements are to be taken and dealt with to meet each critical issue and specific T&E objective. It provides a scheme for collecting the right balance of unbiased quantitative and qualitative data essential to understand the processes taking place and to support conclusions and recommendations. The degree to which this formal structure can be based on the scientific method has a major impact on the ability to meaningfully collect sufficient data, analyze the results, and validly address the critical issues and objectives of the test. This investigative process involves establishing two kinds of test significance—1) operational and 2) statistical. Both are essential to good testing. Operational significance establishes the fact that the value measured for a given test parameter or parameters is operationally important and does indeed make an operational difference in the outcome of the military mission. Statistical significance establishes a confidence level that the operational differences measured did not occur merely by chance. To illustrate operational and statistical significance, consider the following example. Suppose we have collected a sufficient data sample to show with a 99 percent statistical confidence that system A has a better range-to-target prediction accuracy than system B. Further, suppose system A is measuring range to target within ± 10 meters and system B makes the same measurement within ± 15 meters. Also, suppose this range prediction information is being fed into an overall defense system that requires ± 12 meters accuracy for successful execution. This situation is indicative of one where the measured values are both operationally and statistically significant. That is, system A is within the operational requirement of ± 12 meters and system B is not, and we are 99 percent confident that these measurements did not occur merely by chance. On the other hand, suppose the overall defense system required only ± 15 meters for successful execution. The situation then would become one where the differences measured were still statistically significant but were not operationally significant. That is, although we are highly confident that the differences measured did not occur by chance, these differences do not make a difference from an operational point of view because both systems demonstrated the capability to meet the operational requirement of ± 15 meters. Obviously the tester desires to draw conclusions and make recommendations based on T&E data that are both operationally and statistically significant.

A good T&E design furnishes the required information with minimum resources and testing effort. Measurements collected in one way may be relevant to the critical issues and objectives under investigation, whereas measurements collected in another way may be useless, or of minimal value. Thus, the ultimate success or failure of a T&E depends heavily upon the adequacy of the design. To accomplish good T&E design, three things are essential: first, the critical issues and specific objectives must be correctly formulated; second, a correct choice of T&E method must be made in the light of operational realism, measurements, and accuracy required, and the various experimental pitfalls that are likely to be encountered; and, third, the general structure of the T&E (i.e., the number of replications, spacing, and interrelation of the individual measurements) must be correctly chosen.

6.2 Test and Evaluation Design Process

A good T&E design will allow the necessary system measurements and the quantitative extrapolation of measured system performance into a realistic combat environment. The process by which one arrives at a T&E design must have the flexibility to treat a broad variation in T&E requirements ranging from the relatively uncontrolled complex military operations to the highly controlled laboratory and field experiments. If T&E produces relatively uncontrolled results in an extremely large number of variables and, if some variables are not controlled (i.e., set at a fixed level or levels), so much variability is introduced that extremely large and impractical sample sizes are essential to draw conclusions at a reasonable level of statistical confidence. On the other hand, efforts to artificially reduce a complex military operation into the formal squares of an agriculture-type statistical experimental design often are not practical. This is because most operators are not likely to agree to a T&E design that does not realistically follow how they would operate in combat and, even if they did agree, the tester must be careful that the results are descriptive of the projected combat operations—not an abstract and unrealistic experiment.

In order to accommodate this wide range of testing requirements, two formal approaches to the T&E design process are defined and addressed in this book. They are 1) Procedural T&E Design, and 2) Experimental T&E Design. During the discussion of these two approaches, a number of important quantitative methods will be presented. Appendix 1 will also address example problems and solutions that can be used in support of the two design approaches.

6.3 Procedural Test and Evaluation Design

The first T&E design approach, procedural T&E design, offers a great deal of flexibility and is most commonly applied in the operational field/flight testing of military weapons systems. It is depicted by the iterative process shown in Figure 6-1.

6.3.1 Critical Issues and Test Objectives

Selection of the right critical issues and test objectives is vital to the success of the T&E. The wrong critical issues and test objectives mean that the wrong test is being conducted. No matter how well the T&E is designed and executed, it will yield the wrong answers.

TEST AND EVALUATION DESIGN 115

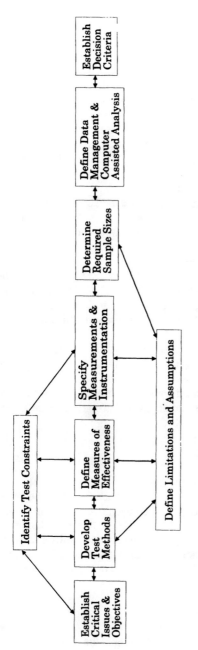

Figure 6-1. Nominal Steps of Procedural T&E Design

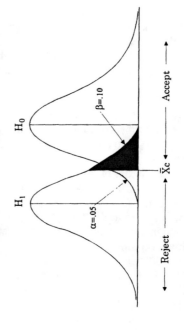

Figure 6-2. Determining Sample Size for a Test of a Hypothesis

T&E designers should approach each T&E task thoughtfully and with an open mind. It is important to look for new ideas and employment concepts, as well as facts and relationships that explain those already known. It is important not to exclude methods, tactics, and concepts merely because they run contrary to past practice or superficially seem impractical.

Ideally, the T&E design should attempt to exercise the system under T&E over its full range of operational use in the environment in which it is expected to be employed. Where possible, emphasis should be placed on the simultaneous consideration of all relevant factors as applied to the specific military tasks. Instead of limiting a T&E to component parts by deliberately neglecting their interactions, T&E designers should make every effort to extend the T&E boundaries, where possible, to determine which interdependencies are important to the effectiveness of the system in total. This is an ideal, of course, and will be limited by economic and practical constraints. Thus, "simultaneous" and "all" refer to an ideal situation. Even though partial testing and approximate criteria must usually be employed, it is desirable for the T&E designers to think about and examine the operational unit as a system and in terms of the total structure of military operations when attempting to determine where the partial testing should occur and to arrive at and apply limiting criteria to objectives.

The critical issues and associated test objectives provided in the test concept are used to form the basis for the T&E design. If there is no test concept when the design effort begins, then the test design effort itself should go through the processes described in Chapters 2 and 3 (i.e., flowcharts, dendritic method, and research of system problem areas) to derive the appropriate critical issues and objectives.

Once the list of critical issues and associated objectives is finalized, the test team must define one or more measures of effectiveness (MOEs) or measures of performance (MOPs) for each critical issue and test objective. These measures, in turn, are used to evaluate the utility or the degree of output of the operational system in terms of the issues and objectives and the overall accomplishment of the military mission.

6.3.2 Test and Evaluation Constraints

Test constraints offer a way to help control and manage the T&E. There is never an unlimited amount of money, tanks, ships, airspace, bombs, and other resources available for testing. Consequently, the T&E design must be constrained to operate within the available resources. On the other hand, the possible impact of each T&E constraint on the test data and objectives must be carefully examined during the design process. The T&E designers must verify that the test constraints imposed have not destroyed the operational combat realism of the test and its ability to produce meaningful results.

6.3.3 Test and Evaluation Method

An extremely important part of procedural T&E design is the method. In procedural design, realistic operational application of the test item in its projected combat environment is paramount to the evaluation. The T&E method can generally be thought of or addressed in two parts—1) a general method covering how the overall T&E will be conducted, and 2) a collection of specific T&E methods covering the unique requirements of each critical issue or objective that cannot be sufficiently addressed by the general method. The choice of a T&E method is the determination of how the specific equipment, personnel, and resources will be exercised and of the simulated combat and climatic/environmental conditions under which the testing will occur. This determination requires a broad, multidisciplined approach that can best be achieved by a test design team consisting of expert operators, maintainers, operations research analysts, intelligence officers, logisticians, testers, and other supporting personnel. In operational T&E, there is no equivalent substitute for operational experience and for live field and flight testing with real operators, real equipment, and/or good surrogate systems under realistic simulated combat conditions.

Historically, testing early in the development cycle has been, for the most part, based on some form of simulation or surrogate systems—often both friendly and enemy. A surrogate system is a substitute system which technical and operational experts agree functions enough like the real system to produce meaningful and applicable test results. The tester takes the best available intelligence information, picks (or builds) a surrogate system to emulate that information, does the same thing for the friendly system, and conducts the test. The validity of the surrogate testing depends directly upon the operational and engineering validity of the surrogates. The same detailed knowledge and intelligence required for good surrogate testing and analysis are also required for good digital modeling and analysis. Working with intelligence agencies, the tester is responsible for knowing what differences exist between the surrogate and the real system and for projecting those differences into the results of the evaluation. In almost every case, surrogate testing has revealed valuable field results not obtainable by pure analytical modeling, especially the first order effects that an operator would experience if he tried to employ his system in combat.

In choosing the T&E method, the designer is primarily concerned with selecting a realistic simulated operational environment in which measurements can be made with sufficient accuracy and frequency to fulfill each critical issue and specific objective. This involves a determination of what the appropriate scenario and simulated environment are for each specific objective of the T&E as well as the overall T&E. Ideally, a common method will exist that can be used to address all of the objectives on each trial of the T&E. If this is not the case, then a separate and unique method may be necessary for selected objectives. Additionally, the designer must establish the level of accuracy required on each measurement made in this environment to quantify or qualify each measure as related to each objective. Also, it is essential that the T&E design team think through each part of the test to make sure that it is operationally realistic and to develop contingencies that can be used to offset the test pitfalls that are likely to be encountered. For example, if a specific T&E objective is to evaluate the military maintainability of a given device, the use of highly skilled contractor maintenance during this part of the test is hardly indicative of a realistic combat environment. Moreover, in many cases, measurement of the operational performance of the device with contractor maintenance may be biased to the degree that it is inconsistent with combat results. In development T&E, analytical M&S is a key tool and often is substituted for field/flight testing. In operational T&E, the use of laboratory tests, hybrid models, and digital models and simulations is sometimes necessary, but is limited to evaluations where field/flight testing is not possible because of the lack of real systems, adequate surrogates, safety, and/or realistic quantities and environments. In all cases, data produced by methods other than field/flight testing should be maintained separately and used to augment and supplement judgments made on the field/flight test data, not to replace them. For these reasons, the T&E method used to address each specific objective in testing should be selected on the basis of providing (as nearly as possible within allotted test resources) a total and realistic environment that takes into account all important factors and that can be used to predict the results that will occur under actual combat conditions.

6.3.4 Measures of Effectiveness and Measures of Performance

The final selection of appropriate, practical, and efficient MOEs and MOPs is extremely important in T&E design. The MOEs and MOPs must be capable of being defined in terms of obtainable data, and they must relate directly or indirectly to the accomplishment of one or more T&E issues or objectives. The process of selecting such measures is iterative. Ideally, a dominant MOE or MOP that is quantifiable can be established for each critical issue or objective (that is, a single MOE or MOP that, if considered alone, will lead to the same conclusions that would be reached if multiple measures were considered). The critical issue or objective is first examined to define the potential MOEs and MOPs that are relevant to it. It is then examined to reveal quantifiable (or qualifiable) data that can be used to evaluate the MOEs and MOPs. These data are examined to ensure that their acquisition is feasible with the planned instrumentation and within the measurement accuracy attainable. The likely impact of such data on the objective is then examined. The repetition of this process lends increased insight until the test team is convinced that satisfactory measures have been selected and that the required data to quantify (or qualify) these measures can be collected within the resources, instrumentation, and environmental constraints of the T&E.

In any T&E, the appropriate measure or measures for each specific objective should have two basic characteristics: first, and most important, they must be highly relevant (that is, they must describe how well the objective is being achieved); and, second, they must be capable of being measured (i.e., either quantitatively or qualitatively). These desired characteristics are often conflicting, with the most relevant being very difficult to measure and vice versa. For this reason, practical MOEs and MOPs normally can best be established by a T&E team which has sufficient operational experience and knowledge in the area of the systems and the operations being employed. For a typical T&E, this overall team knowledge often requires coverage of a four-part evaluation dealing with system maintainability, compatibility, reliability, and performance.

Under all conditions, it is imperative that the measures be selected for each critical issue and specific test objective prior to the beginning of the T&E. This is required to ensure that the data sample collected during T&E will be pertinent to the objectives, and that the data can be adequately expressed in terms of the system measures that have been selected for addressing those objectives.

To illustrate the procedure just described, consider the specific T&E objective of evaluating performance in terms of the free-fall bombing accuracy of a tactical weapon system. To meet this objective, the T&E team might select circular error probable* (CEP) under various sets of operational conditions as the dominant measure. The value of CEP estimated from sample data obtained during the test would then be compared with desired or standard values (i.e., criteria) during the subsequent evaluation. Because CEP is estimated by an analytical technique that depends upon a series of measurements, a sufficient sample of the basic system measurements (e.g., x and y miss distances from the aim point on the ground) must be planned for and data collected for these measurements during the test. These basic test data are then used to compute an estimation of system CEP (that is, to measure the system's effectiveness under the various sets of operational conditions being investigated). Example problems and calculations for these types of test design problems are presented in Appendix 1.

6.3.5 Measurements and Instrumentation

In a well-designed T&E, the test team identifies the type and amount of data needed, the frequency and accuracy required on each measurement, the method of instrumentation to be used, and other measurement details required to carry out the test. Measurement and instrumentation, in turn, must provide a valid representation of the selected characteristics of the system under consideration; and the results must also be valid, or at least valid to an acceptable degree of usefulness (i.e., they must provide system output in terms translatable to the system MOEs and MOPs—circular error probable, etc.).

Normally, a military field/flight test containing pertinent sets of quantitative measurements made by reliable instrumentation has more credibility than other types of evaluations. However, it is not always obvious why one should make quantitative measurements (i.e., assign numbers to procedures or conditions). Although there is a need for quantitative measuring, given any technical endeavor, it is not effective to perform measurements for the sake of measuring. One is inclined to use instrumentation and make quantitative measurements because it seems more precise. Also, there is a certain comfort in assigning numbers to some set of occurrences. These numbers are then presented as information. Precise information would further imply that the results are repeatable and that one is able to distinguish objects and their properties to some degree of refinement. But again, there is no reason to be precise just for precision's sake. It is not rational to state that a measurement is good or bad based only on whether it is precise or not. Actually, sometimes we do not need to make a fine distinction among objects, and sometimes we do. The degree to which this distinction must be made depends on what values are operationally significant and the measurement requirements necessary to show statistical significance (i.e., what values truly make a difference in success, and the substantiation that they did not occur due to chance).

The overall measurement problem can be specified in terms of the total system resolution, accuracy, and precision as well as the requirements essential to the component parts. This can be approached in at least two ways. First, one can attempt to start with an understanding of what levels are operationally significant based on the a priori knowledge of experts, and then proceed to specify the resolution, accuracy, and precision required in the measurement based on this knowledge. Second, if what is operationally significant is unknown, one can specify

---

*The radius of a circle in which each independently dropped bomb has a .5 probability of impacting.

levels of resolutions, accuracy, and precision based on reasonable costs, and then attempt to establish through T&E the levels that are operationally and statistically significant.

There are at least three notions that apply to these measurements.

1) Measurements provide a language by which we are able to express results more precisely. In general, this is the notion of measurements.

2) The use of measurements causes the user to identify certain system characteristics and the environment in which these characteristics apply. In general, this is specification of data.

3) In establishing measurements, the test team must consider how the results will be used. Specifically, will these data require comparison with other data? In general, this is standardization of data. This area also deals with the broader aspects of transforming measurements into data, into information, and into the decision process. These transformations will usually be effected through the system MOEs and MOPs which, in turn, may be expressed in terms of mathematical relationships or models (e.g., system probabilities, frequency functions, etc.). The models based on measurement data can be descriptive, predictive, or decision models.

Instrumentation, in a general way, provides a physical means of obtaining measurements of greater refinement, or complexity, than are possible by the unaided human faculties or measurements which human faculties would be unable to sense—the latter is particularly true in the fields of modern avionics and weapons delivery.

If we can visualize the domain of a system as containing all of the characteristics of the system, then we can think of subsets of these characteristics. It is usually some portion of the total system or of the subsets that we are able to instrument and measure. Because there are other effects either from unidentified system characteristics, instrumentation anomalies, or model deficiencies, "noise" will be induced from one test measurement to another. Thus the nature of these measurements can be treated as statistical (i.e., they must yield information out of noise). When instrumentation is used, the test team must incorporate safeguards—such as calibration checks, minimum energy transformations, and minimum state changes—into the test design to ensure the required validity of the measurements.

The test design team is also interested in instrumentation from two other points of view. First, it must be sure that the data to be collected during a test on the various variables and parameters are sufficiently valid and accurate to support the analysis required to answer the specific objectives of the test. In many cases, this may require special aircraft and ground instrumentation (e.g., scope cameras, magnetic recorders, and range tracking radars) to enhance data collection and validity. Secondly, it is interested in instrumentation from the point of view of ensuring the best use of scarce test resources. Many types of information can be collected during a test at little or no additional expense, whereas others can be generated only by the use of elaborate and expensive instrumentation (e.g., time-space-position, radar cross section). The important thing to the tester is to be certain that information requiring a costly measurement system, when specified in the test design, is absolutely essential to satisfying the critical issues and objectives of the test.

6.3.6 Selecting the Size of the Sample

Each trial in a T&E contributes a certain amount of information pertinent to the parameter or parameters of interest and, in turn, costs a certain amount of testing resources. Because of the test cost in resources for this information, it is important to be interested in how many times the trial must be replicated in order to draw valid conclusions from the data.

When addressing sample size, two general approaches to the underlying statistics apply: 1) parametric, and 2) nonparametric. The parametric approach assumes that the functional form of the frequency distribution is known and is concerned with testing hypotheses about parameters of the distribution, or estimating the parameters. The nonparametric approach does not assume the form of the frequency distribution (i.e., it applies distribution free statistics).

The quantity of information in any T&E varies with the method of selecting the sample or the design of the T&E. However, once the sampling procedure is specified, the test design team must decide upon the sample size necessary to draw meaningful conclusions from the data. The sample size necessary will depend upon the method of inference—that is, whether the tester wishes to test a hypothesis or estimate the value of a parameter. In both cases, the test design team must specify certain input information to allow the estimation of the required sample size.

6.3.6.1 Determining the sample size for hypothesis testing (See References 1–11).
If the inference is to be a test of a hypothesis, the test design team must state the magnitude of the deviation from the hypothesized value that it wishes to detect and must select the value of α and β (the measures of risk of making an erroneous decision). The sample size must then be chosen to provide the information necessary to satisfy these requirements. For example, suppose a specific objective of the test is to determine whether a new visual aid device combined with the present tactic provides an improvement in the range at which aircrews can locate tanks on the battlefield under clear weather conditions. Further, from previous experience we know that if the tactic is used without the visual aid device, on the average we are able to visually acquire tanks at 4 miles with a variance equal to 2 miles. (If this is not known, it may be the objective of a pretest.) At this point, the test design team must answer the following questions:

1) How large a shift in the average acquisition range do we wish to detect ? For example, this determination is based on what is defined as operational significance, that is, how much of an improvement is necessary for it to be of real operational value. Suppose we wish to detect a shift from 4 to 5 miles.

2) How much variability is present in the population? From previous experience, we have estimated the measure of variability (i.e., the standard deviation, s) to be 2 miles. One practical method of estimating s is to subtract the smallest observation from the largest and divide by 4. This information on the *range* of the values might be obtained from pretest data or based on pretest experience. That is,

$$4s \approx \phi_{max} - \phi_{min}$$

where

s = an estimate of the standard deviation
ϕ_{max} = the maximum value observed
ϕ_{min} = the minimum value observed
$\phi_{max} - \phi_{min}$ = range of the observations

3) How much risk* are we willing to take? (Assume that $\alpha = 0.05$ and $\beta = 0.10$ is sufficient.) Given that the distribution of the data are known (or assumed), the values of α and β can be used to find the corresponding numbers that represent the deviation from the mean (average value) expressed in "standard deviation units (t)." In this example, an $\alpha = 0.05$ corresponds to $t = -1.645$ and $\beta = 0.10$ corresponds to $t = +1.282$, where t for the normal distribution is expressed in standard units (see table for normal distibution contained in most of the referenced statistical texts). From the previous information, the sample size required can now be estimated. Visualize two sampling distributions, one of the average acquisition ranges $(\overline{x_s})$ when the null hypothesis (H_0) is true, $\mu = 5$; and the other when the alternative hypothesis (H_1) is true, $\mu = 4$ (see Figure 6-2). Here $\overline{x_c}$ is called the critical value in that if our $\overline{x}$ computed from the test sample is greater than $\overline{x_c}$, we accept H_0; and if $\overline{x}$ computed is equal to or less than $\overline{x_c}$ we reject H_0. The shaded areas under the curves represent the α and β risks. From the above information, we can now form two simultaneous equations, standardizing $\overline{x_c}$ first with respect to a μ of 5 (the α equation) and, second, with respect to a μ of 4 (the β equation). We then solve for n and $\overline{x_c}$:

α equation: $\dfrac{\overline{x_c} - 5.0}{2/\sqrt{n}} = -1.645$ (based on $\alpha = 0.05$)

β equation: $\dfrac{\overline{x_c} - 4.0}{2/\sqrt{n}} = +1.282$ (based on $\beta = 0.10$)

Subtraction of the second equation from the first and solving yield

$n = 34.2 \approx 35$ observations and $\overline{x_c} = 4.4$ miles

The decision rule then becomes: From a random sample of at least 35, if the computed average acquisition range $\overline{x}$ is greater than 4.4 miles, we accept H_0; and if it is equal to or less than 4.4, we reject H_0. The reject region defined by $\overline{x}$ being equal to or less than 4.4 miles is defined as the critical region. Thus, if we found $\overline{x} > 4.4$ miles, we would have shown that the data obtained using the new device

---

*Often a specific part of a test is addressed by the statement of an initial hypothesis called the "null hypothesis." In the analysis of results of the test, the tester may 1) reject a true hypothesis (type I error), or 2) accept a false hypothesis (type II error). The tester may decide what level of significance he/she wishes to use in testing the null hypothesis. The level of signficance (α) is the probability of rejecting a true hypothesis (type I error). In military testing, an α of 0.05 or 0.10 will probably be adequate for most situations. The probability (β) of accepting a false hypothesis (type II error) is usually expressed in terms of the power of the test (1-β). A power (1-β) of .85 or .90 should also be adequate for most military testing.

combined with our present tactic represent both an *operationally* significant and a *statistically* significant improvement under the conditions previously stated.

6.3.6.2 Determining the sample size for estimation (See References 1, 4–11).
If the inference to be drawn from a test is to be the estimation of a parameter, or parameters, then again the test team must specify certain information pertinent to the design. In this case, a limit (L) or bound must be specified on the error of estimation such that

$$P \text{ (error of estimation} \leq L) = 1 - \alpha$$

Here we are saying that the probability that the error in estimating the value of the parameter is less than or equal to L, which is equal to some specified value (say $1 - \alpha$) that would be acceptable to the test design team.

Two important notions can be applied here in understanding the probability associated with the error of estimation. They are the Empirical Rule and Tchebysheff's theorem. The Empirical Rule states that for a distribution of measurements that is approximately normal, the mean ± one standard deviation will contain approximately 68 percent of the measurements, the mean ± two standard deviations will contain approximately 95 percent of the measurements, and the mean ± three standard deviations will contain approximately 99.7 percent of the measurements. When the distribution is not normal, or it is inappropriate to assume it is normal, Tchebysheff's theorem states that given a number k greater than or equal to 1 and a set of measurements $x_1, x_2, x_3, ..., x_n$, then at least $[1 - (1/k^2)]$ of the measurements will be within k standard deviations of their mean. Thus, the mean ± two standard deviations contain about 75 percent of the measurements, and the mean ± three standard deviations will always contain about 89 pecent of the measurements, regardless of the shape of the distribution.

If the estimator* (Q) possesses a mound-shaped distribution in repeated samplings and is unbiased, then the empirical rule can be used to state that the error of estimating a parameter (q) should be less than $1.645\ s_q$ with probability equal to approximately .90.

Consider the following example. Suppose we wish to estimate the difference in the proportion of successful attacks (perhaps "success" is defined as both killing a target and surviving) achieved by a given type of weapons delivery system when two different tactics are employed. Further, suppose we would like our estimate of the difference to be correct within 0.1 with probability equal to .90. How many observations (trials) of the test are required?

Let n_1 and n_2 represent the number of observations using tactic A and tactic B, respectively. Further, let p_1 and p_2 equal the true proportion of successes using A and B, respectively. The estimator of the difference ($p_1 - p_2$) is

$$(\hat{p}_1 - \hat{p}_2) = \left(\frac{x_1}{n_1}\right) - \left(\frac{x_2}{n_2}\right)$$

---

*An estimator is merely a number obtained from calculations made on the test values, and is used as an estimate of the population parameter.

where

$$(\hat{p}_1 - \hat{p}_2) = \text{unbiased estimator of } (p_1 - p_2)$$

$$\hat{p}_1 = \frac{x_1}{n_1} \; ; \; \hat{p}_2 = \frac{x_2}{n_2}$$

where x_1 = number of successes using tactic A in n_1 trials and x_2 = number of successes using tactic B in n_2 trials.

The variance $s^2_{\hat{p}_1 - \hat{p}_2}$ of the estimator is

$$\frac{p_1 q_1}{n_1} + \frac{p_2 q_2}{n_2}$$

where

$$q_1 = 1 - p_1 \text{ and } q_2 = 1 - p_2$$

Since $(p_1 - p_2)$ will be approximately normally distributed for reasonably large samples, we would expect the error of estimation (0.1) to be less than $1.645 \, s_{\hat{p}_1 - \hat{p}_2}$ with probability equal to .90. (See table for normal distribution contained in most of the referenced statistical texts.)

Therefore, we choose $n = n_1 = n_2$ such that

$$1.645 \left(\frac{p_1 q_1}{n_1} + \frac{p_2 q_2}{n_2} \right)^{1/2} = L$$

If approximate values for p_1 and p_2 are known in advance, they can be used in estimating the sample size required. Again, it is under conditions like these that pretest data should be used to estimate values for p_1 and p_2. On the other hand, under the worst conditions, values for p_1 and p_2 can be guessed with $p_1 = p_2 = .5$ representing the most conservative guess. Then,

$$1.645 \left(\frac{(.5)(.5)}{n_1} + \frac{(.5)(.5)}{n_2} \right)^{1/2} = .1$$

since $n_1 = n_2 = n$,

$$\left(\frac{.25}{n} + \frac{.25}{n} \right)^{1/2} = 0.0607$$

$n = 136$ observations

Thus, in order to meet the stated degree of confidence that the resulting estimate of the difference in the fraction successful for the two tactics $(p_1 - p_2)$ lies within $L = 0.1$ of the true difference, with probability approximately equal to .90, we would require approximately 136 trials using each method. If we were willing to accept a probability of .80, then,

TEST AND EVALUATION DESIGN 125

$$1.282 \, s_{p_1 - p_2} = 0.1$$

and we note that the corresponding calculation yields a reduced sample size of $n = n_1 \approx 83$ observations.

Now, consider an example test objective where our purpose is to compare two maintenance procedures, A and B, respectively. Suppose we have a method of scoring each trial of each procedure with a value ranging from 0 to 100 points. Further, suppose we are interested in estimating the difference in the average score achieved by the two methods correct to within 5 points. Assume that two equal random samples are to be taken using each method and that we expect the scores for each trial to range from 20 to 100 points.

Then, using the range to estimate the standard deviation,
$$4 \, s \approx 100 - 20$$

$$s \approx 20$$

If we are again interested in a probability equal to .9 that the error of estimation will be less than 5 points, and if we assume that the difference in average estimates $(\overline{x_1} - \overline{x_2})$, where $\overline{x_1}$ and $\overline{x_2}$ are the average scores computed from samples A and B, respectively is distributed normally, then,

$$1.645 \, s_{\overline{x_1} - \overline{x_2}} = 5$$

$$1.645 \left(\frac{s_1^2}{n_1} + \frac{s_2^2}{n_2} \right)^{1/2} = 5$$

since $n = n_1 = n_2$, then,

$$1.645 \left(\frac{20^2}{n} + \frac{20^2}{n} \right)^{1/2} = 5$$

$$n \approx 87 \text{ observations}$$

Hence, a sample of approximately 87 observations using each method represents a practical solution to the sample size problem for the conditions stated.

6.3.6.3 Determining the number of bombs that must hit to achieve a given circular error probable.
Suppose the success criterion has been defined such that 50% of the bombs dropped must have miss distances of 60 feet or less to be considered hits [i.e., circular error probable (CEP) 60 feet]. Further, suppose it is desired to be 90% confident that the system meets this criterion. During the test, 24 bombs will be dropped. What is the minimum number of bombs that must score 60 feet or less to meet the criterion?

N = number of bombs to be dropped = 24
r = number of bombs that are successes (i.e., score 60 feet or less)

q = proportion of the population that is not successful (i.e., score greater than 60 feet)
g = level of confidence = 90%

Problem

Find an upper limit q, such that we are 90% confident that no more than q proportion of the population do not meet the criterion, where $q = 1 - p$. In this problem we establish $q = 50\%$, $N = 24$, and look up the corresponding value for r [see binomial tables for one-sided confidence limits (see Reference 12)]. From the table,

$$r = 16 \text{ for } q = 1 - .51551 = .48$$

Therefore, we are 90% confident that the CEP will be 60 feet or less if at least 16 bombs score 60 feet or less out of the 24 drops.

6.3.7 Data Management and Computer-Assisted Analysis (See Reference 13)

Data management is the process of ensuring adequate methods of data collection, the maintenance of data files, the screening and retrieving of selected information, and the preparation of selected data summaries. In T&E it also deals with the problem of providing for a repository of data to allow the establishment of performance standards and, in many cases, computerized mathematical and statistical analysis routines to aid in quality control and the analysis of the T&E data. During many T&Es, a large quantity of data must be collected on a daily basis and numerous data collection formats used. To assist in the management, processing, and handling of the data, in most cases, the tester must make special arrangements for data processing and computerized analytical support. This support is usually of the following kinds:

1) Research and analysis of stored data that may totally or partially satisfy a T&E objective or aid in establishing data collection requirements.

2) Special assistance in data collection techniques and computer-assisted analysis techniques.

3) Machine data edit services to improve the quality of data and to pinpoint data collection flaws and/or areas requiring more closely monitored collection management.

4) Computer support services for rapid building of data files for specific sets of data.

5) Updating of data files through computer programming techniques that permit adding to, deleting from, and changing data items in a data record or file.

6) Data retrieval services permitting rapid retrieval of any or all data in a file, machine-printed to meet the user's requirement in terms of content, format, and timeliness.

7) Standard periodic machine-written reports designed to meet special T&E requirements.

8) Computer programming and data systems services that permit the development and production of new machine-written reports and/or the modification of

existing machine-written reports. In addition, special requirements may exist to develop statistical subroutines or to modify existing routines to fit particular T&E needs.

Normally, these types of capabilities can be used on sets of data that meet one of the general conditions listed below (i.e., data repository services are not restricted to just those applications requiring computer support because of data volume):

1) Where large numbers of observations are to be made or large numbers of data elements must be measured for ultimate collective evaluation.

2) Where large numbers of combinations of observations are to be made, though the total events or observation period may be relatively short.

3) Where a base must be established for new weapon systems coming into the inventory, regardless of the size of the initial test. (Examples: F-22, F-117, F-15X, missile X.)

4) Where complex analysis must be performed that would be too cumbersome to handle manually.

5) Where it has been determined that frequent reference to the T&E data will be necessary in the future, or the information is determined to be vital and of historical value.

Upon completion of a T&E and its corresponding analysis and reporting, it may be desirable to retain all data collected. Under these conditions, the data will normally be stored in a format that allows for computerized data retrieval. In addition to the data collected during a given T&E, it is sometimes desirable to include in the data bank well-specified sets of data from other sources that have been quantified. The data system used for these purposes should allow rapid data retrieval and also be compatible with computer analysis routines.

An important aspect of data management is the timely and orderly flow of data into the data control and collection function. These data should be processed and summarized continually during the T&E to the degree practical to allow "quick look" analyses. These analyses form the basis for providing timely feedback concerning project status, data adequacy, and preliminary evaluation to operating elements within the T&E project. This feedback must be sufficient to support management decisions and to allow the necessary adjustment, refinement, and completion of the T&E in such a manner that the critical issues and objectives are met.

6.3.8 Decision Criteria

There are several types and levels of decision criteria associated with T&E design (Reference 14). These criteria should be identified or established before the beginning of the T&E to the degree practical. Decision criteria are those explicit values of quality or performance that must be reached or exceeded for continuing the T&E process, accepting a data trial into the data base, or stating that an objective of the T&E has been successfully met. Examples of T&E design areas where decision criteria are essential are listed below.

1) Go/no go mission criteria for each mission.
2) Data validity criteria for each trial.
3) Evaluation criteria for each objective.
4) Overall evaluation criteria for areas of performance, such as reliability, bombing accuracy, and survivability.

The primary reason for establishing these criteria before the T&E begins is that they are essential to ensuring a consistent-quality data product for use in the analysis of results. Generally, it will be possible to establish all T&E decision criteria during the design process with one notable exception. This exception most often occurs when little is known about "how much is enough" with regard to some specific objective, and the T&E itself is used to contribute to that body of knowledge for the decision process. Given that this is the case, the T&E designers should acknowledge such during the design, establish the need for these values as a part of the T&E itself, and clearly define the decision criteria as they were applied to those objectives after the T&E when the results are being reported.

6.3.9 Limitations and Assumptions

The limitations and assumptions associated with any T&E can greatly influence the level of confidence that can be placed in the results. When limitations and assumptions are made in the T&E design to the extent that the T&E will most likely produce unrealistic results, the design team must either correct these shortcomings or seek out a new design that eliminates the problems.

The limitations of the T&E must be known, understood, and documented by the tester. For example, if only six test items are available for the T&E, statistical confidence or proof is not likely to be a product of the T&E. On the other hand, if the same result or problem occurs on every trial of the T&E, the probability is significant that the problem requires attention and resolution even though only six test items were expended.

Assumptions usually must be made to allow the T&E design, conduct, and analyses to proceed. These assumptions should be documented during T&E design along with the supporting rationale stating the need for and the reasonableness of the assumptions.

6.4 Experimental Test and Evaluation Design (See References 1–32)

The second approach to T&E design, experimental T&E design, is a formal and structured process that requires a significant degree of control to examine hypotheses and to measure developmental and operational phenomena. Properly conducted and analyzed experimental tests can provide a great insight into how military equipment and operations might be improved. The actual method for an experimental T&E may be driven by the needs of the analysis and the assumptions of the experiment rather than the total realism of the operation. Experimental T&E design is often used to determine the degree to which selected variables and factors can possibly affect the outcome of the experiment. Usually the results revealed from an experimental T&E are fed into a procedural T&E based on a more conventional and realistic military operation for final verification and impact.

6.4.1 Test Variables

Experimental test variables may be classed as dependent or independent. In an experimental test, the effects of controlled changes of the independent variable are measured on the dependent variable. The system or object to which treatment (change of the independent variable) is applied in a single trial of the test is called an experimental unit. Although any number of the proposed variables identified by military and technical experts might be expected to affect operational performance, the tester must choose to deliberately vary only a small number of them in a controlled way if his/her objective is to establish a cause-and-effect relationship. The remainder of the factors that affect the outcome of the experiment can also be controlled by being held constant. In a test, dependent variables usually represent the mechanism by which MOEs and MOPs are evaluated for each specific objective, and it is highly desirable that they be quantitative, whereas independent variables may be either quantitative or qualitative.

6.4.2 Sensitivity of Variables

The tester must determine the relative importance of each candidate independent variable for inclusion as a controlled variable during the experiment. He/she should select as controlled variables those likely to have the greatest influence on developmental and operational performance. This selection will be aided by a thorough knowledge of the military operations and environment, the results of previous studies and tests, and in some cases by the use of mathematical modeling or other pretest analyses of the functions under consideration. To successfully perform this task, the T&E design should be made by a team of persons with a variety of knowledge and skills. This is necessary for two reasons. First, a complex military test is likely to involve many diverse disciplines which cannot be handled by a single specialty. Secondly, a testing problem looks different to a pilot, an operations research analyst, an engineer, a maintenance technician, or a professional tester; and their different views contribute to a broader and more comprehensive design.

6.4.3 Operations Tasks

Any given military operation will consist of a number of tasks that must be performed to some degree simultaneously or sequentially in some pattern to achieve success for the operation. Definition of what constitutes the successful accomplishment of each task will aid in finalizing the objectives of the test. The test participants should be aware of such definitions as constituting criteria by which their performance of the task will be judged. For example, typical objectives that might be selected for the ordnance delivery* task are 1) maintaining guidance tracking to target; and 2) achieving weapon impact or detonation within the specified envelope. To address the stated test objectives, a number of basic measurements must be made on each flight.

Examples of test measurements for the ordnance delivery task are: 1) target and/or sensor position and altitude as a function of time; 2) target-to-weapon range at impact or detonation; and 3) events (weapon release, fire control break-lock,

---

*The task of ordnance delivery is defined here as all activities affecting the trajectory of the bomb, bullet, missile, or other weapon from release to impact or detonation. It includes guidance tracking, if applicable.

fire control reacquisition, weapon acquisition, weapon break-lock, and impact or detonation) as a function of time.

6.4.4 Analysis

The analysis presentation, developed on the basis of the T&E design, is used to describe the analytic process by which the basic data to be collected during the test will be incorporated into meaningful relationships (e.g., MOEs, MOPs) that determine how well the test objectives have been satisfied. For example, in considering the ordnance delivery task, the analytic presentation might consist of 1) miss distances, 2) distribution of miss distances along with the identification of major error sources, 3) probability of impact or detonation within a specified envelope, 4) probability of certain events occurring, or 5) times between significant events.

When other factors are introduced into the test design as variables, these basic analytic functions should also be determined with respect to the new variables. For example, if aircraft dive angle is introduced as a variable factor (two or more dive angles), the probability of impact or detonation within a specified envelope as a function of aircraft dive angle may be calculated.

6.4.5 Factor Categories

Independent variables can be referred to as "factors" and may be further classed as 1) treatment factors, 2) classification factors, or 3) random factors.

6.4.5.1 Treatment factors. Treatment factors may be of three kinds:
1) Factors of direct interest (e.g., dive angle).
2) Factors included to provide a better understanding of the factors of direct interest (e.g., aircraft speed).
3) Factors connected with performing the test (experimental technique).

Treatment factors are varied in a controlled manner during the test, and may be either quantitative or qualitative.

6.4.5.2 Classification factors. Classification factors are qualitative and may be of two kinds:
1) A natural grouping of experimental components (e.g., homogeneous items such as premission briefing factors) that could influence the test results.
2) Deliberately inserted variations in the experimental units.

Classification factors are either varied in a controlled manner or held constant throughout the test.

6.4.5.3 Random factors. Random factors are those factors that cannot be isolated as either treatment factors or classification factors because they are either unknown or not subject to test control. However, the effects of random factors can be reduced by randomizing experimental units or by making supplementary (concomitant) measurements. Examples of random factors might be 1) the differing abilities of test participants and 2) some natural phenomena which are difficult to control, such as weather, visibility, turbulence, insects, and birds.

6.4.6 Design Considerations

Any experimental T&E involving several levels of one or more factors can be designed in such a way that the data obtained can be analyzed by the technique of analysis of variance. This technique enables us to break down the variance of the measured variable into portions caused by the several factors, varied singularly or in combination, and a portion caused by the experimental error. As discussed in earlier chapters, measurements are not precisely reproducible; therefore, to an extent depending upon the size of the experimental error, the test must be replicated in order to draw valid conclusions. Therefore, the most efficient procedure for planning the analysis and for designing the test is to 1) estimate the errors likely to be encountered during the test on the basis of previous experience; 2) estimate the number of replications required; 3) perform the test according to the test design for treatment factors, randomizing or holding constant all other important factors; and 4) analyze the results according to the plan [e.g., analysis of variance (See Reference 15), and regression analysis (see Reference 17)].

Initially, all pertinent factors should be identified and listed. This total factor list is reduced by selecting factors that apply to the specific equipment, operations, and tasks chosen for consideration in the experiment. The factor reduction criteria applicable to the particular test should be applied (e.g., sensitivity, relative importance, operational significance, and degree to which satisfied by previous research or testing) to identify which factors are to be chosen as variables.

The factor list should include all factors that must be measured or described, introduced as independent variables, or held constant throughout the test. This list can then be divided into the three factor categories (treatment, classification, and random). This determination is made by the test design team only after careful study of the critical issues and specific objectives to be addressed by the test.

The number of levels of control of a qualitative treatment factor or a classification factor is dictated by the nature of the factor and the objectives of the test. *Treatment* factors should have a minimum of two levels, whereas classification factors may be held constant at a single level. *Quantitative treatment factors* may be considered at only two levels if an objective of the test is to see whether the factor has an effect, and, if so, to determine its direction. At least three levels of a quantitative factor are necessary to investigate nonlinear response. If only two levels of a quantitative factor are to be used, the levels should cover the test range of interest.

The factors within the factor groups are categorized, and levels of treatment and classification factors are specified in accordance with the objectives of the test. Most factors for any one investigation will fall within the classification and random factor categories, to be either controlled at one level or randomized. The example of the sequence discussed here for establishing the factor list and categorizing the various factors for test design is summarized in Figure 6-3.

6.4.7 Example of Categorizing Factors for a Multiple-Factor Test

For the purpose of illustration, let the following list represent independent variables selected from a more comprehensive list for a test involving the task of ordnance delivery for the F-15X aircraft.

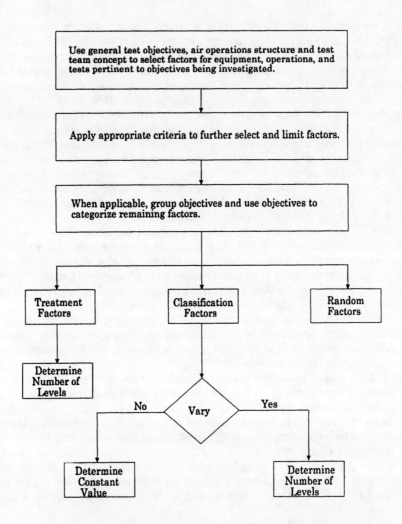

Figure 6-3. Example Sequence for Determining Factor Categories

6.4.7.1 *Example list of test variables for ordnance delivery.*

Aircraft types—1 level
   F-15X
Target types—3 levels
   Oil tank
SAM site
   Bridge

Aircraft speed—3 levels
  430 knots
  330 knots
  230 knots
Aircraft dive angle—3 levels
  10–20 deg
  25–35 deg
  40–50 deg
Aircraft altitude—2 levels
  1000–5000 feet
  15,000–18,000 feet
Terrain/vegetation—1 level
  Hilly; forested
Illumination/visibility—1 level
  Nighttime; Vertical Flight Rules (VFR) flying conditions
Intelligence/briefing—1 level
  Good intelligence; preplanned targets
Experience/training—1 level
  Average combat-ready
Distance from Initial Point (IP) to target—1 level
  2 miles
Target contrast—1 level
    Target painted natural color in clear area

Factors should then be grouped as treatment, classification, or random factors as shown below.

6.4.7.2 Example grouping of test factors for ordnance delivery.

Treatment factors
  Factors of direct interest
    Aircraft speed—3 levels
    Aircraft altitude—2 levels
    Aircraft dive angle—3 levels
Classification factors
  Natural groupings
    Aircrew aids—1 level
    Intelligence—1 level
    Doctrine—1 level
    Topography—1 level
    Ground cover—1 level
    Checkpoints—1 level
  Deliberately inserted variation
    Target characteristics—3 levels
Random factors
  Human factors
  Heading
  Maneuvers

Meteorological conditions
Illumination/visibility

Within the ordnance delivery task, suppose we are interested in the effects of aircraft altitude, speed, and dive angle upon the dependent variables [task objectives, such as the achievement of a specified circular error probable (CEP)]. These factors should be considered as treatment factors of direct interest. The remaining factors should be considered either as classification factors controlled at one level or as random factors that might be measured as concomitant variables (i.e., variables which accompany the primary variables and are measured in a subordinate or incidental way). Factors concerning primary systems are usually treatment or classification factors, depending upon the objectives of the tests. Intelligence and aircrew aids are generally considered standard and are, therefore, classification factors at one level. Factors affecting natural environment can be classified as treatment, classification, or random, but are generally selected and as such are classification or random factors at one level. Target characteristics are normally considered as classification factors and can be used to increase the generality of a test. Individual pilots or aircrews are considered the experimental unit characterized by human factors and will usually, but not always, be considered as random factors.

6.4.8 Analysis of Variance (See References 1–11, 15–31)

One of the most useful techniques for increasing the amount of information derived from a test is designing it in such a way that the total variation of the variable being studied can be separated into components that are of interest or importance to the test. Splitting up the total variation in this manner allows the test team to use statistical methods to eliminate the effects of certain interfering variables and thus increases the sensitivity of the test. The analysis of variance is a technique for carrying out the analysis of a test from this point of view. The analysis of variance technique allows the test team to design formal tests that apply a greater amount of statistical sensitivity to better ascertain the true effects of selected variables.

Two well-known mathematical models for analysis of variance are available for application to tests of this type. One is called the linear hypothesis model, and the other is called the components-of-variance model. The basic difference between the two models lies in the assumptions made concerning the population of tests of which the given test is considered to be a random sample. Essentially, the linear hypothesis model makes assumptions about the means of the basic variables and draws inferences on the basis of a comparison of these means, whereas the components-of-variance model makes a linearity assumption about the basic variables themselves. For a more detailed treatment of the underlying theory of these models, the reader is referred to References 1–11 and 15–31.

6.4.8.1 Example analysis of variance problem. Suppose each of three pilots delivers weapons during a test with four different weapons delivery modes and achieves a different circular error probable (CEP) as listed in the matrix below:

TEST AND EVALUATION DESIGN

Delivery Mode	Pilots 1	2	3	$\overline{Y}_{i.}$
1	50	80	75	68.3
2	98	90	105	97.7
3	68	63	88	73
4	50	55	55	53.3
$\overline{Y}_{.j}$	66.5	72	80.8	73.1

We may wish to apply analysis of variance to ascertain whether the skills of the individual pilots or the weapons delivery modes were significant in producing the different CEPs. We assume that each Y_{ij} represents a random sample value from a normal population and that a different normally distributed population is associated with each cell in the matrix. We also assume that the standard deviations of all the normal populations from which the Y_{ij} are drawn are equal to the same value σ, which is estimated by $\hat{\sigma}$. Based on these assumptions, the following calculations can be made:

Analysis of Variance (Two-Way Classification)

Variation	Sum of Squares	Degrees of Freedom	Mean Square
Rows	$S_R = s\sum_{i=1}^{r}(\overline{Y}_{i.} - \overline{Y})^2$	$r-1$	$\hat{\sigma}_R^2 = \dfrac{S_R}{(r-1)}$
Columns	$S_c = r\sum_{j=1}^{s}(\overline{Y}_{.j} - \overline{Y})^2$	$s-1$	$\hat{\sigma}_c^2 = \dfrac{S_c}{(s-1)}$
Errors	$S_E = r\sum_{i=1}^{r}\sum_{j=1}^{s}H_{ij}$	$G=(r-1)(s-1)$	$\hat{\sigma}_E^2 = \dfrac{S_E}{G}$
Total	$S = r\sum_{i=1}^{r}\sum_{j=1}^{s}(Y_{ij}-\overline{Y})^2$	$rs-1$	$\dfrac{S}{(rs-1)}$

$$S = S_R + S_c + S_E$$

$$H_{ij} = \left(Y_{ij} - \overline{Y}_{i.} - \overline{Y}_{.j} + \overline{Y}\right)^2$$

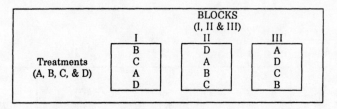

Figure 6-4. Example of a Randomized Block Design

r = number of rows (i.e., weapons delivery modes) = 4

s = number of columns (i.e., pilots) = 3

$\overline{Y} = 73.1$

$\overline{Y_{1.}} = 68.3,\ \overline{Y_{2.}} = 97.7,\ \overline{Y_{3.}} = 73,\ \overline{Y_{4.}} = 53.3$

$\overline{Y_{.1}} = 66.5,\ \overline{Y_{.2}} = 72,\ \overline{Y_{.3}} = 80.8$

$S_R = 3060.8,\ S_c = 416.4,\ S_E = 569.8,\ S = 4047$

Test Hypothesis 1 (No significant difference in pilots). This test is based on use of the F distribution and associated tables cited in the statistical references.

$$F \text{ calculated } = F_c = (r-1)\frac{S_c}{S_E} = (4-1)\frac{416.4}{569.8} = 2.19$$

For a significance level $\alpha = .05$, F table shows $F_\varepsilon = 5.14$ for $m_1 = s - 1 = 2$ and $m_2 = (r-1)(s-1) = 6$.

Since F_c is less than F_ε (i.e., 2.19 < 5.14), we conclude that there is not a significant statistical difference in the delivery accuracy of the pilots.

Test Hypothesis 2 (No significant difference in weapons delivery modes). This test is also based on the F distribution and associated tables cited in the statistical references.

$$F \text{ calculated } = F_s = (s-1)\frac{S_R}{S_E} = (3-1)\frac{3060.8}{569.8} = 10.7$$

For a significance level $\alpha = .05$, then F table shows $F_\varepsilon = 4.76$ for $m_1 = (r-1) = 3$ and $m_2 = (r-1)(s-1) = 6$.

Since F_c is greater than F_ε (i.e., 10.7 > 4.76), we reject the hypothesis and conclude that there is indeed a significant statistical difference in the weapons delivery accuracy achieved by the different weapons delivery modes. Operators and analysts must further examine these differences to ascertain whether they are operationally significant.

6.4.8.2 Some standard designs based on analysis of variance.

Numerous statistical techniques and experimental T&E designs are available in the referenced texts and can be readily applied in military developmental and operational T&Es. The advantages and disadvantages of each of these designs must be considered when one is attempting to apply them to a particular situation. When these theories were developed, certain assumptions were necessary (e.g., "errors" are purely random and normally distributed and samples are unbiased). These assumptions were made by the statisticians to simplify the mathematics and enable them to present the results in a concise and usable form. The tester, when employing the theory, should accept the statisticians' good intentions by planning and conducting the T&E in such a way that the assumptions in the theories are met. This section presents some of the more important features of these designs and cites references where the appropriate analysis procedures are documented in greater detail.

6.4.8.2.1 Randomized blocks (See References 6-11, 16). Theoretically, to obtain the highest precision in comparative trials, the trials should be carried out under identical conditions apart from the conditions being compared. In practice, particularly when the number of trials required is quite large, it is usually difficult to ensure such uniformity because of the natural variability of the functions and processes involved. Even if the variability could be eliminated, it is doubtful whether it would be wise to try, since such variability broadens the basis of comparison and thus renders the results more generally applicable. It is often possible, however, to split up a set of trials into smaller subsets within which the variation is likely to be less than in the set as a whole. Thus, aircraft of the same configuration are expected to be more alike in their properties than aircraft of different configurations, and samples taken over a short period of time can be expected to vary less than those taken over a longer period. Where these conditions hold, the precision of the test can be increased by dividing it into *blocks* within each of which random variations are expected to be smaller than in the test as a whole. Figure 6-4 illustrates an example of a layout of a randomized block design. The dependent variable in the experiment is a function of two qualitative variables, "blocks" and "treatments." This example consists of four treatments (A, B, C, and D) and three blocks (I, II, and III).

Suppose in the test design (Figure 6-4) the blocks represent tactics, and the letters within the blocks (treatments) represent sets of aircrews with four different levels of combat training. Further, suppose that an equal number of observations are made for the treatment factor combinations shown (cells) and that the corresponding number of successful missions achieved is listed in the appropriate cells. Then, using an F statistic, it is possible to test the hypothesis that there is a statistically significant difference in the number of successes achieved that can be attributed to the type of tactic employed. Also, it is possible to test the hypothesis that there is a statistically significant difference which can be attributed to the different levels of aircrew training. The test team would have to examine the amount of difference in success resulting from the tactics and aircrew training to ascertain whether it is operationally significant. In tests designed to meet objectives where all treatments are applied to each block, several blocks can be used to ensure that the number of observations is sufficient to give the required precision to the test as a whole. Where the block is too small to accommodate all the treatments, it may be necessary to use a more complex *incomplete block* design. The methodology used for analysis of both of these types of test designs is discussed in detail in Chapter 4 of Reference 6.

Although the variability within any block in this type of design is likely to be less than that in the test as a whole, there may be a systematic variation within the individual blocks. Thus, aircraft of a given configuration selected at the beginning

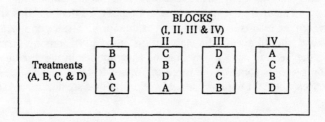

Figure 6-5. Example of a Latin Square Design

of production may differ systematically from those selected at the end of production. If, therefore, the treatments are introduced in the same relative positions in successive blocks, spurious effects caused by the systematic variations associated with position within a block are likely to be introduced into the results. To overcome this, the arrangement of treatments should be different in each block, with the actual positions of the treatments within any block being chosen by an adequate random process.

6.4.8.2.2 Latin, Graeco-Latin, and Youden squares (See References 4, 6–11). In the special case where the number of blocks is equal to, or is a multiple of, the number of treatments, it is possible to improve on the random arrangement by using a design in which each treatment appears an equal number of times in each position in a block. For example, suppose our previous sample problem is expanded to include four different tactics (blocks) to be exercised during the test by the four sets of aircrews (treatments). The treatments would then be distributed randomly among the various sets of positions. Such designs are known as Latin square designs.

Figure 6-5 illustrates a Latin square design with four blocks (I, II, III, and IV) and four treatments (A, B, C, and D).

The square can be of any size, and it should be noted that each letter (treatment) occurs once in each row and once in each column, but always in a different posi-

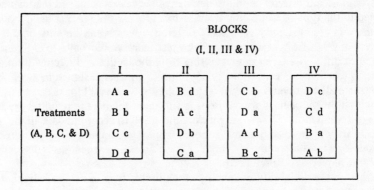

Figure 6-6. Example of a Graeco-Latin Square Design

tion in each. The analysis for this design is similar to that discussed for randomized blocks and can be found on page 66 of Reference 6.

Another design of interest involves the combining of two orthogonal* Latin squares and is called the Graeco-Latin square. This design is illustrated in Figure 6-6. Notice in this design that a third restriction is at levels a, b, c, and d; and not only do these appear once and only once in each row and each column, but they appear once and only once with each level of treatment (A, B, C, and D). To illustrate how this design might be applied to our previous example, suppose the four treatments still represent four sets of aircrews and the blocks still represent four different tactics. The third restriction (a, b, c, and d) could then be used to represent four different types of targets. The analysis technique for this design is very similar to that for a Latin square and is discussed in considerable detail beginning on page 79 of Reference 6. In some cases, it may be impossible to use the same number of levels for all factors as is provided by the Latin square. In these cases, a special kind of incomplete Latin square design called Youden squares may be applicable. The analysis of variance for this design is discussed in Chapter 16 of Reference 6.

6.4.8.2.3 Factorial and fractional factorial designs (See References 6–11, 16).
The classical idea of testing is to have all the independent variables (factors) but one constant. It is frequently not recognized that this may sometimes be far from ideal, because, in order to get a fair assessment of the effects of varying the particular variable, one must allow the others to vary over their full range as well. If they were held constant, they likely would have to be held constant at arbitrary values. Thus, varying factor A from a value of A_1 to some other value A_2 may produce a given change in the output when B is at a value of B_1, but a different change in the output when B is at a value of B_2. The technique of factorial testing developed by R.A. Fisher and his colleagues (References 25, 28, and 31) for use in the science of agriculture is designed to detect this type of effect.

Assuming that the experimental T&E has been correctly designed, it is possible to determine not only the effect of each individual factor, but also the way in which each effect depends on the other factors (i.e., the interactions). The factorial design is most useful in cases where the experimental errors are usually large compared with the effects being investigated and the cost of the individual observations is not very great. Elaborate designs with much replication are therefore frequently used.

In certain types of testing, the test errors may be smaller in proportion to the effects sought, and the cost of the individual test trials may be considerable. Moreover, it is sometimes possible to state on theoretical grounds which factors are likely to interact and which may safely be assumed not to interact. Under these conditions it may be desirable to use a *fractional factorial* design, which enables the size of a test to be reduced to a fraction of that of a full factorial test and still yield the important results. These designs are generally thought of as volume-increasing designs because, by proper choice of the combinations of levels of variables, it is possible to shift information derived from the test to focus on the spe-

---

*Two designs are said to be orthogonal if the estimators of the parameters associated with any one factor are uncorrelated with those of another (i.e., if, when they are combined, the same pair of symbols occurs no more than once in the composite square).

cific parameters of interest. Factorial tests are covered in detail in References 4, 6, and 7.

There are a few other items that should be noted concerning experimental T&E designs. For the randomized block designs, it should be noted that, as the number of treatments is increased, the size of the block must increase correspondingly. On the other hand, as the size of the block is increased, the within-block variability usually increases and thereby reduces the advantages of blocking; however, in many cases, this difficulty may be overcome by use of an incomplete block design. It is also important to note the assumptions implied by block designs. First, it is assumed that the effect of any row or column of the Latin square remains the same, regardless of the combination of row, column, and treatment. In other words, if a trend exists within columns (caused by a difference between rows), it is assumed that this trend will be identical in all columns, without regard to experimental error. Also, it is assumed that the trend within rows (caused by a difference between columns) is identical for all rows. From this, it can be seen that the Latin square design is not constructed to detect row-column, column-treatment, or row-treatment interactions if they exist. When such interactions exist, blocking will fail to remove them and, therefore, will be ineffective in reducing experimental error. Under these conditions, other designs like factorial and fractional factorial may be used.

6.4.8.2.4 Missing values (See References 6, 8). No matter how well an experiment is planned, there are unusual cases where experimental observations could not be made or are missing from the data. In these situations it is sometimes possible to estimate the missing values with appropriate analytical techniques. Some of these techniques for the various designs are discussed beginning on page 70 of Reference 16.

6.4.8.2.5 Randomization (See References 4, 6, 15). In any experimental T&E design it is important that the various experimental units be randomized. The term "random" refers to the assignment of treatments to the experimental units in such a manner that there is equal probability of any particular assignment. Most experimental designs require specific schemes of randomization. Randomization is necessary for a hypothesis testing procedure that relies on the underlying assumption of independent errors and allows the experimenter to proceed as if the errors were indeed independent. Sometimes this is randomization in space, as the layout of targets in an area, or in time, where a series of block tests are carried out on a single aircraft. Where there are several aircraft of the same type in use, it may be appropriate to address the question of which experiment is done on which aircraft by randomization.

In a single factorial experiment, the position of any experimental unit can be randomized by giving each unit a serial number and then choosing these numbers randomly. If the experiment is confounded,* we should randomize the order of the blocks and then randomize the order of the units within the blocks. When a Latin square design is used, the treatments should be allocated to the letters randomly, and the rows and columns should be allocated to their respective factors.

---

*The term "confounded" describes a certain phenomenon which occurs in experimentation and is simply a synonym for "mixed together"; that is, two (or more) effects are said to be confounded in an experiment if it is impossible to separate the effects when the subsequent analysis is performed.

With balanced incomplete blocks, the treatments should be allocated to the letters randomly, and the blocks then arranged in random order.

6.4.8.2.6 Limitations. In practice, three major circumstances may seriously affect the validity of the results of military experimental T&Es. One is the collection of an insufficient quantity of usable data to answer the specific objectives of the test. The statistical aspects of the experimental test design are directed at minimizing this possibility given that it can be afforded. The second circumstance deals with violation of the assumptions that underlie the analysis procedure. Often in complex military T&E, it is not possible to obtain the control and randomization required to apply experimental T&E design. The last circumstance is the unreliability of the test data collected because of the lack of test environment realism. The experimental test design procedures discussed herein are not intended to imply that military T&E should be controlled to the extent that operational realism is lost. Because a trade-off must always be made between test control and realism, it is advocated that the trade-off be made only after careful consideration, and that the experiment be controlled only to the degree necessary to produce valid data that are amenable to subsequent analytical evaluation.

6.5 Sequential Methods (See References 15, 19, 32)

In a sequential test of a statistical hypothesis, certain calculations are made after each observation (or group of observations) and, on the basis of the calculated results, a decision is made as whether 1) to accept the hypothesis under test; 2) to accept a selected alternative hypothesis; or 3) to withhold judgment until more data have been examined or have become available.

The decisions are so governed that the hypothesis is accepted or rejected as soon as it appears that the available data are adequate for making a decision with preselected degrees of reliability. In a sequential testing process, three quantities play important roles: first, the number of observations (n); second, as discussed earlier under hypothesis testing, the probability (α) of erroneously rejecting the hypothesis under test; and third, the probability (β) of erroneously accepting this hypothesis. In theory, the number of observations considered necessary to reach a decision is minimized while fixing the α and β risks. This procedure is somewhat in contrast to that used in nonsequential tests where n is likely to be fixed, and one of the risks is minimized while fixing the other. Sequential tests are frequently more economical than nonsequential tests, especially when n is readily changed, whereas nonsequential tests are more likely to be more economical when n is not easily changed. Often the number of observations (n) required in a sequential test of a given reliability is considerably less than the number of observations required by the corresponding nonsequential test. Sequential analysis and testing methods are treated in considerable detail, beginning on pages 352 and 221 of References 15 and 19, respectively.

6.6 Nonparametric Example (See References 33–36)

Most of the testing techniques previously described are based on parametric statistics, that is, statistical methods which rest on particular assumptions about distribution forms and their parameters. Often information about the distribution forms is not available, or the tester does not choose to make such assumptions. Under these conditions it is sometimes desirable to apply methods called nonparametric which do not require particular assumptions about the form of population distributions. An excellent example is the contingency test in the 2×2 table shown below.

		Options		
		I	II	Total
Result	Successes	A	B	A + B
	Failures	C	D	C + D
	Total	A + C	B + D	N

This table can be used to classify and statistically compare two options (I and II) in terms of their successes and failures during the test. The Chi Square (x^2) statistic is used with $(2-1)(2-1) = 1$ degree of freedom. There should be no less in each cell than 5, unless N is greater than 40. However, in cases where this condition is not met, the comparison can still be made using Fisher's Exact Probability Test. The value of x^2 can be calculated as follows:

$$x_c^2 = \frac{N\left(|BC - AD| - \frac{N}{2}\right)^2}{(A+B)(C+D)(A+C)(B+D)}$$

Suppose we wish to compare the success of two electronic jammer systems in protecting an aircraft. Both systems are scored in terms of their capability to successfully cause missiles fired at the aircraft to miss beyond their lethal range. The hits and misses for systems #1 and #2 are shown in the matrix below.

	Electronic Jammer #1	Electronic Jammer #2	Total
Hits	3	15	18
Misses	53	65	118
Total	56	80	136

Test Hypothesis (There is no statistical significant difference in hits against jammer #1 and jammer #2.)

Alternative Hypothesis (Hits against jammer #1 are significantly fewer than hits against jammer #2.)

We calculate

$$x_c^2 = \frac{136\left([(3)(65) - (53)(15)] - \frac{135}{2}\right)^2}{(18)(118)(56)(80)} = 4.05$$

For a significance level $E = .05$, x^2 table shows $x_\varepsilon^2 = 3.84$ for a one-sided test with 1 degree of freedom.

Since χ_c^2 is greater than x_c^2 (i.e., $4.05 > 3.84$), we reject the null hypothesis and accept the alternative—that is, hits against jammer #1 are significantly fewer than hits against jammer #2 at the .05 level. Again, the operators, analysts, and testers must make the additional determination of whether this statistically significant difference is indeed operationally significant.

References

[1] Mace, A.E., *Sample-Size Determination*. New York: Reinhold Publishing Corporation, 1964.

[2] Tang, P.C., "The Power Function of the Analysis of Variance." Tests with tables and illustrations of their use. *Statistical Research Memoirs*, Vol. 2, 1938.

[3] Pearson, E.S., and Hartley, H.O., *Biometrika*, Vol. 39, 1951.

[4] Sheffe, H., *The Analysis of Variance*. New York: John Wiley and Sons, Inc., 1959.

[5] Dixon, W.J., and Massey, F.J., Jr., *An Introduction to Statistical Analysis* (2nd ed.). New York: McGraw-Hill Book Company, 1957.

[6] Hicks, C.R., *Fundamental Concepts in the Design of Experiments* (3rd ed.). New York: Holt, Rinehart and Winston, 1982.

[7] Mendenhall, W., *Introduction to Linear Models and Design and Analysis of Experiments*. Belmont, California: Wadsworth Publishing Company, Inc., 1968.

[8] Davies, O.L. (ed.), *The Design and Analysis of Industrial Experiments*. New York: Hafner Publishing Co., 1956.

[9] Ostle, B., *Statistics in Research*. Ames, Iowa: Iowa State University Press, 1963.

[10] Kempthorne, O., *The Design and Analysis of Experiments*. New York: John Wiley and Sons, Inc., Second Printing, 1960.

[11] Brownlee, K.A., *Industrial Experimentation*. New York: Chemical Publishing Co., Inc., 1948.

[12] Cooke, J.R., Lee, M.T., and Vanderbeck, *Binomial Reliability Table*, NAVWEPS REPORT 8090. Springfield, Virginia: National Technical Information Service.

[13] Green, W.R., *Computer Aided Data Analysis*. New York: John Wiley and Sons, Inc., 1985.

[14] DoD Instruction 5000.2, "Defense Acquisition Management Policies and Procedures," Office of the Secretary of Defense, Under Secretary of Defense for Acquisition, 23 February 1991.

[15] Hoel, P.G., *Introduction to Mathematical Statistics*. New York: John Wiley and Sons, Inc., 1962.

[16] Anderson, V.L., and McLean, R.A., *Design of Experiments*. New York: Marcel Dekker, Inc., 1974.

[17] Draper, N., and Smith, H., *Applied Regression Analysis* (2nd ed.). New York: John Wiley and Sons, Inc., 1981.

[18] Johnson, N.L., and Leone, F.C., *Statistical and Experimental Design*. New York: John Wiley and Sons, Inc., 1964.

[19] Burington, R., and May, D., *Handbook of Probability and Statistics With Tables* (2nd ed.). New York: McGraw-Hill Book Company, 1970.

[20] Brownlee, K.A., *Statistical Theory and Methodology in Science and Engineering*. New York: John Wiley and Sons, Inc., 1960.

[21] Johnson, P.O., *Statistical Methods in Research*. New York: Prentice-Hall, Inc., 1950.

[22] Winer, B.J., *Statistical Principles in Experimental Design*. New York: McGraw-Hill Book Company, 1962.

[23] Chew, V. (ed.), *Experimental Designs in Industry*. New York: John Wiley and Sons, Inc., 1958.

[24] Cochran, W.G., and Cox, G.M., *Experimental Designs*. New York: John Wiley and Sons, Inc., 1957.

[25] Fisher, R.A., *The Design of Experiments*. New York: Hafner Publishing, 1960.

[26] Kempthorne, O., *Design and Analysis of Experiments*. New York: John Wiley and Sons, Inc., 1962.

[27] Villars, D.S., *Statistical Design and Analysis of Experiments for Development Research*. Dubuque, Iowa: William C. Brown, 1951.

[28] Fisher, R.A., *Statistical Methods for Research Workers*. New York: Hafner Publishing, 1958.

[29] U.S. Army Materiel Command, *Experimental Statistics*, Sections 1-5 (pamphlets 110-114). Washington, D.C.: USAMC.

[30] Young, H.D., *Statistical Treatment of Experimental Data*. New York: McGraw-Hill Book Company, 1962.

[31] Fisher, and Yates, *Statistical Tables for Biological, Agricultural, and Medical Research*. New York: Hafner Publishing, 1963.

[32] Wald, A., *Sequential Analysis*. New York: John Wiley and Sons, Inc., 1948.

[33] Tate, M.W., and Clelland, R.C., *Nonparametric and Shortcut Statistics*. Danville, Illinois: Interstate Printers and Publishers, Inc., 1957.

[34] Siegel, S., *Nonparametric Statistics for the Behavioral Sciences*. New York: McGraw-Hill Book Company, 1956.

[35] Conover, W.J., *Practical Nonparametric Statistics*. New York: John Wiley and Sons, Inc., 1971.

[36] Hollander, M., and Wolfe, D.A., *Nonparametric Statistical Methods*. New York: John Wiley and Sons, Inc., 1973.

7
Test and Evaluation Planning

7.1 General

Successful Test and Evaluation (T&E) planning requires the development, production, and execution of a series of planning activities and documents throughout a system's or component's life cycle. T&E planning should begin very early in the life cycle of a program and be pursued in a deliberate, comprehensive, and structured manner throughout acquisition and operation. Successful T&E planning results in the right information being injected into the tasking and planning documents as the program matures and at the correct points in time to provide an appropriate basis for the development of the T&E concept, the T&E design, and the detailed T&E plan (see References 1–4 for Department of Defense guidance).

Successful planning requires a full understanding of the operational user's concepts and requirements, the mission equipment, the employment environments, and the applicable tactics, strategies, doctrine, and procedures. The methodology used in T&E planning should rely heavily on a balanced combination of the scientific method and operational expertise. T&E planning culminates in the publication of the T&E plan and the final preparations to begin the test. The T&E plan depicts how, when, where, how much, and by whom T&E will be accomplished to execute the T&E concept and design. It should contain the detailed procedures to be used in conducting the T&E on a mission-by-mission and event-by-event basis. The T&E plan is prepared primarily for use by project personnel in the daily conduct of T&E. It also serves as the document by which higher level organizations and independent authorities can review, critique, and approve the planned T&E activities. It should include experimental layout, order, sequence, and control procedures; data collection and synthesis procedures; and preliminary analysis planning, along with other information deemed necessary to carry out the T&E. The T&E plan provides the basis for day-to-day test scheduling, T&E conduct, data collection, analysis, and reporting. The T&E plan should relate critical issues (CIs) and objectives, appropriate evaluation techniques, T&E methodology, and evaluation criteria. It should define the T&E scope by identifying the purpose, resources, and timing of events, describe the detailed T&E design, and provide information to the test team and support personnel from which they may develop other required detailed procedures to carry out their portion of the test. The T&E plan should ensure that all responsible participants clearly understand their roles in the test execution, data collection, analysis, evaluation, and reporting of the T&E results.

Any number of formats and documentation approaches could be used to present the detailed information needed in the T&E plan. An example table of contents for a T&E plan is presented in Figure 7-1. The format approach chosen for use is

> **Table of Contents**
>
> Executive Summary
> Table of Contents
> List of Figures
> List of Tables
> List of Abbreviations/Acronyms/Symbols
> Introduction
>   General
>   System Information
>   Operational Environment
>   Program Structure
> T&E Concept
>   CIs and Objectives
>   Scope and Test Approach
>   Planning Considerations and Limitations
>   Contractor Involvement
>   Schedule and Readiness Requirements
>   Resource Summary
>   User Requirements
> Methodology
>   General
>   CIs/Objectives Matrix
>   Addressing of Individual Objectives
>   Scope
>   Measures and Criteria
>   Scenarios
>   Method of Evaluation
> Survivability Assessment
> Administration
>   T&E Management
>   Tasking
>   Training
>   Safety
>   Security
>   Environmental Impact
> Reporting
>   Activity, Status, and Events
>   Final
>   Interim
>   Briefing and Service Reports
> Supplements
> Glossary of Terms
> References
> Distribution

Figure 7-1. Example Table of Contents for a T&E Plan

primarily concerned with illustrating the appropriate subject matter, content, and level of information necessary for effective T&E planning.

7.2 Advanced Planning (See References 1-17)

The planning for new or modified military systems begins with a statement of the combat user's operational requirement. This statement of requirement identifies an operational deficiency that cannot be satisfied through changed tactics, strategies, doctrine, or training; improved technological opportunities; or expanded missions. This requirement forms the basis for the development program to obtain the needed new or modified system and further planning documentation essential to satisfy the requirement. The T&E master plan (TEMP), as required by Department of Defense Directive (DoDD) 5000.3, serves as the major control mechanism for planning, reviewing, and approving all military T&E. It defines and integrates both developmental and operational T&E. It relates the program schedule, decision milestones, T&E management structure, and T&E resources to the critical technical characteristics, critical operational issues, evaluation criteria, and procedures. It is used as a tool for oversight, review, and approval of the total T&E effort by the Office of the Secretary of Defense and all Department of Defense components. It defines the specific requirements for a system to proceed to the next acquisition milestone. The initial TEMP must be completed and submitted prior to program Milestone I (beginning of system development) and updated at least annually thereafter. In summary, the TEMP is a living document throughout the acquisition cycle, outlining the roles of developmental and operational T&E. It and other program management documents provide the framework, basic T&E philosophy, and guidance on which the further, more detailed T&E planning is based.

7.3 Developing the Test and Evaluation Plan (See References 18–33)

The T&E plan is a formal document with sufficient detailed information and instructions to enhance successful completion of the T&E. It is based on the advanced planning information, the T&E concept, and the detailed T&E design. The T&E plan should include chapters covering introduction, T&E concept, methodology, administration, reporting, and appropriate supplements.

7.3.1 Introduction

The introduction of the plan should provide general information as to why the T&E is being conducted (for instance, to support a production buy, to support a product improvement, and so forth), who will conduct the T&E, and when and where it will occur. It should provide introductory information about the system or component under test, the operational environment, and the overall program structure.

7.3.1.1 System. Information should be presented that describes the system, along with its previous testing results, and other relevant program documentation. It should trace the development of new systems to the current model and describe the proposed system, how it works, what the major interfaces are, the level of threat environment within which the system must operate, how many systems are needed for testing, which command and what types of personnel will operate it, any known unique reliability and maintainability issues, and, most importantly, what its primary mission or missions will be. Enough information is required to enable personnel not familiar with the system to understand the CIs and objectives of the test. If appropriate, a description of how the system or component interoperates with and relates to other systems should be provided. Where applicable, the different types of systems in which a subsystem will be installed along with important interfaces should be identified. Include the names of the organizations doing the primary concept exploration and/or designs, the major trade-off studies that are being conducted on the program, the assessment of the program's technical risk, the management strategy for system development, and photographs, drawings, or sketches of the primary system components. If appropriate, address program competition and the sensitivity of contractor source selection. Describe the flight profiles, operations, existing procedures being modified, and other pertinent information to give a clear understanding of the T&E item. List significant characteristics of the T&E item, such as performance, size, and capability. Identify any known differences between the T&E item and anticipated production hardware, and describe how the T&E design will accommodate those differences. Identify T&E locations and facilities and any planned test deployments. Provide the rationale for selection of the particular T&E areas, laboratory environments, models/simulations, and/or test ranges. Describe how data taken from the various T&E facilities and environments will address the overall CIs and objectives.

7.3.1.2 Operational environment. The threat, operations concept, maintenance concept, and general training requirements developed by the operating command are all part of the operational environment and should be addressed. The software support concept can probably best be addressed as an integral part of the maintenance support concept. It is important to identify the current and postulated threats the system will encounter from initial operational capability (IOC) through IOC plus 10 years and the actual systems against which the system will be tested. The T&E plan should substantiate how testing against the available threats will yield results appropriate for extrapolation against the current and potential threats. When surrogate threat systems are used, the T&E plan should provide (or refer to other documents which provide) authentication of the representation of the real threat. When analytical modeling and simulation (M&S) are used to evaluate or expand upon test data, information on the validation and credibility of the M&S should be included, or referenced in the plan. In conjunction with the threat summary, some inputs will be applicable to the "Planning Considerations and Limitations" chapter of the plan if conditions exist that would affect the realism of the T&E. However, the emphasis in the plan should be on what is being tested and the justification that such testing is sufficient to meet the CIs and objectives. Threat simulation shortfalls (fidelity, density, and so forth), range limitations (for instance, size, airspace control, and instrumentation accuracy), time constraints, and test item configuration shortfalls are important, but more important is how the T&E will make up for the shortfalls. The operational concept should describe the primary mission scenarios and wartime use of the system. This may include primary and alternate missions, system interfaces, day/night sortie rates, surge rates, accuracy, system and weapons numbers, altitudes, profiles, duration, hours per day, numbers of operators, dormancy, shelf life, durability, typical environment, and so forth. Drawings to graphically portray the way in which the system will be used in an operational environment are often beneficial. The maintenance concept should briefly describe the maintenance philosophy for both peacetime and wartime environments (if different). It should address who will maintain the system (contractors, military, or a combination of the two), what specialty codes will be needed, which maintenance levels will be used (organizational, intermediate, and/or depot), what adverse environments (chemical gear, etc.) the maintainers will experience, what type of diagnostic concept will be used (such as built-in test equipment, external test sets, and manual procedures), what mobility or deployment concept/requirements exist, and what system-specific support equipment requirements exist. A subset of the maintenance concept should briefly describe depot-level support of the system. It should identify how the depot support concept will be employed (Will the system be supported organically, by a contractor, or supported commercially?), what the depot software support concept will be (Will the software be maintained in-house, and by what organization?), what additional facilities are needed, if an Inter-Service Support Agreement is being used, what the planned supply support concept is, and how the support equipment will be maintained (see References 5 and 6 for information on joint Service support and defense system software development). The training concept should describe who will define and develop the training necessary for the personnel who will operate and maintain the system. It should review the numbers of personnel, skill levels, and training required. This training should include any special training required to support the T&E, as well as training required to support the system once it is introduced into the operational inventory. It should also address training requirements which are not presently available and what is needed to ensure that all training requirements are fulfilled in accordance with the program schedule.

7.3.1.3 Program structure. The developer's integration strategy from concept exploration through production and deployment should be addressed. Discuss how program integration will impact T&E. If applicable, address when source selection will occur and whether it will restrict information release and program meetings. Examine the status of the assets and determine whether the system is a developmental prototype or a production representative system. Identify program risks and how they are being mitigated. Highlight those areas where inadequate T&E would affect system certification. Discuss whether the required contractual provisions to support T&E are in the statement of work and the contractor's test plans. Describe the intermediate levels of performance and capability that will be used to document the development process (that is, the growth process). Include pertinent items from the TEMP program summary. It is also desirable to include a tabular schedule or a milestone chart with as much information as is available at the time. Figure 7-2 is a sample T&E milestone chart. Known or projected program, test, and production events, and meetings of high-level, decision-making bodies, should be depicted and described.

7.3.2 Test and Evaluation Concept

The T&E concept described in Chapter 5 and documented in the detailed test plan should identify the major CIs and objectives that will be addressed; the overall measures of effectiveness (MOEs) and measures of performance (MOPs) that will be applied; the general criteria that must be met for success; the key data to be collected, analyzed, and evaluated; the primary test scenarios under which the system will be evaluated; and the overall approach (i.e., overall T&E structure or model) by which the T&E will be conducted.

7.3.2.1 Critical issues. CIs must identify those aspects of a system's capability that have to be addressed in testing to allow its overall worth to be estimated. As stated earlier, they are of primary importance to the decision authority in deciding to allow a system to advance into the next phase of acquisition. CIs are associated with key system characteristics relating to effectiveness and suitability and are addressed as "show-stoppers" in T&E. Once identified, the CIs are used to

Event	Date
T&E Plan Signed	Nov 93
Delivery of Validated Technical Data	Nov 93
Training (maintenance and aircrew) Complete	Dec 93
Support Equipment Delivered	Dec 93
Aircraft F-113 Delivered to Test Organization	Feb 94
Class II, Instrumentation Modification Complete	45 Days
Aircraft Checkout and Electromagnetic Compatibility Testing	15 Days
Contractor Sorties (3) and T&E Maintenance Demonstration	25 Days
T&E Manual Mode Test (3 sorties)	15 Days
T&E Manual Mode Message Report Preparation	15 Days
Lead Time from Message Report to Low Rate Initial Production Decision	30 Days
End of Test[1]	1 Jun 94
Milestone IIIA	30 Jun 94
T&E Final Report Published[1]	1 Aug 94
Milestone IIIB	1 Oct 94

Note 1: Required date to meet Milestone IIIB as scheduled in the TEMP

Figure 7-2. Example T&E Milestone Chart

develop the more specific T&E objectives. The responsible agencies must identify CIs as early as possible and relay them to program managers, decision makers, and participating agencies. This gives the participants an opportunity to address major problems or shortfalls before they are elevated to the decision maker. Often many problems relating to CIs can be corrected before testing. DoD Directive 5000.3 ties CIs to the decision process. For these issues to be meaningful as decision tools, the test planner must ensure that they are adequately defined to allow T&E and that they can be reported in meaningful terms. This implies the definition of some aspect of performance as opposed to general terms like "effectiveness" and "suitability" (for example, "the reliability of the missile is critical to the program; its failure to meet the success criterion would result in an inability to meet the intended wartime commitment"). A discussion should be provided that identifies how the CIs will be addressed by the specific T&E objectives. State which phase of testing will address each CI. If appropriate, list the issues in relative order of importance. Explain why these particular issues are critical to this program. Explain how they were derived and tie them back to the user requirement. New CIs may surface at any time—even during testing. The T&E agency should validate and update the CIs during the planning and conduct of the test. Ideally, CIs are developed by the operating command (user) with inputs by other developing and supporting organizations early in the planning process. They describe the effectiveness and suitability issues most critical to the system's ability to perform its mission in its intended environment. CIs may be established at a "global" or "mission" level, in which case they would take the form of: "How well does the F-113 perform the offensive counterair mission?"; or "How susceptible is the GBU-16 laser guided bomb to countermeasures?" The "dendritic or tree method" described in Chapter 3 can be used to trace the global CIs to more specific CIs and objectives. For example, an alternative could be to establish CIs at the "mission segment" level, which tends to deal with subsystems. CIs of this nature will take the form of: "Will the F-113's inertial navigation system support navigation accuracy required to find the target?"; or "Is the GBU-16's data link sufficiently jam-resistant?" The level of application selected for the CIs will dictate the formation of specific test objectives to support the CIs. The scope of CIs may change during the development of a system, and/or during its operational life cycle. Early CIs which adequately addressed developmental testing, for instance, will likely have less direct application during operational testing. Conversely, some developmental results may suggest additional or amended operational issues. In all cases, T&E should be structured to carefully address the issues most relevant at the time. Once the CIs are identified, the T&E objectives must be structured so the CIs may be addressed as completely as possible. Objectives should partition the CIs into parts that can be effectively tested in the planned test facilities and environments. The objectives are supported by MOEs and MOPs that can have specific evaluation criteria. An important factor in T&E is maintaining the integrity of objectives throughout the T&E process. The T&E manager should ensure that T&E planning will satisfy the objectives and the objectives remain focused on the CIs they support. When adequate program documentation is not available and the T&E planner does not have the option of waiting for published documents, the planner must take the lead by working with experts from the operating, implementing, and supporting commands to make valid assumptions about system employment and maintenance. These assumptions should address the threat, tactics, and mission scenarios planned or expected for the new system. They also should be clearly documented as assumptions and coordinated with the operating command as soon as possible. When CIs are not readily available in program documentation, the T&E planner should request that the operational user research three basic sources of background information and assist by developing them. These sources are mission area analysis (MAA), available program documentation, and knowledge regarding the intended operational concept and environment.

The military combat commands are responsible for reviewing their capability to conduct the war fighting missions assigned to them. The command's current capability is balanced against the current threat for a specific mission. Shortcomings in capability are identified and documented to initiate a weapon system acquisition or modification program to meet the need. The threat documents, user's requirements, developer's specifications, and the operational command's acceptance criteria are used to describe what a system must do to meet this need. Evaluating each of these characteristics against the identifiable risk will determine which issues are, or have a potential of becoming, critical.

As a program matures, the amount of reference documentation increases. The program management document (PMD), acquisition decision memorandum (ADM), decision coordinating paper (DCP), and threat documents must be reviewed by the T&E planners to fully understand the requirement for the system, the capabilities it must have, and the threat-stressed environment in which it must operate. The most useful information in developing CIs is the using command's statement of mission needs and a thorough knowledge of available technologies, technical challenges, and the state of system accomplishments. The statement of mission needs lists the operational user's requirements and describes the projected operational environment. As such, it is a source of preliminary information on tactics, doctrine, threats, weather, terrain, training, and maintenance concepts. Additionally, the acquisition program baseline (APB) is a key management tool for establishing commitment and accountability within the acquisition community. The APB includes cost, schedule, and performance parameters as approved by the Defense Acquisition Board (DAB), and other management agencies.

There are three elements to the operational environment: the threat, the natural environment, and operational tactics/procedures. Threat information is available from appropriate military intelligence agencies. The natural environment includes terrain, vegetation, and weather. Knowledge about the capability of previous systems to perform effectively, survive, and be supportable in all types of natural environments may well help identify candidates for CIs. Operational tactics/procedures are the specific techniques to be employed against certain targets and threat systems.

7.3.2.2 Scope and overall Test and Evaluation approach. The scope and overall T&E approach should provide an overview of the test and how the test program will be structured to collect the information needed to address the CIs. It must answer the following: 1) How will the overall T&E facilities be used to realistically test weapon system utility in the operational environment? 2) Are there differences between the projected operational use and the way the system will be tested? 3) What steps will be taken to minimize the impact of the differences? This chapter should provide an overview of the methodology to collect logistics data, highlighting whether there will be distinct logistics test events or whether logistics data will be collected in conjunction with ongoing effectiveness test events. If appropriate, it should address the pertinent aspects of separate or combined tests; actual tests versus simulations; use of surrogates; use of modeling and its accreditation; prioritization of test events; and test planning assumptions (validated and those that are not currently validated). When laboratory tests, ground tests, flight tests, and analytical M&S are planned, the tester should develop and document a management scheme by which the performance demanded in each of these environments will be synthesized and evaluated in a manner consistent with the user's requirements. If the T&E program is part of a combined or joint test program, it should discuss the appropriate parts of that effort. It may include a test concept summary table which groups such information as the test phase, existing test capabilities, the outcome expected for that phase of testing, and the CIs which will be addressed. Additionally, a test summation matrix is desirable. This matrix should provide the reader with an overview of the relationships between CIs/objectives and user requirements, test objectives and events, test events and scenarios, and test scenarios and resources. It can then be used by the T&E manager and other personnel as part of the internal briefing information to assist in T&E execution.

7.3.2.3 Planning considerations and limitations. The assumptions or circumstances that significantly affect the test concept and scope should be identified and discussed. Areas should be addressed that will require "workarounds" to allow

TEST AND EVALUATION PLANNING

the test to be successfully completed. Also, areas that require additional management attention to ensure a successful T&E should be included. Circumstances that limit the ability of the test to address user requirements should be stated. These are areas for which no suitable workaround exists and which will result in undetermined system capabilities. When such areas exist, state their impact and identify the need for follow-on T&E or risk reduction associated with the undetermined capability.

7.3.2.4 Contractor involvement. The system contractor's involvement in the T&E should be only that which is appropriate. The level of contractor involvement in the T&E should be clearly stated in the plan. If contractors must be involved in data collection or analysis, state why. For example, a proprietary data analysis package from the system contractor may be the only feasible method of collecting and reducing the test data. If so, justify in terms of cost and schedule impact why it is necessary to use the system contractor's package or people. Briefly describe how the T&E team will ensure that participation by the system contractor and any other support contractors will not influence the integrity or outcome of the test. Describe any system contractor maintenance during T&E. If the role of contractors during T&E is different from the planned operational and maintenance concepts for the system, explain in terms of cost or schedule impact why it is necessary to allow such contractor support. Explain the controls that will be applied to ensure that the T&E results are not improperly influenced or biased because of contractor involvement.

7.3.2.5 Schedule and readiness requirements. The T&E chapter containing the Milestone Chart should show the projected planning versus the testing completed to date. Discuss dates of key T&E planning events, briefings, T&E readiness events, internal reviews, joint reliability and maintainability reviews, and other relevant events. Highlight those events, actions, and items that must be finished or be in place before T&E begins, including significant T&E events, certifications for operational T&E, training, support equipment, software and technical data verifications and validations, accreditation of M&S, facility completion, and so forth. Improve overall program planning by communicating this list early in the planning process and as often as necessary.

7.3.2.6 Resource summary. The T&E plan should summarize major resource requirements for the test organization, the implementing command, the supporting command, the operating command, the contractor, and other agencies participating in the T&E. It should identify the major items of support required (personnel, ranges, automated data processing, analytical modeling/simulation, training, aircraft, equipment, and so forth). Reference can be made to the current test resources plan for operating and supporting command resource requirements, to the current TEMP for implementing command resource requirements, and to the statement of work for contractor resource requirements.

7.3.2.7 User requirements. The requirements of the operational combat user should be clearly identified as specified in the statement of operational needs documentation. The test plan should show how the T&E will establish the degree to which the user's operational requirements are met by the system under test.

CIs/SPECIFIC OBJECTIVES

Objectives	CI 1	CI 2	CI 3	CI 4	CI 5	CI 6
1	X	X	X	X		
2	X	X				
3	X			X	X	
4	X	X				
5	X					
6					X	X
7			X			
8			X	X		
9		X				
10			X		X	X
11				X		X
12			X	X		
13			X		X	X
14			X	X	X	
15				X		X
16			X		X	X

Figure 7-3. Example Matrix Relating CIs to Specific Objectives

7.3.3 Methodology

The methodology should provide an objective-by-objective description of how the test is to be conducted and illustrate how the objectives support the CIs. It should briefly explain test and analysis events planned to provide data at appropriate confidence levels and how they are to be integrated. It should identify and describe the end products expected from the test. Flowcharts, block diagrams, and matrices are very helpful in depicting the relationships between the various T&E events and the major inputs and expected outputs of each.

7.3.3.1 Critical issue/objective matrix. It is often informative to include a matrix that shows the relationship between the CIs and the objectives (see example in Figure 7-3). Each CI should be addressed by one or more objectives. List all objectives and the survivability assessment. The survivability assessment in military plans should always be of high interest and be included as a CI or objective. Cross-referencing the MOEs and MOPs to the objectives and/or CIs is also recommended.

7.3.3.2 Objectives. The objectives should be listed and addressed in a logical order driven by specific program priorities. They should be written clearly and concisely, using simple, common wording. For some systems, objectives may best be written by addressing effectiveness and suitability separately. When this is true, it may be desirable to state them in 1) priority order, stating first those that are most important; 2) mission phase order, such as for aircraft with preflight planning, ground activities, takeoff, cruise, recovery, and landing phases; 3) functional activities, such as testing of a ground segment, airborne segment, shipborne segment, and so forth; or 4) by time sequence, such as when test activities are driven by availability/delivery of hardware or software. It is helpful to include a representative mission scenario depiction in the discussion of each objective. This depiction should give the reader a graphic portrayal of the types and numbers of missions or events to be performed to support the objective. These scenarios may be critical to the evaluation of a single objective, or common to more than one of the objectives.

7.3.3.3 Example for addressing individual objectives. The example described in the following subsections demonstrates how to address individual objectives by stating the objective.

7.3.3.3.1 Scope. Describe, in general terms, the testing required (number of missile firings, and so forth) to satisfy the objective. State areas to be examined, such as electronic counter-countermeasures capability, to gather the data. To identify specific pieces of information needed to evaluate the objective, the discussion should: 1) quantify what effort will be expended to meet the objective (testing in ground test facilities, number of flights, launches, faults induced, and so forth); 2) outline the test activities, exercises, or operations that will provide the means for acquiring data for analysis; 3) specify whether test activities are combined events, dedicated events, or observation of normal operations/exercises; 4) when possible, specify the length of testing periods, number of test events, and so forth, required to gather the appropriate amount of data to satisfy the MOE or MOP; 5) distinguish between field/flight test activities, contractor facility tests, and pure analytical M&S activities; and 6) explain how much M&S will contribute to the evaluation, who will build any required M&S, who will accredit the M&S, and so forth.

7.3.3.3.2 MOEs/MOPs and evaluation criteria. An MOE or MOP is a quantitative or qualitative measure of a system's performance or a characteristic which indicates the degree to which it performs a task or fulfills a requirement under specified conditions. CIs are normally addressed by multiple objectives. Each objective is addressed by one or more MOEs or MOPs. An MOE or MOP may be a single parameter (for example, miss distance) or may be a combination of several parameters or characteristics, such as weapon system delivery accuracy expressed in terms of circular error probable. Usually more than one MOE or MOP is essential to address a complex objective where several specific major indicators must be examined. The evaluation of CIs also usually requires addressing the MOEs and MOPs associated with multiple objectives. Generally, the analysis of these complex objectives and CIs requires synthesis and judgment by experts to arrive at conclusions and recommendations. Consequently, the operational experience and technical knowledge of system operators, testers, operations research analysts are extremely important and should not be underestimated.

Evaluation criteria are the standards by which achievement of required operational effectiveness/suitability characteristics or resolution of technical or operational issues may be judged. They are established by the operating command (i.e., the bottom-line customer or combat user). The tester may provide some assistance during the planning phases. The requirements established in the military command's statement of need and carried throughout other system documentation provide the foundation for evaluation criteria. The formulation of evaluation criteria should not be undertaken without an understanding of the complicated process which precedes formal requirements definition. Each parameter with its associated numerical/quantitative value represents the user's compromise of several conflicting viewpoints. Parameters and associated values may be useful when actual performance can be measured against them, and evaluation criteria are linked to the specific performance to be measured. The early development of the T&E concept, however limited, can be an invaluable aid to the user requirements development process. Whether the eventual criteria are fixed values which do not change or "grow" over time or are interim values which can be expected to change or mature over time is an important additional consideration. When requirements have been identified, the operational user and tester must ensure that the evaluation criteria applied in the T&E adequately reflect those needs. If not stated, the tester must request the user to identify pertinent values. This will require the tester to work closely with the operational user so that the tester understands the criteria. In this case, it is absolutely essential that the criteria be provided by the operating command and coordinated on by the implementing and supporting commands at an appropriate level. The most important considerations when identifying evaluation criteria are that values be based on the user's operational needs and that the T&E be capable of determining whether those values are met.

7.3.3.3.3 Mission scenarios. Representative mission scenarios should be included along with the proposed outline of test events. Also, for each objective, identify whether the tests events are to be live firing missions, captive-carry flights, simulated missile shots, or combined events.

7.3.3.3.4 Method of evaluation. A description of how the data will be used to address the objective should be provided. If possible, present a sequence of events that logically leads to whether the objective has been met.

7.3.3.4 Survivability assessment. The survivability of military systems is usually a major program concern and is addressed as a CI and/or objective. Discuss the appropriate MOEs and MOPs that will be used in this evaluation.

Identify the type of information to be reported to the user/decision maker. State where the data will be collected in support of the survivability assessment. Explain how the assessment will provide an examination of the impact of system characteristics and performance on survivability.

Outline the T&E activities that will provide the survivability assessment. This assessment should use all available data, including specifically designed field/flight testing, laboratory testing, intelligence estimates, vulnerability analyses, and analytical M&S results, as appropriate. For assessment, it is normally essential to conduct an analysis prior to physical testing that helps identify the primary factors that should be addressed in the physical test.

TEST AND EVALUATION PLANNING

7.3.3.5 Example problem relating a critical issue to objectives, measures of effectiveness, and evaluation criteria. Critical Issue: Can the F-113 aircraft effectively perform the close air support (CAS) mission in a high-threat environment?

Objectives:

1) Evaluate the capability of the F-113 to receive and exchange the appropriate CAS mission information in a high-threat environment.

2) Evaluate the capability of the F-113 to provide timely response to mission tasking and to kill enemy CAS targets in a high-threat environment.

3) Evaluate the capability of the F-113 to survive while performing the CAS mission in a high-threat environment.

MOEs:

1) Objective 1. MOE 1-1. The number of CAS targets on which successful information exchange and tasking occurred versus the number of targets attempted (use separate breakout for "preplanned" and "immediate" types of requests).

2) Objective 2. MOE 2-1. The number of CAS targets successfully attacked within x minutes after tasking versus the number of targets tasked (use separate breakouts for "preplanned" and "immediate" types of requests).

3) Objective 3. MOE 3-1. The number of F-113 aircraft that survive while performing the CAS mission in a high-threat environment versus the number of aircraft attempting such missions.

Evaluation Criteria:

1) Objective 1. At least 95 percent of the "preplanned" tasking and 90 percent of the "immediate" tasking will result in successful information exchange.

2) Objective 2. At least 95 percent of the CAS targets will be successfully attacked within the time response criteria (for instance, within x minutes of "preplanned" time over target, or "immediate" request).

3) Objective 3. Survival estimates will indicate that at least 95 percent of the F-113 aircraft performing CAS in a high-threat environment would survive.

7.3.4 Administration

The administration portion of the T&E plan should cover things such as test management, tasking, training requirements, safety, security, environmental impact, release of information, and disclosure of T&E information to foreign governments.

7.3.4.1 Test and Evaluation management. The T&E management chapter should provide an overview of the responsibilities and the relationships among the agencies supporting the T&E program. Often an organizational chart showing the relationships of the responsible organizations is beneficial. Figure 7-4 is a sample chart depicting the external organizational relationships.

7.3.4.2 Tasking. Identify the principal agencies responsible for planning and conducting the T&E and identify specific responsibilities and relationships. The implementing, operating, participating, and supporting organizations who have a direct input to the test planning/conduct and the T&E manager and his key staff should be identified by organization and office symbol. See the sample list of key personnel in Figure 7-5.

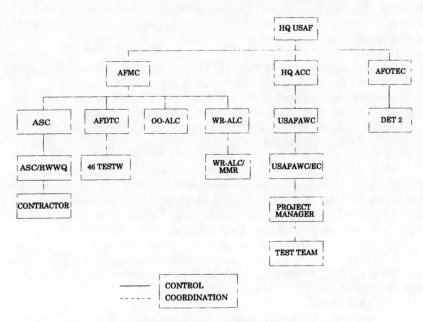

Figure 7-4. Example of External T&E Organizational Chart

Include an organizational chart of the T&E team. Figure 7-6 is a sample T&E team organizational chart. Briefly outline the test team control authority and the responsibility for scheduling and programming T&E resources.

7.3.4.3 Training requirements. Identify training requirements to include operator, maintenance, and specialized training (analyst, data base, computer programming, contractor provided systems training, and so forth) and how and by whom these requirements will be satisfied. Also include a schedule to accomplish any specialized training for T&E prior to the initiation of T&E.

7.3.4.4 Safety. It is important to ensure that required statements contained in Military Standard (MIL-STD) 882B (Reference 7) or other applicable MIL-STDs or documents are included. Also include a safety responsibility statement, required safety support, required safety reviews and documentation, and specific safety concerns and limitations.

7.3.4.5 Security. Classification: The security classification of the system and the test program in accordance with the program security classification guide or by the original classification authority should be identified. Defense documents are marked in accordance with DoD 5200.1R. It is important that the correct security classification be determined at the outset of a program to preclude any violations in the handling of classified information. Before a technical document is released for distribution, its contents should be reviewed and the appropriate limited distribution statements applied.

TEST AND EVALUATION PLANNING

KEY PERSONNEL

Name	Position	DSN	Office
Lt Col Jones	JPO Program Manager	785-3199	ASC/RWWQ
Maj Smith	JPO Dep Manager	785-3198	ASC/RWWQ
Ms Bown	JPO Proj Engineer	785-2575	ASC/RWEAW
Capt Davidson	F-16 SPO EW Dvlpmt	785-8953	ASC/YPD
Maj Tweet	ACC Prog Manager	574-7440	ACC/DRW
2Lt Flaek	Embedded Computers	574-3422	ACC/SCXE
Lt Col Dawton	AFOTEC Prog Manager	872-8535	Det 2 AFOTEC/TEM
Mr Smith	DT&E Prog Manager	872-3716	46 TW/TZPR
Mr Jones	F-16 EW Engineer	872-8916	46 TW/TZWD
Mr Lawton	DT&E Test Engineer	872-8916	46 TW/TZWD
Maj Cash*	IOT&E Manager	872-4042	USAFAWC/ECEX
Mr Davey*	IOT&E Manager	872-4042	USAFAWC/ECEX
Mr Randy*	Ops Analyst	872-2088	USAFAWC/ECED
Capt Dickey*	Ops Project Officer	872-2425	4485 TESTS/DO
Capt Smelt*	Deputy Ops Proj Off	872-2425	4485 TESTS/DO
1Lt Jaken*	Computer Support	872-4125	USAFAWC/SCA
SSgt Lackey*	Intelligence	872-4043	USAFAWC/INW
Mr Money*	IOT&E Support Equip & Maintenance Analyst	872-4350	USAFAWC/ECESA
SSgt Nimble*	MET Chief	872-8533	USAFAWC/ECEST
Mr Howe	EW Systems Engineer	872-8181	USAFAWC/ECE
Mr Cooley	AN/ALR-XX Manager	872-8888	USAFAWC/ECED

*Indicates IOT&E test team member.

Figure 7-5. Example List of Key Personnel

Source Selection: For programs where source selection is involved, refer to the appropriate military regulations for safeguarding procedures. Any deviations or exceptions to these procedures should be identified and discussed.

Communications Security (COMSEC): The T&E plan should identify those aspects of the program which require COMSEC protection. If it is determined that COMSEC protection is not required, a statement to that effect should be included.

Control of Compromising Emanations (TEMPEST): The T&E plan should identify those aspects of the program which require TEMPEST protection. If it is determined that TEMPEST protection is not required, a statement to that effect should be included.

Physical Security Requirements: Identify and document the physical security requirements of the T&E program.

Operations Security (OPSEC): OPSEC is an important consideration in planning. Depending on the outcome of the OPSEC review, the T&E plan must contain the appropriate OPSEC statement.

Security Test and Evaluation (ST&E): Although not strictly a part of T&E, the provisions of Computer Security Policy should be applied. If ST&E is required

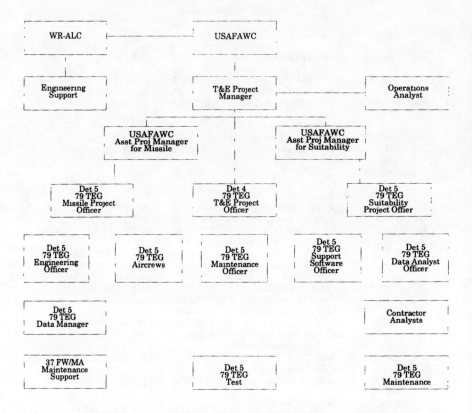

Figure 7-6. Example T&E Team Organizational Chart

and has not been accomplished prior to the start of T&E, the T&E team should coordinate with the risk assessment team to ensure that ST&E requirements are accounted for, where possible, during T&E.

7.3.4.6 Environmental impact (See Reference 8). The implementing agency is normally responsible for filing an environmental impact statement for T&E. Depending on the location and the planned operational environment, this statement should be referenced and used as required. Where T&E locations are incompatible, a separate statement should be included to describe the differences and impact. The local environmental coordinator for the T&E location should be contacted when a separate statement is needed or when the environmental impact statement for the T&E program either is not included or is inadequate. The environmental coordinator is normally a part of the military civil engineering community. This office should be able to provide the expertise to help determine the impact of the program, if any, on the environment. It is important that the T&E plan contain the appropriate environmental impact statement.

7.3.4.7 Release of information. Instructions should be provided for publicly releasing information about the system and T&E results.

TEST AND EVALUATION PLANNING

7.3.4.8 Foreign disclosure. Any release of T&E information to foreign nationals, governments, or agencies is normally approved through the appropriate military foreign disclosure office.

7.3.5 Reporting

Reporting is one of the T&E team's most important responsibilities. Many reports receive high-level attention and must be carefully planned if they are to effectively communicate the needed information. T&E teams should examine their program early in the planning phase to determine the reporting requirements to ensure that the proper information is disseminated to the right people in a timely manner. It is helpful during planning to devise a milestone chart identifying known report requirements and to use this chart to specify the requirements for various types of reports. When planning, specifically identify the timing of reports, writing responsibilities, addressees, formats, contents, and transmission means and include the appropriate specifics in the T&E plan. Special consideration should be given to distribution of T&E reports to ensure that the related information is released only to agencies with a valid need to know. Additional information on reports and briefings is contained in DoDD 5000.3 and ANSI Z39.18-198X, References 4 and 9, respectively.

7.3.5.1 Activity report. Activity reports are used to provide immediate feedback on major test events, such as missile launches, live firings, and so forth.

7.3.5.2 Status report. Status reports relay to the managing staff offices pertinent information regarding progress on the T&E phase. They are usually submitted monthly or at the interval specified in the T&E plan.

7.3.5.3 Significant event report. These reports address specific test events, such as missile launches or live firings, identified as significant in the TEMP and normally reported within 24 hours of the event.

7.3.5.4 Final report. The final report is the most detailed and important T&E report. It should be an in-depth, succinct document that clearly states and supports the results of the test. Extreme care must be used in planning for and constructing this report, since it stands as a permanent historical record of the system's performance. The important characteristics of how to develop a final report are discussed in greater detail in Chapter 8.

7.3.5.5 Interim summary report. If the final report cannot be made available at a major decision point, an interim summary report may be required to support the decision. The T&E manager should determine the requirement for an interim summary report for this or other reasons and submit the report at the appropriate time.

7.3.5.6 Briefings. Briefings are normally required before, during, and after T&E. A short narrative on the format, content, and intended audience of planned briefing reports should be included in the T&E plan.

7.3.5.7 Service report. This portion of the plan should describe how deficiencies and proposed enhancements discovered during T&E will be reported and managed. At a minimum, the description should discuss the duties and responsibilities relating to service reports, to include attendance at boards, the overall management of the service report process, and the procedures for prioritizing service reports.

7.3.6 Test and Evaluation Plan Supplements

Supplements to the T&E plan should be used as necessary to provide more detailed planning information. The number and type of supplements used are program-dependent. Supplements also can be used to "extract" classified material from the body of the test plan to simplify handling procedures. The exact format of each supplement usually varies depending on program specifics, such as the number of objectives, MOEs, MOPs, and so forth.

References

[1] DoD Directive 5000.1, "Defense Acquisition," Office of the Secretary of Defense, Under Secretary of Defense for Acquisition, 23 February 1991.

[2] DoD Instruction 5000.2, "Defense Acquisition Management Policies and Procedures," Office of the Secretary of Defense, Under Secretary of Defense for Acquisition, 23 February 1991.

[3] DoD Manual 5000.2-M, "Defense Acquisition Management Documentation and Reports," Office of the Secretary of Defense, Under Secretary of Defense for Acquisition, 23 February 1991.

[4] DoDD 5000.3, Test and Evaluation. Establishes policy and guidance for the conduct of test and evaluation in the Department of Defense, provides guidance for the preparation and submission of a Test and Evaluation Master Plan and test reports, and outlines the respective responsibilities of the Director, Operational Test and Evaluation (DOT&E), and Deputy Director, Defense Research and Engineering [Test and Evaluation (DDDR&E (T&E)].

[5] DoD 4000.19-R, Defense Regional Interservice Support Regulation (DRIS). Establishes procedures for formulating DD Forms 1144, Interservice Support Agreements.

[6] DoD STD 2167, Defense System Software Development.

[7] MIL-STD-882B, System Safety Program Requirements. Provides uniform requirements for developing and implementing a comprehensive system safety program to identify the hazards of a system and to ensure that adequate measures are taken to eliminate or control the hazards.

[8] Heer, J.E., Jr., and Hagerty, J.D., *Environmental Assessments and Statements*. New York: Van Nostrand Reinhold Company, 1977.

[9] ANSI Z39.18-198X, Scientific and Technical Reports: Organization, Preparation, and Production. (American National Standards Institute, 1430 Broadway, New York, NY 10018, Dec., 1986.) The accepted reports standard for DoD as well as a part of academia for the private sector.

[10] LeBreton, P.P., and Henning, D.A., *Planning Theory*. Englewood Cliffs, New Jersey: Prentice-Hall, Inc., 1961.

[11] Rudwick, B.H., *Systems Analysis for Effective Planning*. New York: John

Wiley and Sons, Inc., 1969.

[12]Novick, D., *New Tools for Planners and Programmers*. The Rand Corporation, P-2222, February 14, 1961.

[13]Kahn, H., and Mann, I., *Techniques of Systems Analysis*. The Rand Corporation, Research Memorandum RM-1829-1, rev., June 1957.

[14]Enthoven, A., and Rowen, H., *Defense Planning and Organization*. The Rand Corporation, P-1640, rev. July 28, 1959.

[15]Hitch, C.J., and McKean, R.N., *The Economics of Defense in the Nuclear Age*. The RAND Corporation, Report R-346. Cambridge, Massachusetts: Harvard University Press, March 1960.

[16]McKean, R.N., *Efficiency in Government Through Systems Analysis*. New York: John Wiley and Sons, Inc., 1958.

[17]Novick, D., *System and Total Force Cost Analysis*. The Rand Corporation, Research Memorandum, RM-2695, April 15, 1961.

[18]Arrow, K.J., Karlin, S., and Scarf, H., *Studies in the Mathematical Theory of Inventory Production*. Stanford, California: Stanford University Press, 1958.

[19]Moffat, R.J., "Using Uncertainty Analysis in the Planning of an Experiment." *Journal of Fluids Engineering*, Vol. 107, June 1875, pp. 173-178.

[20]Gottfried, B.S., and Weisman, J., *Introduction to Optimization Theory*. Englewood Cliffs, New Jersey: Prentice-Hall, Inc., 1973.

[21]Bellman, R.E., and Dreyfus, S.E., *Applied Dynamic Programming*. Princeton, New Jersey: Princeton University Press, 1962.

[22]Brown, B., and Helmer, O., *Improving the Reliability of Estimate Obtained from a Consensus of Experts*. The Rand Corporation, P-2986, September 1964.

[23]Chestnut, H., *Systems Engineering Methods*. New York: John Wiley and Sons, Inc., 1967.

[24]Enke, S. (ed.), *Defense Management*. Englewood Cliffs, New Jersey: Prentice-Hall, 1967.

[25]Forrester, J.W., *Industrial Dynamics*. Boston, Massachusetts: MIT Press, 1961.

[26]Hitch, C.J., *Decision-Making for Defense*. Berkeley, California: University of California Press, 1965.

[27]Mandel, J., *The Statistical Analysis of Experimental Data*. New York: John Wiley and Sons, Inc., 1964.

[28]Kline, S.J., and McClintock, F.A., "Describing Uncertainties in Single-Sample Experiments." *Mechanical Engineering*, Vol. 75, January 1953, pp. 3-8.

[29]Schenck, H., *Theories of Engineering Experimentation* (3rd ed.). New York: McGraw-Hill Book Company, 1979.

[30]Taylor, J.R., *An Introduction to Error Analysis: The Study of Uncertainties in Physical Measurements*. Mill Valley, California: University Science Books, 1982.

[31]Moffat, R.J., "Contributions to the Theory of Single-Sample Uncertainty Analysis." *Journal of Fluids Engineering*, Vol. 104, June 1982, pp. 250-260.

[32]Hildebrand, F.B., *Advanced Calculus for Applications*. Englewood Cliffs, New Jersey: Prentice-Hall, 1962.

[33]Montgomery, D.C., *Design and Analysis of Experiments* (2nd ed.). New York: John Wiley and Sons, Inc., 1984.

8
Test and Evaluation Conduct, Analysis, and Reporting

8.1 General

Once a Test and Evaluation (T&E) project has been planned and all the necessary resources assembled, there remain the all-important tasks of carrying it out effectively, analyzing the results, and reporting the findings. Although the success of a T&E greatly depends on its planning and the overall support provided for it, T&E rarely occurs exactly according to the plan. Thus, in T&E as in combat, the best results will be obtained by the organizations that have thought through the various possible contingencies and that have the knowledge and versatility to make adjustments for unplanned and undesirable events.

T&E is always affected by hardware and software, human failures, priorities, weather, and other factors that can be only partially forecast and controlled. Because T&E often involves complex systems, with humans interacting as a part of the system as well as being the evaluators of its capability, it is likely to require some adjustment from the original plan. For complex T&Es, findings often include the discovery of unforeseen capabilities and limitations which require adjustments and may make the original design of the test less than optimum.

Some adjustments are almost always required during the process of testing. Often it seems that good T&E managers acquire an extra sense that enables them to foresee difficulties and correct small problems before they become large ones. These changes will sometimes have a profound effect on system operators and maintainers and upon the requirements levied upon support agencies, such as range operations; instrumentation; and data collection, reduction, and analysis. In some instances, desirable changes will not be possible under the constraints of the planned T&E support and priorities, and compromise adjustments will be necessary.

8.2 Test Conduct

Since T&Es vary widely in terms of critical issues, objectives, scope, complexity, and priority for completion, it is not possible to describe precisely all the challenges involved in getting every T&E accomplished. However, there are certain characteristics that appear to have broad applications in the successful conduct of T&E.

8.2.1 Professional Leadership and Test Team Performance
(See References 1 and 2)

The T&E manager and his team are critical to the success of the test. The T&E manager must be capable of leading a diverse group of experts whose interests and priorities vary widely. Teamwork is essential in T&E, and most complex T&Es require a large test team, with each professional bringing a specialized talent to the group. Because disagreements and misunderstandings will occur during the conduct of any T&E, the team must establish both formal and informal methods of achieving their timely resolution and proceeding with the T&E. The T&E manager must exercise his or her position of leadership to take full advantage of the broad range of expert advice in rendering important decisions regarding the test.

8.2.2 Systematic Approach

The T&E manager and test team must do their best to fully understand the total system or equipment being tested, the design of the test, and the test support which can be called upon. This includes the capabilities and limitations of the instrumentation (to include threats, facilities, and environments), data collection and reduction procedures and problems, and the techniques and goals of the analysis planned for the T&E.

The T&E manager and team must determine the correct level at which to make demands on those people and organizations assisting them to get the T&E accomplished on schedule. This will require achieving the right balance to avoid making unreasonable demands, while at the same time sensing and reacting to unjustified resistance to reasonable requests for planned support.

The testing of military systems and equipment is a dynamic process. The T&E manager and team must continuously monitor the results from the data collection and feedback system and be alert to the desirability, and especially the necessity, of making changes or adjustments in the T&E conduct. These changes could include suspending testing for a period, ending the T&E early because the objectives have been met, or extending the T&E so that objectives can be met. On the other hand, before anyone makes any change in the T&E plan and conduct, he or she should be aware that the original T&E design (and, therefore, the comparability and validity of T&E results) may be significantly affected. Prior to making any changes, it should be demonstrated that the changes are essential, that valid results have been (or will be) obtained, and that useful conclusions can be drawn from the modified test.

8.2.2.1 Objectivity.
The T&E manager and team should develop and maintain an unbiased attitude toward the results, and they should resist any actions that may cause actual or perceived bias in the results. There will always be preconceived notions about how a T&E will progress and what its outcome should be. In fact, without such pretest predictions, T&E design would be almost impossible and certainly inefficient. The T&E itself is usually designed to ascertain whether a given system or piece of equipment is capable of performing at some desired level. However, once testing begins, the T&E manager and team must maintain their integrity and deal only in the facts produced by the T&E. Above all, it should be the T&E results that drive the conclusions and recommendations of any T&E, and not the preconceived notions of what the results should be.

8.2.2.2 Pretest or dry runs.
In any complex T&E, pretest or dry runs are necessary to make sure all parts of the T&E execution work. The preparation of the detailed T&E plan will have involved numerous meetings and discussions,

many of which resulted in compromises to arrive at an understanding of who does what, when, where, and how. Often this higher level of detail is reflected in subplans such as Flight Operations; Data Collection, Reduction, and Analysis; and Instrumentation. When documented and published, these understandings usually represent the "final" compromise versions of planning deliberations. Dry runs should be designed to rehearse the T&E down to the last detail before valuable test items and resources are expended. No matter how much discussion and planning have taken place, pretest runs usually reveal unforeseen problems that can and should be eliminated or accommodated prior to the beginning of T&E. Sufficient time and funds should be planned and budgeted for the present effort.

8.2.2.3 Test and evaluation records and control. Overall records of the T&E should be maintained by the T&E manager in a project case file and a project data base. These include such items as the T&E initiating documents, T&E plan, analysis plan, and other pertinent material affecting the conduct and control of the test. A considerable amount of planning and coordination effort is necessary to successfully arrange and carry out each test trial. Using the overall T&E plan as the controlling document, the T&E team plans each test trial or mission. From the mission planning effort, weapon systems operations and ordnance are selected and range support is scheduled.

Airborne, ground, and range instrumentation data must be collected and adequately recorded. No matter how simple the T&E objectives may be, testers cannot assume that useful data will be collected as a routine result of good design. T&E design, both procedural and experimental, creates the need for detailed data control by designating what variables are to be measured during each phase of the T&E and to what accuracy these measurements must be made. Data control, in turn, dictates how these measurements must be made to ensure valid implementation of the T&E design. Data control should specifically provide the mechanisms, operational procedures, and rules for the collection of useful data in a form amenable to the planned data processing methods and overall evaluation criteria.

Data control mechanisms can be generally classified into two broad categories: 1) instrumentation, and 2) data collection methods and procedures. In a sense, both forms of data control can be considered as test instrumentation in that each requires the design and fabrication of a sensing and recording function (or scheme) capable of being used to accumulate the required data in usable form. The mechanisms differ only in the nature of the collection devices and the amount of human participation required to perform the recording function. Although this interpretation may be considered a valid comparison of the two categories of data control, the more familiar concept of the relationship between instrumentation and data collection is that instrumentation is used where required to ensure accuracy limits defined by the T&E design are achieved. Data collection incorporates those mechanisms used for the same purpose, but is applied where the use of extensive instrumentation is too expensive, not available, unnecessary or would destroy the realism of the T&E data.

T&E information may be collected in many ways to include electronic, mechanical, film, video, direct observation, and so forth. When processed for analysis, the information usually consists of numbers (quantitative) and words (qualitative). It is extremely important that the information be correlated (that is, that observations relating to an event, or interval, be properly identified with the event or interval and conveniently maintained for subsequent analysis). Identification will require a data management system which correctly labels each bit of information. Operating procedures must be developed and applied to ensure that the data control techniques selected (or designed) will be used properly and that valid data, within the required accuracy limits, are obtained. These procedures should indicate any deviations that might be necessary to satisfy data requirements of each T&E facility or phase. The operating procedures for the complete data control system (instrumentation and data collection) for any given T&E will probably be unique. Conversely, where instrumentation support is required and the instrumentation system selected is standard or composed of standardized components that have been used for similar T&Es, the operating procedures have probably also been standardized and may be reusable with very little modification. Experienced data management personnel can help implement systems and methods that have worked well in previous similar T&Es or can recommend special methods based on the specific needs of a particular T&E.

Aircrews, ground crews, and test conduct personnel must be briefed and debriefed for each mission; and they must accurately complete data records for each T&E trial or observation. It is essential that the collection of T&E data be in accordance with the preconceived data collection plan. Human data collected as a primary source or to augment instrumented data should be documented as soon as possible during or after each observation or mission since human memory is notoriously unreliable (details that seem perfectly clear and obvious at one minute have a way of being completely lost in an hour, a day, or a month). Moreover, all records should be made with the attitude that they must be easily understood by someone other than the one who witnessed the event, interpreted the instrumentation readout, or recorded the measurement. Inexperienced testers too often make the mistake of recording only a few measured values during the operation, usually in a nonstandard way, and of trusting to memory to fill in details of equipment and performance at a later time. This practice is certain to lead to mistakes and to introduce uncertainties that may eventually render the entire test worthless. The only safe way is to make a complete record of operations, equipment, personnel, and other data on well-designed collection forms and to use instrumentation where possible and appropriate. It is better to record an occasional superfluous item than to discover at a later date that some extremely important information has been omitted. Data should be recorded in such a way that an experienced analyst could duplicate the test and/or results.

In some cases, data may be adequately obtained through normal ongoing data collection systems (for instance, standard data collection systems employed by the military Services, and so forth). In others, special data collection teams using rigid control and collection procedures or specialized data collection systems may be required. In any case, when T&E conditions exist that are not specified in the T&E plan, records must indicate these variations so that the analysis of results can be conducted accordingly.

8.2.2.4 Periodic evaluation and feedback.

Only in the most simple T&E does a tester perform the complete test and then examine and evaluate the data collected. When T&Es involve multiple critical issues and objectives and considerable replication, it is essential to periodically examine and evaluate what has been done and consider the impact of testing to date upon the testing that remains.

The places for such periodic evaluations are usually self-evident and can be prescheduled and put into the T&E plan. For a T&E involving flying aircraft, for example, evaluations could occur in real time for some important information, but would normally occur after each mission for most information. A mission would ordinarily be planned to fulfill more than one objective, and the information relating to each objective should be examined carefully to see whether subsequent missions require modification. Collecting data and assembling them in a form for quick evaluation takes time and may cause impatience on the part of test operations personnel. In the interests of practicality, it may be necessary to wait to conduct periodic evaluation until after two missions, or even three. There is a point, however, beyond which it is bad management to collect more data without conducting evaluation and obtaining feedback.

During T&E, a large amount of data may be collected on a daily basis and numerous data collection formats used. In most cases, to assist in the management, processing, and handling of the data, the project team must make special arrangements for data processing and computerized analytical support. Practically every sizable data processing effort is keyed to the efficient use of one or more electronic computers. As more and more routine tasks are assigned to automated computers, technology responds by providing computers which are physically smaller, faster, and more versatile. Rough approximate computations are replaced by more elaborate and sophisticated procedures. Larger sets of data are fed into the computer to yield better results. More information is obtained from each set of data. Data processing becomes less prone to computational errors. The speed and reliability of electronic computers allow them to be used in real time data processing to control the T&E operation and collection of data. Solutions are attempted for more complicated problems (for instance, real time kill assessment and removal). Much of the burden of T&E analysis can be alleviated by assigning the routine work to the computer. T&E results reflect a higher degree of objectivity as the result of assigning simple logic problems to the computer. Simulation methods alleviate much of the burden of testing by predicting and extending the results of tests. The development of computerized data collection and processing has led to automated support of the following kinds (see Reference 3):

1) Research and analysis of stored data that may totally or partially satisfy a critical issue or T&E objective, or aid in establishing data collection requirements.

2) Special assistance in data collection techniques and computer-assisted analysis techniques.

3) Machine data edit services to improve the quality of data and to pinpoint collection flaws and/or areas requiring more closely monitored collection management.

4) Computer support services for rapid building of data files for specific sets of data.

5) Updating of data files through computer programming techniques that permit adding to, deleting from, and changing data items in a data record or file.

6) Data retrieval services permitting rapid retrieval of any or all data in a file, machine-printed to meet the user's requirement in terms of content, format, and timeliness.

7) Standard periodic machine-written reports designed to meet special test requirements.

8) Computer programming and data systems services that permit the devel-

opment and production of new machine-written reports and/or the modification of existing machine-written reports. In addition, special requirements may exist to develop statistical subroutines or to modify existing ones to fit particular T&E needs.

Normally, computerized data management capabilities are used on sets of data that meet one of the general conditions outlined below (that is, data repository services are not restricted to just those applications requiring computer support because of data volume):

1) Where large numbers of observations are to be made, or large numbers of data elements must be measured for ultimate collective evaluation.

2) Where large numbers of combinations of observations are to be made, though the total events or observations may be relatively short.

3) Where a base must be established for new weapon systems coming into the inventory, regardless of the size of the initial test. (Examples: FX aircraft, missile X.)

4) Where complex analysis must be performed that would be too cumbersome to handle manually.

5) Where it has been determined that frequent reference to the T&E data will be necessary in the future, or the information is determined to be vital and of historical value.

From the previous discussion, it is obvious that a highly important aspect of T&E conduct is the timely and orderly flow of data into the data control and collection function. To take full advantage of available specialized technical assistance, the T&E manager and team should not only be familiar with the T&E system and facilities, but also should have sufficient knowledge of basic instrumentation and data collection concepts. This will enable them to evaluate the general application of the systems using these concepts to their specific data control needs. T&E information should be processed and summarized continually during the T&E to support the periodic "quick look" analyses. These analyses help provide timely feedback concerning project status, data adequacy, and preliminary T&E evaluation to operating elements within the project. This feedback must be sufficient to support management decisions and to allow the necessary adjustment, refinement, and completion of the T&E in such a manner that the critical issues are addressed and the objectives are met.

8.3 Analysis

Analysis is the process by which the tester treats the T&E data to infer conclusions and recommendations regarding the combat application of the weapons system. The data gathered in the course of a T&E are seldom of interest in themselves. Almost their only use is to draw inferences about what would happen in a wider, or at least a different, environment (i.e., the combat environment). One typical case in military T&E is a test conducted to make a forecast of what an item would accomplish in combat; another is testing a few items with the goal of predicting what would happen if a large number of similar items were used throughout the operational establishment. In these examples, the logical step of inference is the translation of the T&E data into a conclusion and recommendation concerning the "real world" situation in which the decision maker is actually interested. The ability to successfully make this translation depends greatly on the effectiveness of the T&E design, its proper execution, and the resulting analyses.

The T&E design itself will drive the analysis to a large degree. Once the T&E data have been satisfactorily collected and processed according to the plan, the analysis should also proceed according to the plan. In procedural T&E designs, this will usually involve calculating values for the measures of effectiveness and measures of performance of each objective and comparing those values with previously established decision criteria. (See Chapter 6 for a discussion of "procedural" and "experimental" T&E designs.) In experimental T&E designs, it will likely involve applying analysis of variance to identify the relationships and the significance of the relationships. In either case, deviations from the original T&E design, plan, and assumptions should have been recorded during testing and must be taken into account during the analysis.

A tester often learns significant new and unplanned information about the test item during T&E. This information generally leads to the need to examine potential analytical relationships and to conduct a variety of comparative analyses. The T&E team must decide which analytical comparisons are needed and which analytical relationships are likely to be meaningful. Arrangements can then be made for the available data to be tabulated or graphed to help in this process. Some computerized statistical techniques, such as contingency table comparisons and regression analysis, may also be used to assist in these examinations.

8.4 Reporting (See References 4–34)

When a T&E has been completed, the data processed and analyzed, and conclusions reached concerning what was learned, there remains the very important task of communicating the results of the T&E to those who need the information. This phase of the T&E is crucial because the best-planned, most capably conducted, and thoroughly analyzed T&E is of little value until it is reported to those in positions to translate its findings and recommendations into action.

8.4.1 General

T&E reports have two primary purposes: 1) to convey information useful to operators and decisions makers, and 2) to document T&E results for future reference. The first purpose may be fulfilled by an oral report, a written report, or both. This purpose is driven by the need both to be timely and to communicate the essential information. There is definitely a trade-off between completeness and timeliness, since, in a dynamic environment, the value of T&E information may decay rapidly with time. All T&E agencies face this problem of timeliness, and, in the past, the value of many excellent T&Es was much diminished because reports were slow to reach those who needed the information. The second purpose (documentation) always requires a formal written report. In some cases, the length of time required to produce a final report may dictate the need for a preliminary written report. This preliminary report may be considerably less comprehensive and detailed than the final report to communicate the most important test results in a timely manner.

Formal technical reports by various T&E organizations are often quite different in format and appearance. Nevertheless, all formal reports of T&Es have a number of characteristics in common. They all contain sections relating to what was to

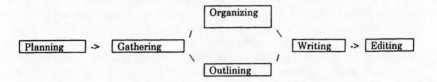

Figure 8-1. Functional Steps in Report Preparation

be accomplished, how the T&E was conducted, how the results were obtained and what they were, how the results were analyzed, the conclusions reached on the basis of the analysis, and, except in rare instances, recommendations for action.

Technical reports on T&E can also be viewed in terms of six generic functional steps involved in their production. Figure 8-1 depicts those functional steps.

8.4.2 Planning

Physical testing is usually expensive and can result in the expenditure of millions of dollars and valuable resources. Indeed, the value of the T&E results should transcend the actual money which the test costs. Since the T&E report is the method by which these results are made available, it should be planned and started early in the T&E process. The audience(s) for which the report will be written must be determined early. Not only will this influence the distribution and style of the report, but it will also influence the amount of detail, justification, and explanation included. The decisions reached in the planning stage will be continuously subject to examination, improvement, and revision. They will be made in accordance with the best information available at the time so that report preparation can start early and proceed, to the degree possible, as the T&E proceeds.

8.4.3 Gathering Data

For the most part, the information used in the final report will have been gathered during the T&E. Usually, this information will become available incrementally throughout the T&E. Trends in testing will be observable, and a good idea of the content and emphasis of the final report can ordinarily be obtained before physical testing is concluded. Those responsible for presentation of the report should take advantage of this early information and have the report draft substantially formulated and partially written before the T&E is actually completed.

Continual report preparation during T&E offers great advantages. Probably the most obvious one is that it should make it possible to issue the report promptly. Also very important, however, is the immediacy of the data and the circumstances surrounding the testing which produced them. Getting on paper at the time exactly what happened while it is fresh in the minds of T&E personnel is infinitely better than attempting to reconstruct events at a considerably later date. It is entirely possible that what is written immediately may not actually appear in the final version of the report, but it will likely have had an important influence on what does appear.

Finally, there is an aspect of gathering data which often is accorded insufficient emphasis and sometimes is neglected completely. This is research devoted to obtaining pertinent and related information not generated during the T&E being reported on. It is especially important to have the research done early—preferably before the T&E commences. Information from previous tests, analytical studies, war games, combat experience, and so forth, may corroborate T&E data. Such external data may make a good T&E better (more valid, of higher confidence). These sources of information can be mutually reinforcing. External data may improve a T&E, and, on the other hand, a T&E may also add to the credibility of the external data. When the T&E data do not corroborate with previous research, it is even more important that the T&E be operationally and statistically sufficient to support the findings in light of the differences with previous research.

8.4.4 Organizing

The organization of a report can have a significant impact on the ability to effectively communicate vital information to the reader. An executive summary is a must in just about any kind of T&E report or lengthy document. Decision makers and other officials are busy people and often do not have the time to wade through a mass of information to arrive at the significant findings, conclusions, and recommendations. It is important to those writing reports to make it easy for the reader if they wish to successfully communicate the important information.

The particular T&E organization originating the report will normally establish its general framework (that is, format). However, within that framework, the tester has a great deal of latitude in how to present the material. This is the key reason why, although the formats are the same, reports within an organization may vary from excellent to catastrophic.

There is another aspect of organization besides structuring the report in a logical manner: selecting the material to be included and deciding how to use it. As the T&E progresses, there will be a tendency to accumulate more information than can or should be used in the report. This information must be evaluated and classified. In general, it may be separated into three classes: 1) vital material that must be in the body of the report; 2) corroborative or tangential information which is worth saving and making available, but which should not be in the body of the report; and 3) material which can now be evaluated as having little or no worth in preparing the report.

Sometimes it is difficult to know what to do with the second class of information. Probably the most general practice is to include it in appendices to the report. In this manner, it does not interfere with the smooth flow of the body of the report, and yet it is available in the event the reader wishes to consult it. The third class of information should be retained for a sufficient period of time until it is established that it is of no value and can then be destroyed.

8.4.5 Outlining

The outlining of a report is a process of arranging the principal features of the material and presentation into a logical sequence. It is the single most important beginning point in the writing of a report. Outlining not only identifies all subjects to be covered in the report, but also relates each to the others and orders them in sequence and importance of presentation. As Figure 8-1 illustrates, there is a close relationship between outlining and organization, and both assist in the logical and convenient presentation of the material.

Basically, the outline is a tool for laying out the report in condensed form. As such, it warrants sufficient thought. The outline makes it possible for the tester/author to shift things around, expand one subject, and condense another; and it helps him/her decide whether material belongs in the main body of the report or should be relegated to an appendix. It allows the author to examine the material in condensed form and check it for unity, coherence, and logic (organization). Conscientious use of the outline also permits the author to view the report as an organic whole. Careful study of the outline will reveal gaps or omissions, over- or underemphasis, inclusion of irrelevant and unnecessary material, illogical sequence, and so forth. Also, when a report is being written jointly by a number of people, an outline is an excellent way to communicate to them how each person must contribute to the final report as a whole.

As a first step in the outlining process, the author should concentrate on the content of the information to be presented; that is, what the major findings are that must be presented in the report, and how they logically lead to and support the conclusions and recommendations being made.

No one method of outlining can be singled out as best; the method depends on the particular reporting situation. A good rule, however, is to *keep it simple*. A complex outline will result in a complex report. If an outline consisting of key topics, phrases, or sentences is unclear, the report based on it will likely be equally unclear.

8.4.6 Writing

Technical report writing is a form of descriptive writing that generally follows an iterative process. Rarely is the tester/author capable of going directly from an outline to a finished manuscript in one step. Writing usually begins by expanding the outline into a meaningful text (a draft), followed by several reviews and improvements to the draft. Writing the report consists essentially of translating the outline into prose, and including the tables, graphs, photographs, and other expository material selected to help communicate this essential information to the reader. The outlining process can be highly beneficial in the assignment of writing workloads to team members for complex T&Es. Frequently, dictation proves to be a good method of writing a report. More often, however, the chore is done by sitting in front of a computerized word processor or taking pencil in hand. This is the difficult stage for many T&E personnel. It is suggested that those who find the actual writing difficult make an attempt to dictate in accordance with the outline, topic by topic. In the first draft, emphasize descriptive communication of accurate information rather than grammatical correctness or style. One of the challenges in good writing is to have thought through the subject matter to the extent that you clearly understand the substantial and relevant material you wish to communicate. Remember that the purpose of a final report is to communicate important information, and that anything which does not contribute to effective communication does not belong in the report. It is also suggested that one or two persons with a talent for writing be assigned the task of reviewing the overall report and ensuring that it has the proper emphasis on content as well as sufficient continuity and flow.

8.4.7 Editing

Editing of technical writing is usually painful to the author, but it is essential. The first draft of a technical report always needs editing by the author. Subsequently, there may be second and third drafts and editing by others before the report is ready for publication. The object of editing is to improve the report so that it will be as perfect a vehicle as possible for conveying the selected information to a given audience. The ultimate effectiveness of the communication process depends in large measure on editing and the time available to do it. The process of editing can be discussed in terms of what the author personally can do and what can be done for the author by a professional editor.

The author should always try to improve his/her work through self-editing, a very difficult job. Usually this is because authors tend to read what was meant to be written rather than what is actually written. Understandably, authors have great difficulty standing back and viewing their own work objectively or assuming the role of the intended audience.

Consider first what authors can do to edit their own work. When they have actually produced a first draft of the report, victory has not been achieved, but is in sight. There is now a body of expository material to review, expand, revise, and improve. There is a manuscript to edit. Even when the authors plan to turn the manuscript over to a professional editor—in fact, especially when they are going to do so—they should attempt to edit their own work first. They must make certain that the professional editor has all the materials needed; and the better the author's writing is, the less likely it will be that changes in meaning or missed points will occur as a result of the editing.

It is always wise to wait before tackling the self-editing problem. Even 2 or 3 days will give authors a chance to forget what they wrote and help them adopt an objective reader's viewpoint. This, of course, is an important benefit of writing the first draft at an early date—even before the T&E is concluded, if possible. Authors should begin by reviewing both the purpose of the report and the outline used in preparing the first draft. Then they should consider the reader's needs—actually try to place themselves in the reader's position. They should read the report for coverage and organization and make notes on sections that need more work. In this reading, they should be concerned primarily with the essential information, conclusions, and recommendations, and be less concerned with style, grammar, or mechanical difficulties with the format. These other aspects can be addressed in subsequent readings.

The professional editor's job is to make the report better. This can be done only with the author's help and complete cooperation. The professional editor will undoubtedly read the report many times during the editing process. Routinely, the editor will polish the style—correct grammatical errors, check references, and so forth. The professional editor should then discuss the changes he/she has made or would like to make with the author. Most of the suggested changes will be good and will help to make the report more effective. Some changes, however, may introduce unintended changes in the meaning which would be harmful to the report. It is the writer's responsibility to identify when and where such unintended changes occur and to work with the editor for their resolution. The author and the editor must cooperate so that their collective talents are pooled to produce the best report possible within the available time schedule.

References

[1] Stewart, R.D., and Stewart, A.L., *Managing Millions*. New York: John Wiley and Sons, Inc., 1988.

[2] Roberts, G.W., *Quality Assurance in Research and Development*. New York: Marcel Dekker, Inc., 1983.

[3] Green, W.R., *Computer Aided Data Analysis*. New York: John Wiley and Sons, Inc., 1985.

[4] ANSI Z39.18-198X, Scientific and Technical Reports: Organization, Preparation, and Production. (America National Standards Institute, 1430 Broadway, New York, NY 10018, Dec. 1986) The accepted reports standard for DoD as well as academia and the private sector.

[5] Hoshousky, A.G., *Author's Guide for Technical Reporting*. USAF Office of Aerospace Research, 1964.

[6] Rathbone, R., and Stone, J., *A Writer's Guide for Engineers and Scientists*. Englewood Cliffs, New Jersey: Prentice-Hall, 1961.

[7] Reisman, S.J. (ed.), *A Style Manual for Technical Writers and Editors*. New York: The Macmillan Company, 1962.

[8] Freeman, L.H., and Bacon, T.R., *Shipley Associates Style Guide* (rev. ed.). Bountiful, Utah: Shipley Associates, 1990.

[9] AFR 80-44, Defense Technical Information Center. Prescribes DTIC and outlines Air Force support responsibilities.

[10] DoDD 5230.25, Distribution Statements on Technical Documents. Establishes policies and procedures for marking technical documents to show that they are either releasable to the public or that their distribution must be controlled within the United States Government. This regulation applies to all newly created technical data.

[11] Jordan, S. (ed.), *Handbook of Technical Writing Practices*. New York: John Wiley and Sons, Inc., 1971.

[12] Meredith, J.R., and Mantel, S.J., Jr., *Project Management: A Managerial Approach*, (2nd ed.). New York: John Wiley and Sons, Inc., 1989.

[13] Kezsbron, D.S., Schilling, D.L., and Edward, K.A., *Dynamic Project Management: A Practical Guide for Managers and Engineers*. New York: John Wiley and Sons, Inc., 1989.

[14] Moffat, R.J., "Describing the Uncertainties in Experimental Results." *Thermal and Fluid Science*, Vol. 1, pp. 3-17, January 1988.

[15] Hicks, C.R., *Fundamental Concepts in the Design of Experiments* (3rd ed.). New York: Holt, Rinehart, and Winston, 1982.

[16] Dally, J.W., Riley, W.F., and McConnell, K.G., *Instrumentation for Engineering Measurements*. New York: John Wiley and Sons, Inc., 1984.

[17] Chenery, H.B., and Clark, P.G., *Interindustry Economics*. New York: John Wiley and Sons, Inc., 1959.

[18] Lindgren, B.W., *Statistical Theory* (2nd ed.). New York: The Macmillan Company, 1968.

[19] Loeve, M., *Probability Theory* (3rd ed.). Princeton, New Jersey: D. Van Nostrand Company, 1963.

[20] Mangasarian, O.L., *Nonlinear Programming*. New York: McGraw-Hill Book Company, 1968.

[21] Martin, J.J., *Bayesian Decision Problems and Markov Chains*. New York:

John Wiley and Sons, Inc., 1967.

[22]Owen, G., *Game Theory*. Philadelphia, Pennsylvania: W.B. Saunders Company, 1968.

[23]Doebelin, E.O., *Measurement Systems Application and Design* (3rd ed.). New York: McGraw-Hill Book Company, 1983.

[24]Holman, J.P., *Experimental Methods for Engineers* (4th ed.). New York: McGraw-Hill Book Company, 1984.

[25]*Measurement Uncertainty*. ANSI/ASME PTC 19.1-1985, Part 1, 1986.

[26]Tucker, H.G., *A Graduate Course in Probability*. New York: Academic Press, 1967.

[27]Weiss, L., *Statistical Decision Theory*. New York: McGraw-Hill Book Company, 1961.

[28]White, D.J., *Dynamic Programming*. San Francisco, California: Holden-Day, 1969.

[29]Wilde, D.J., *Optimum Seeking Methods*. Englewood Cliffs, New Jersey: Prentice-Hall, 1964.

[30]Choquet, G., *Topology*. New York: Academic Press, 1966.

[31]Derman, C., *Finite State Markovian Decision Process*. New York: Academic Press, 1970.

[32]Ferguson, T.S., *Mathematical Statistics: A Decision-Theoretic Approach*. New York: Academic Press, 1967.

[33]Hays, W.L., and Winkler, R.L., *Statistics: Probability, Inference, and Decision*. New York: Holt, Rinehart, and Winston, 1970.

[34]Freeman, L.H., and Bacon, T.R., *Style Guide*. Bountiful, Utah: Shipley Associates, 1990.

9
Software Test and Evaluation

9.1 General

Although final Test and Evaluation (T&E) of any part of a system should be conducted at the total system level, there are many parts of a complex system to include the software that benefit from progressive T&E. In many cases, the software design for a complex system is approached by use of a well-modularized top-down approach. This requires that both the problem being addressed by the software and its solution be well understood. Success in achieving a good design for a new complex system on the first try is difficult, if not impossible. Consequently, a progressive approach is generally applied whereby a simpler subset of the problem and solution is addressed, and then it is iteratively enhanced until the full system solution is implemented. Each step allows the system to be refined and to become more effective as it is better understood through the iterative process. The emphasis of the T&E effort during this developmental testing period is on both the debugging (i.e., correction of errors) and the enhancement of the solution. After systems are fielded, software products often tend to become obsolete because of such things as threat and mission changes, and emphasis must be placed on replacing or upgrading the software to meet the evolving conditions.

Software T&E can be discussed in terms of the activities that must be achieved during the design and production of a system and the activities required to enable the continued application and use of the system throughout its life cycle. The term "maintenance" is used in the software community to describe the correction of errors, upgrading to maintain compatibility with new operating systems/compilers, and the introduction of small improvements to a computer program. Consequently, the distinction between the "development" of new capabilities, and "maintenance" of the system is a qualitative judgment depending on what constitutes a small change or improvement versus a major developmental change to the software.

Good software is vital to the success of any modern weapon system. Above all else, it is important that it work properly in the system for which it was designed (i.e., in accordance with design specifications). It is also important that software errors be corrected prior to large-scale fielding of the system (i.e., it has been debugged) and that the software execute quickly and make efficient use of the computer memory. Finally, it is desirable that the software be flexible, well documented, and delivered on a schedule that meets the user's needs.

9.2 Important Terms and Concepts

Computer software, like most other areas, has some basic terms that are used in the description of activities associated with software development and maintenance. A "system" is normally considered to be made up of "subsystems" (sometimes called "suites") and the subsystems are made up of "modules." A module (sometimes called a "subroutine") is a collection of instructions sufficient to accomplish some logical function. A module is also sometimes defined in terms of physical constraints, and generally must be combined with other modules to be useful. A program is a collection of instructions which, when combined with data and control information, is sufficient to accomplish some well-defined function. Software reliability is the probability that the software operates for a specified period of time without an error, on the machine for which it was designed, given that it is operated within design limits. Software reliability (like hardware) is also described in terms of mean time between failure (MTBF) and mean time to repair (MTTR). A "bug" in the software is an error which is consistent and repeatable. On the other hand, a "glitch" is some unforeseen feature that renders the software inefficient, inelegant, or clumsy (i.e., a glitch implies a problem with the specification itself, or some property that may be outside of the scope of the specification). Testing of software is a set of procedures and activities to demonstrate that the program performs correctly in its intended environment. "Debugging" is the process of finding and removing errors from the software. "Maintenance" is all of the activities necessary to correct errors, upgrade with new operating systems/compilers, and the introduction of minor enhancements. Some of the other very important terms like verification, validation, and accreditation have already been described under analytical modeling and simulation in Chapter 4.

9.3 Software Maintenance (See References 1–11)

Military weapons systems require a high level of software maintenance throughout their life cycle. New software features and enhancements must be periodically installed, and the software must be adapted to new data and hardware in response to increasing enemy threats and capabilities.

Software maintenance is often performed by someone other than the person who wrote the original program. When this occurs, the software maintainer must first become knowledgeable about the existing program by consulting the written documentation prepared by the original programmer. This documentation normally includes flowcharts at various levels of detail, narrative descriptions and, ultimately, program symbolics. Often this documentation is lacking, and the correction of errors and the evaluation of proposed changes based only on these materials are generally a challenge for the software maintainer.

9.4 Modifying Software (See References 12–23)

Modifying software requires understanding the reason for modification (i.e., the change), specifying the change, developing the software change, testing the change, and installing the change (see Figure 9-1).

The operational user generally communicates the need for the change to the software maintainer who is responsible for knowing how the software and system work before the change, and how they are supposed to work after the change. The activity and emphasis of the change are concentrated on the difference between how the system now works versus how it is desired to work. The modification activity (sometimes referred to as cut-line and patch) involves identifying the ex-

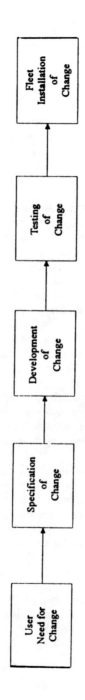

Figure 9-1. Simplified Software Change Process

isting code, data, and procedures to be modified, and developing the patch (i.e., new code, data, and procedures). This activity should be accomplished in a manner that minimizes interactions between the existing system and the patch to reduce the risk of introducing unintentional changes in the existing functions. The change is tested to ensure that it functions properly and that appropriate system actions occur. A process called *regression testing* is normally applied whereby the system is tested by carrying out tests saved from previous configurations, changes, and modifications. Only after the change has been properly verified and system tested, should it be released for large-scale fleet installation. When software is released without adequate testing, it generally results in the system not working properly and expensive and time consuming additional changes being required.

9.4.1 Reducing Debugging Problems

Testing and debugging computer software generally requires about 1/3 to 1/2 of the total time expended on a software project. Therefore, anything that can make the program simpler, more logical, straightforward, easy to maintain, better documented, easy to upgrade, and easy to test is important. Programmers often do not provide sufficient comments and documentation on code. The code tells what the program is doing, but the comments are necessary to understand why the programmer has used those specific instructions. Comments that are incorrect or misleading about program statements are generally worse than no comments at all. Assembly language programs are difficult to list and should not be used when a high-level language will do just as well. Multitasking (or a synchronous execution) allows the programmer to initiate and control the execution of several different tasks within one application. However, this approach also allows the programmer to take on some of the functions normally performed by the operating system and thus makes program testing much more difficult. Shared files also can lead to difficulty if the operating system allows one program to read a file while another is updating it. For example, if one program is reading two or more logically related files while another program is updating them, the first program may receive information that is not consistent.

Sometimes programmers can misuse instructions on a given machine to work for a specific application. Often this misuse can destroy the overall program logic. Sometimes programs are written in a way to modify themselves. Such programs are difficult to debug. Sharing variables and temporary storage among many subroutines can also lead to problems. Even when the two subroutines are logically independent, there sometimes can be a temporary storage conflict. It is normally good programming practice not to share storage between different routines unless large amounts of memory can be saved (e.g., several hundred elements in an array). Programs with an excessive number of GO-TO statements are difficult to debug because they tend to lack recognizable structure or flow of control. An alternative to the GO-TO statements are decision statements, subroutines, or some other modular structure.

9.4.2 Software Errors

Software errors can occur because of such things as faulty logic, incorrect documentation, overloads, timing problems, throughput and capacity problems, fallback and recover difficulties, hardware and software system problems, and nonadherence to standards. Logic errors are generally the most common type of errors found in software, and most testing efforts are expended to reveal these problems. When a given test input exposes the presence of an error, then the same input presented to the system a second time should expose the same error. Documentation errors can result in the operational user not knowing how to properly operate, or interpret operation of, the system. Overload errors can occur when various interval tables, buffers, queues, and other storage areas are filled up beyond their capacity. Timing errors in real time systems can cause problems where logic errors occur that cannot be easily repeated because they represent situations that are a function of timing conditions or coincidental events in the software. These types of errors are not uncommon in complex embedded systems and they can lead to the inability to successfully accomplish the mission. Throughput and capacity errors can affect the performance of the software by having an adverse impact on processing time, excessive use of memory, and so forth. Testing in the area of fallback and recovery should ensure that the system can recover in a reasonable amount of time, and that operators are not confused in the recovery process. For example, if a Radar Warning Receiver (RWR) system or a navigational system has automatically reset itself because of system overload, it is important that the operator know that the system outputs warnings during the overload and reset period may not be accurate or reliable. Hardware and system software errors can also result in major problems if not properly tested. Finally, violations of the standards for attributes of software, such as modularity, program comments, and being free of nonstandard programming statements, all can result in increased testing and operational problems.

9.4.3 Top-Down Programming

Top-down programming (sometimes called systematic programming, hierarchical programming, and design by explosion) attempts to identify the key functions that must be accomplished, and then proceeds to lesser functions from the major functions. The process begins by addressing the total programming problem, and then breaks it down into lesser parts (i.e., "primitives") until the primitives are small and simple enough to be addressed without great difficulty. Each primitive in a program or system should be rigorously examined in terms of the function, input, and output. The problem should also be addressed in terms of the design of the interfaces among the modules, and the interfacing among modules should follow a formal hierarchical input/output processing approach. During the top-down programming approach, professionals generally try to put priority on the higher level modules and avoid becoming too immersed in the details of the lower level problems until the higher level solution is obtained.

The size of each module should be limited to situations where the inputs are known, the functions are known, and the required outputs are known. Any given module should be implemented by a relatively small number of submodules. The top-down approach applies to the design of data as well as algorithms. The output is usually specified in terms of the system or user needs; however, programmers are generally free to design the internal files, tables, and other variables as they see necessary to efficiently produce the required output.

9.4.4 Top-Down Testing

Top-down software testing also approaches the problem from an overall system viewpoint; that is, it attempts to first test the main program and, to some degree, lesser levels of subroutines as a system framework. Often this results in many of the lower level modules being implemented initially as dummy routines until the total program has been developed. With top-down testing, the tester adds modules and code in a manner that allows the source of the problem to be readily apparent when something goes wrong. Bottom-up testing adds increasing numbers of modules during each step of the test process, making debugging more difficult. Top-down testing identifies most major interface, logic, and control errors early in the process, rather than at the end. Top-down testing is generally applied in concert with top-down design and top-down programming.

9.4.5 Bottom-Up Testing

The classical method of software testing is normally referred to as bottom-up testing. Here, testing begins with the lowest components of the system hierarchy (i.e., module testing), and proceeds through subsystem testing and then on to system testing. Module testing is also sometimes referred to as unit testing, single-threat testing, and program testing. It involves some form of stand-alone testing of each module and should reveal basic logic and coding errors. However, in order to do this, the tester generally has to have a test data base and various other elements of the environment within which the module will eventually reside. Testers believe that thorough and comprehensive testing of each module is the key to success when using this approach. The degree of success that can be achieved here is highly dependent upon how well the interfaces and environment created for the module represent the subsystem and the final overall system. Subsystem testing attempts to address the interfaces among the various modules in the software. Testers hope to reveal logic and interface errors as opposed to basic coding errors during this level of testing. System testing should reveal complex control and logic errors, throughput and capacity errors, timing errors, and recovery errors as well as interface errors. This type of testing attempts to examine the system under realistic operational conditions and to compare its output with known situations, or the output of other systems.

9.5 Assessment of Software (See References 24–44)

Software T&E must address the total utility of the software in terms of both system effectiveness and system suitability. The total systems approach for software evaluation is depicted in Figure 9-2. The ultimate success or failure of the software package depends upon its end performance within the system for which it is designed to work (i.e., system effectiveness) and the sustained capability to provide adequate support for the software over the system's life cycle (i.e., system suitability*). As such, the T&E of software is a subset of the T&E of the total system.

---

*Software maintainability is a vital part of system suitability.

SOFTWARE TEST AND EVALUATION

```
                    ┌─────────────────┐
                    │  WEAPON SYSTEM  │
                    └─────────────────┘
┌──────────────────┐                      ┌──────────────────┐
│   OPERATIONAL    │                      │   OPERATIONAL    │
│  EFFECTIVENESS   │                      │   SUITABILITY    │
└──────────────────┘                      └──────────────────┘
```

Software Evaluation

Does the total system perform as advertised?

If not, what problems are attributed to the software?

Is the software effectively partitioned to the degree advertised?

Can the software be supported at the wing/squadron level under the planned software support concept (i.e., personnel, training, support equipment, Technical Orders, documentation, etc.)?

Does the planned depot and higher headquarters software support concept work satisfactorily?

What should be done to resolve any identified software effectiveness and suitability problems?

Are there any software changes that need to be made for improvements in operations and/or support?

Figure 9-2. Systems Approach for Software Evaluation

The T&E of software is carried out throughout the acquisition and life cycle of a system and must be capable of addressing a wide range of issues to include software design concepts and attributes, performance measures such as the accuracy and timeliness of the solution, and system support measures such as what it takes to keep the system performing on a mission-by-mission basis as well as to make any required changes. The software evaluation process includes paper evaluations of the source code listings and documentation; simulations, prototyping, and laboratory evaluations; system integration testing; and total system operational testing evaluations. The final T&E of any software version should be conducted as an integral part of the system for which it was designed. The major thrust of the evaluation should be based on the software's contribution (or lack of contribution) to total system performance.

9.5.1 Design Concepts and Attributes

Structural or design T&E should be conducted on the software architecture/engineering and the associated source code listings and documentation. During design T&E, it is important to address software attributes such as modularity, descriptiveness, consistency, simplicity, expandability, and instrumentation. These attributes can be evaluated by answering questions such as those discussed in the following paragraphs.

9.5.1.1 Modularity. Has the software been partitioned into parts logically with few connections between modules, are processes within each module functionally related, and are documentation volumes essentially self-contained? Is the partitioning adequate to provide the degree of subsystem isolation essential to the operational/support concept as well as system safety?

9.5.1.2 Descriptiveness. Does information about the software exist, and is it useful? Does the information contain the objectives and assumptions, inputs, processing, outputs, revision status, and is it accurate, clear, and easily located?

9.5.1.3 Consistency. Do the software documentation and source code listings correlate? Is there uniform notation, terminology, and symbology containing organization, names, and comments?

9.5.1.4 Simplicity. Are there few fundamental elements with good terminology, structure, and techniques?

9.5.1.5 Expandability. Can physical change be easily accomplished once the requirement for the change is understood? Is there adequate margin for growth, numbering schemes, and independence of structure from logic?

9.5.1.6 Instrumentation. Does the software contain aids which enhance testing (i.e., embedded tests, test tools, and commonly used code)?

9.5.2 Performance Measurement

Functional or requirements T&E should be conducted to ensure that the software is effectively accomplishing its intended design functions. These evaluation measurements should begin early in the design and development process and proceed through simulations, prototyping, and laboratory evaluations, with end result performance being measured on the total system. Performance measurements generally deal with the accuracy, timeliness, and overall adequacy of controls, displays, computations, solutions, and other system end results. It is extremely important that the operators of embedded computer systems know when the software (and system) is functioning in a manner that degrades, or precludes, mission accomplishment. Catastrophic results can occur when operators are not alerted to highly degraded situations and continue their mission and procedures as if things are normal.

9.5.3 Suitability Measurement

Software suitability measurement involves measurement of the activities necessary to ensure that during a computer system's life cycle, the implemented and fielded software fulfills its original mission, any subsequent modifications to that mission, and any requirements for mission enhancements. These measurements can involve support personnel and systems at several organizational levels. For example, day-to-day scheduled and unscheduled maintenance actions are normally required at wing/squadron level, as well as periodic support actions at depot and higher headquarters levels. The factors of software maturity and software supportability are discussed below and they must be addressed during software T&E because they affect the performance of the software at all levels.

9.5.3.1 Software maturity. Software maturity is a measure of the software's progress in its evolution toward full satisfaction of the user requirements. Primary indicators of this evolution are the rate, number, and severity of the required software changes. Two conditions generate software changes: the correction of faults (in-scope deficiencies) and the incorporation of enhancements (out-of-scope deficiencies). Observables include fault trends, cumulative trends (as a function of time and of program size), and cumulative enhancements (as a function of time and of program size). Problem reports, engineering change proposals (ECPs), and the usability of the software are important indicators of maturity. However, the ultimate measure is whether the system can accomplish the mission for which it was procured and do it at the success rate specified as essential for combat.

9.5.3.2 Software supportability. Software supportability is a measure of the capability of the system software to be changed by typical software experts according to the planned software support concept. Three essential elements are required to adequately support software: a well-designed software product exemplified by well-engineered and useful system source code and the information pertaining to that code contained in the associated documentation (i.e., user's or supporter's manual, engineering notebook, etc.); the existence and adequacy of the software support resources (personnel, equipment, and facilities); and a well-defined process exemplified by the existence and adequacy of the procedures needed to manage and control all aspects of a software change from its inception and fielding throughout the system's life cycle. For some systems, software supportability will be impacted by the memory size, the processing time reserves, and the remaining growth capabilities.

9.6 Test Limitations

Even when the final test of the software is conducted in the system for which it is designed, some errors will likely not be detected. At best, developmental testing will have occurred at all appropriate levels, and most of the major errors will have been corrected prior to operational testing of the system. During operational testing, the system should be exercised over enough of the most demanding missions so that any major errors can be discovered and corrected prior to large-scale fielding of the software. During this testing, the interfaces of the system are extremely important. For example, if the system prepares missiles for launch, it is imperative that either missiles be prepared and launched by the system or some acceptable form of testing be substituted if it is not physically or economically feasible to conduct the launch activity. Operational testing must verify and confirm, to the degree practical, the full operational capabilities of the system. The consequences of not discovering errors prior to large-scale fielding of software for weapons systems are that problems will reveal themselves during daily operations, or possibly in combat. If revealed during daily operations, the expense could be enormous in terms of the potential for causing accidents or disrupting training and mission capability, not to mention the loss in readiness of the forces and the ability to win in combat should a conflict occur while the software problems exist.

References

[1] DoD STD 2167A, Defense System Software Development.
[2] AFR 800-14, Volume I, Management of Computer Resources in Systems. Establishes policy for acquiring and supporting computer equipment and programs

developed or acquired under the management concept of AFR 800-2.

[3]AFOTECP 800-2, Software Operational Test and Evaluation Guidelines. A five-volume pamphlet which describes the activities associated with planning, conducting, analyzing, and reporting software OT&E assessments. Volumes are:

Volume 1, Management of Software Operational Test and Evaluation. A guide for the headquarters software evaluation manager (SEM) and the test team's deputy for software evaluation (DSE).

Volume 2, Software Support Life Cycle Process Evaluation Guide. Provides the software evaluation manager and the deputy for software evaluation information needed to evaluate mission critical computer software life cycle processes as they influence software supportability. In this pamphlet are the means to track the processes affecting mission critical computer software supportability, beginning as early as necessary to provide insight into the quality of the evolving software products, software support resources, and operation support life cycle procedures themselves.

Volume 3, Software Maintainability Evaluator's Guide. Provides the software evaluator the information needed to participate in the AFOTEC's software maintainability evaluation process. It provides an overview of the process for assessing software design and documentation, detailed instructions for using AFOTEC's standard software maintainability questionnaires and answer sheets, and the questionnaires themselves.

Volume 4, Software Operator-Machine Interface—Evaluator's Guide. Provides the operator the information needed to participate in AFOTEC's software operator-machine interface evaluation. This evaluation methodology provides a standardized approach to determine the adequacy of the software action between a computer-driven system and its operator. In addition to describing the method, the guide includes a questionnaire and explanatory details.

Volume 5, Software Support Resources Evaluation—User's Guide. Describes the method and procedures used by the AFOTEC for evaluating the software support resources for an embedded computer system acquisition. The evaluation of resource capabilities, required to provide an overall evaluation of a system's software supportability, is outlined in general, and detailed guidance is provided for the software evaluation manager, the test team deputy for software evaluation, and the software evaluator members of the test team to support the evaluation.

[4]Adey, R.A. (ed.), *Engineering Software*. London, England: Pentch Press, 1979.

[5]Parikh, G., and Zvegintzor, N., *Tutorial on Software Maintenance*. Silver Spring, Maryland: IEEE Computer Society Press, 1983.

[6]Glass, R.L., and Noiseux, R.A., *Software Maintenance Guidebook*. Englewood Cliffs, New Jersey: Prentice-Hall, Inc., 1981.

[7]Lientz, B.P., and Swanson, E.B., *Software Maintenance Management—A Study of the Maintenance of Computer Application Software in 487 Data Processing Organizations*. Reading, Massachusetts: Addison-Wesley Publishing Company, Inc., 1980.

[8]Parikh, G., *Techniques of Program and System Maintenance*. Boston, Massachusetts: Little, Brown, and Company, 1982.

[9]McClure, C.L., *Managing Software Development and Maintenance*. New York: Van Nostrand Reinhold Company, 1981.

[10]Martin, J., and McClure, C.L., *Maintenance of Computer Programming*. Carnforth, England: Savant Institute, 1982.

[11] Kernighan, B.W., and Plauger, P.J., *Software Tools*. Reading, Massachusetts: Addison-Wesley Publishing Company, Inc., 1976.

[12] Jackson, M.A., *Principles of Program Design*. New York: Academic Press, 1975.

[13] Dijkstra, E.W., *A Discipline of Programming*. Englewood Cliffs, New Jersey: Prentice-Hall, 1976.

[14] Swaum, G.H., *Top-Down Structured Design Techniques*. New York: Petrocelli Books, 1978.

[15] Dahl, O.J., Dijkstra, E.W., and Hoare, C.A.R., *Structured Programming*. New York: Academic Press, 1972.

[16] Shneiderman, B., *Software Psychology: Human Factors in Computer and Information Systems*. Cambridge, Massachusetts: Winthrop Publishers, Inc., 1980.

[17] Weinberg, G.M., *The Psychology of Computer Programming*. New York: Van Nostrand Reinhold Company, 1971.

[18] Yourdon, E., *Design of On-Line Computer Systems*. Englewood Cliffs, New Jersey: Prentice-Hall, Inc., 1972.

[19] Markov, A.A., *The Theory of Algorithms* (translated from Russian). U.S. Department of Commerce, Office of Technical Service, No. OTS 60-51085.

[20] Linger, R.C., Mills, H.D., and Witt, B.I., *Structured Programming: Theory and Practice*. Reading, Massachusetts: Addison-Wesley Publishing Company, 1979.

[21] Yourdon, E., *Techniques of Program Structure and Design*. Englewood Cliffs, New Jersey: Prentice-Hall, Inc., 1975.

[22] Constantine, L.L., *Concepts in Program Design*. Cambridge, Massachusetts: Information and Systems Press, 1967.

[23] Siewiorek, D.P., and Swarz, R.S., *The Theory and Practice of Reliability System Design*. Bedford, Massachusetts: Digital Press, 1982.

[24] Rustin, R. (ed.), *Debugging Techniques in Large Systems*. Englewood Cliffs, New Jersey: Prentice-Hall, Inc., 1971.

[25] Lala, P.K., *Fault Tolerant and Fault Testable Hardware Design*. London, England: Prentice-Hall International, 1985.

[26] Anderson, T., and Lee, P.A., *Fault Tolerance Principles and Practices*. London, England: Prentice-Hall International, 1981.

[27] Kohavi, Z., *Switching and Finite Automata Theory*. New York: McGraw-Hill Book Company, 1978.

[28] Pradhan, D.K., *Fault-Tolerant Computing Theory and Techniques*. Englewood Cliffs, New Jersey, 1986.

[29] Johnson, B.W., *Design and Analysis of Fault Tolerant Digital Systems*. Reading, Massachusetts: Addison-Wesley Publishing Company, 1989.

[30] Beer, S., *Decisions and Control—The Meaning of Operational Research and Management Cybernetics*. New York: John Wiley and Sons, Inc., 1966.

[31] Athans, M., and Falb, P.L., *Optimal Control*. New York: McGraw-Hill Book Company, 1966.

[32] Sage, A., *Optimum Systems Control*. Englewood Cliffs, New Jersey: Prentice-Hall, Inc., 1968.

[33] Goodwin, G.C., and Payne, R.L., *Dynamic System Identification: Experiment Design and Data Analysis*. New York: Academic Press, 1977.

[34] Chandy, K.M., and Sauer, C.H., *Computer Systems Performance Analysis*. Englewood Cliffs, New Jersey: Prentice-Hall, 1981.

[35] Davis, G.B., and Hoffmann, T.R., *FORTRAN 77: A Structured, Disciplined Style* (3rd ed.). New York: McGraw-Hill Book Company, 1988.

[36] Jensen, K., and Wirth, N., (Revised by Mickel, A.B., and Miner, J.F.), *Pascal Users Manual and Report* (3rd ed.). New York: Springer-Verlag, 1985.

[37] Kernighan, B.W., and Ritchie, D.M., *The C Programming Language* (2nd ed.). Englewood Cliffs, New Jersey: Prentice-Hall, 1988.

[38] Banks, J., Carson, J.S., and Sy, J.N., *Getting Started with GPSS/H*. Annandale, Virginia: Wolverine Software Corporation, 1989.

[39] Law, A.M., *Statistical Analysis of Simulation Output Data with SIMSCRIPT II.5*. La Jolla, California: CACI Products Company, 1979.

[40] National Bureau of Standards, *Computer Model Documentation Guide*. Washington, D.C.: National Bureau of Standards, Special Publication 500-73, 1981.

[41] Knuth, D.E., *The Art of Computer Progamming*, Volume 2 (2nd ed.). Reading, Massachusetts: Addison-Wesley Publishing Company, Inc., 1981.

[42] Kobayaski, H., *Modeling and Analysis: An Introduction to System Performance Evaluation Methodology*. Reading, Massachusetts: Addison-Wesley Publishing Company, Inc., 1978.

[43] Kennedy, W.J., Jr., and Gentle, J.E., *Statistical Computing*. New York: Marcel Dekker, Inc., 1980.

[44] Knuth, D.E., *The Art of Computer Programming, Volume 3, Sorting and Searching*. Reading, Massachusetts: Addison-Wesley Publishing Company, Inc., 1973.

10
Human Factors Evaluations

10.1 General

The concept of most civilian and military system design and development includes the ever-important person-machine interface. The performance of any individual equipment is partially a function of the performance of other equipment to include the humans who must interface with the system to accomplish the mission. This includes the human beings who operate the system as well as those who must maintain and support the system. Although many systems have been and are being designed to rely less upon human intervention, people are still vital parts of most systems. People must decide when and how to apply the systems, and act and react to inputs and outputs from the systems. Systems work well only if the humans who must interface with them can perform their roles successfully, for it is the humans that the machines must satisfy. Consequently, when addressing the system as a total entity, the evauation must include both of the vital parts: the humans and the machines.

The definition of a system is somewhat arbitrary, and depends upon the purposes for which it is designed and the mission it is to accomplish. A small system with limited person-machine interface is most often part of a larger system, and the performance of one to some degree depends upon the performance of the other. The definition and understanding of these interfaces are of vital importance when one is addressing the scope of what and how much to consider in human factors evaluations.

The human being plays a major role in both the performance and Test and Evaluation (T&E) of any system. Effective person-machine interfaces are a critical part of the design of a system and usually have an indispensable impact on total system performance. Like software evaluations, the final proof of human factors capabilities is best achieved as a part of the total system evaluation. Human factors evaluations are generally directed at examining how system performance is affected as a function of certain human inputs and interfaces, assessing the effects of one or more variables on the various aspects of human behavior, and/or describing a population in terms of selected attributes. Often, the variability introduced into T&E measurements by the presence of the human element is significantly greater than that caused by other sources. Therefore, the selection of sample personnel to be involved in a T&E can have a major impact on the results. For example, if all equipment operators in a given T&E have engineering degrees and the equipment is to be operated in combat by high school graduates, the performance results achieved during T&E likely will not be representative of combat. Many times the human being is also used as an observer and recorder of data which, as a consequence, are subject to human biases, inaccuracies, and other fallacies. Finally, both the system design and the evaluation of T&E results depend directly upon the intelligence, initiative, integrity, experience, industriousness, and other attributes and capabilities of the human participants.

The human engineer is a vital member of any team in system design and application. He/she is the users' champion with respect to safety, comfort, ease of operating and maintaining the equipment, and a multitude of other important human considerations. He/she plays an important role in evaluating how the human as an essential component contributes to the overall system and mission accomplishment.

10.2 System Analysis and Evaluation
(See References 1–31)

System analysis and system evaluation are viewed as two distinctly different efforts in human engineering. A system analysis is conducted to give an accurate picture of the structure and functions of a system. It is a prerequisite to any form of system evaluation or comparison. System analysis addresses how the system will be put together and what processes it will employ to accomplish the mission. The system analysis describes in detail the components of the system, the operating characteristics of the components (both humans and machines), and the required interfaces necessary to successfully accomplish the mission. It provides an analytical statement of the capabilities, limitations, and interdependencies of the components in a manner that is relevant to the system's capability to accomplish the mission. On the other hand, system evaluation provides a measure (or set of measures) that establishes how well the system can accomplish its mission. Often, system evaluation must apply a combination of methods, such as operations analysis, modeling, simulation, physical testing, and comparative techniques, to arrive at measures of how well the "human system" and its interfaces can actually perform their portion of the mission.

10.2.1 System Analysis

The system analysis should form a conceptual structure that helps clarify the relationships and interdependencies of human engineering to the total system engineering process. It includes the definition of requirements and constraints, the description of system functions, the detailed descriptions of operational event sequences (including environmental considerations), and the detailed descriptions of the various components and associated processes. It addresses scheduling, limiting factors, system performance criteria, design options, analysis of functions, tasks, system characteristics, human interactions, and all other factors that affect performance.

Generally, the first step in a system analysis is to construct a block diagram (or pictorial representation) of the overall system. Figure 10-1 is a simplified representation of a pilot mission planning system. Pilots use the system to aid in the planning of their mission and to facilitate the transfer of important planning data from the planning system to the aircraft and weapons. The input to the planning system is the knowledge and expertise of the pilot with regard to target selection, profile, tactics, etc., and the output allows the pilot to manipulate a multitude of options to arrive at an acceptable approach to the mission. Obviously, destroying

HUMAN FACTORS EVALUATIONS

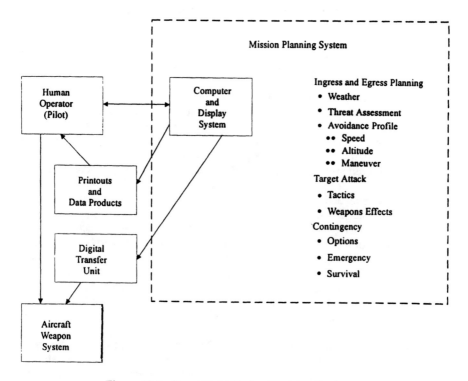

Figure 10-1. Simplified Mission Planning System

the target and surviving are key concerns of the pilot. Once an acceptable planning solution has been achieved, the planning system provides printouts and data products as well as a digital transfer unit solution for use by the pilot in preparing himself/herself and the aircraft for the mission. The block diagram (or pictorial representation) provides a depiction of the components, both man and machine, and the conditions under which the system operations will occur. The overall mission requirement for the system is a key driver in this process. It should include such things as what the system is to accomplish, the environment in which it is to operate, operating and maintenance concepts, design parameters and limits, emergency and safety considerations, and interfaces and interdependencies with other systems. Both the internal and external operating environments of the system should be addressed. A detailed accounting of the system components to include sensors, communication and navigation devices, and operator displays are important. Obviously, the detailed specification of the human component of the system is vitally important. The number of personnel, jobs they are to perform, control devices, information displays, and other interface methods are important.

10.2.1.1 Analyzing system functions. The functional analysis of a system addresses how each specific requirement of a system will be met. From the requirement, system engineers must identify the specific functions that the system must perform to carry out its mission and fulfill the requirement. One function might be maintaining the track of a target during attack. Functional analysis must be performed to ascertain what specific capabilities are needed in terms of human skills and equipment. These determinations include what capabilities should be automated and what capabilities should be left to the human operator. Functional analysis translates system requirements into an organized program for design implementation. System requirements are defined in detail and the system functions that must be performed to accomplish those requirements are identified. Emphasis is placed on describing the functions that must be accomplished during mission operations rather than the equipment or personnel. For example, the requirement might be for a system to accomplish the mission of attacking ground targets. One important function of the system might be to locate the ground targets in order for them to be considered for attack. Once the function is identified and described, the function itself can be used to help determine how best it can be achieved in terms of man and machines relative to the environment. Candidate components to accomplish the target location function might include ground intelligence agents, space surveillance systems, aircraft surveillance systems, aircraft search and attack systems, or some combination thereof. Each function should be analyzed in terms of establishing the kinds of capabilities needed to meet the system performance requirements, possible combinations of man-machine interfaces, and the total system cost effectiveness of the overall design. The potential effect of each option should be addressed in terms of the overall system with the goal of providing the most realistic and cost-effective design.

10.2.1.2 Analyzing system tasks.

Once the system functions have been identified, described, and specified, the important performance characteristics of the tasks that must be performed by humans can be addressed. This examination includes the psychological aspects, the actions required, skills and knowledge, and the probable human and equipment errors and malfunctions that could occur. The relative importance and sequencing of each task should be established. The required speed, accuracy, and special abilities should be identified along with the human characteristics and training needed to successfully accomplish each task. Each task should be examined in light of whether humans can actually meet the requirements of the task in terms of available equipment, operations, mission, and environment.

The analysis of tasks provides a foundation on which the overall human engineering effort can be based. It establishes information on the location at which the tasks will be performed and their relationship to other tasks. It helps identify which skills and knowledge are required to successfully achieve all steps and parts. It should result in a clear definition of the psychological demands of the task, the decisions and discriminations that must be made, and the motor and other skilled responses needed. Human factors within the tasks are concerned with the knowledge and insight relating to human beings that are relevant to the design and operation of the man-made features. Human factors, like operations research analysis, are not a separate discipline, but are interdisciplinary in nature and represent an area of overlapping interest of specialists in a variety of social and engineering sciences. The implications of various perceptual, mental, and physical concepts and procedures as they might affect or influence human factors generally can best be characterized by the results of scientific research investigations and documented operational evaluations.

The tasks that must be accomplished can be examined in terms of a time-line analysis where critical functions are reviewed to ascertain whether situation and time constraints warrant a reallocation of functions. The examination of sequential and parallel functions versus time can reveal periods of peak workload, conflicting demands, and sometimes unnecessary as well as additional requirements. Time-line analysis helps depict the number and duration of functional activities assigned to each part of the system. It allows the examination of system operator's tasking versus time for the functional areas of the system. When overloads or task saturation exist, they should be eliminated by adding the appropriate mix of machines and personnel, dividing workload according to ability and performance, eliminating workload, or some other method. Undertasked situations can also be revealed and can lead to the elimination or combining of workload and operator positions.

Task analysis should be conducted with a view toward predicting the occurrence of operator errors, and the likely behavior of the operators under situations where machine design or malfunctions can lead to human errors. Each potential adverse situation should be examined in terms of the contingencies that can be applied to alleviate the situation and allow successful mission completion. Contingencies are defined as action plans that can be implemented to successfully deal with the nonroutine situations that occur during the mission. Contingency analysis should address such things as potential malfunctions, accidents, failures, extreme climatic conditions, environmental conditions, and enemy actions.

Activity analysis, flow analysis, and decision analysis all play an important role in human engineering. Activity analysis approaches the human involvement from the aspect of the activities carried out by the human in the person-machine system. It examines human activity over time to judge the relative importance of the various operator activities and the demands placed on the operator. It provides information for establishing operational procedures; redesigning interfaces, displays, and equipment; unburdening operators; and redistributing workload and responsibilities. Flow analysis identifies and portrays the sequence of events in a system. It is useful for indicating the functional relationships between system components, tracking the flow of data and information, showing the assignment and distribution of workload, displaying inputs and outputs, and depicting essential component interfaces. Decision analysis involves data and information flow and the resulting decisions and actions that must take place to successfully accomplish the task or mission. It is most applicable to situations where the operator receives and fuses data, reacts to a situation (perceived or real), and takes action to initiate or complete a task.

10.2.1.3 Analyzing equipment characteristics. During system design, the human interface and equipment characteristics must be optimized in a manner to support successful mission accomplishment. This translates to identifying those equipment characteristics that are essential in the design for successful task performance by the operator, and in turn, accomplishment of the mission. These essential characteristics can best be addressed in terms of system displays and other interfaces, operator controls, workplaces, and such things as space, environment, safety, and operator selection/training.

The human sensing system consists of receptors responsive to a large variety of sources and ranges of physical energy. It processes and acts upon stored and sensed inputs, discriminates signals from noise, recognizes situations from among information patterns, makes decisions, and determines appropriate responses for situations. Machines are extremely efficient at storing large amounts of data, rapidly processing it, disseminating the results, and essentially doing anything humans are smart enough to program them to do. Consequently, humans need machine devices to help gather the needed information and to translate it into inputs that they can perceive. These devices are called displays. The essential characteristics of the displays should be selected to suit the particular conditions under which they will be used, the method of use, and the purposes they are to serve. Viewing distance, illumination, presence of other displays, and compatibility with related controls are important characteristics that must be considered in the design. Displays are used for a variety of purposes to include such things as quantitative reading, qualitative reading, check reading, setting, tracking, and spatial orientation. Displays can be symbolic (i.e., have no pictorial resemblance to the conditions represented) or pictorial.

The characteristics of operator controls play a vital role in mission accomplishment. In general, controls should be designed based on the function of the control and the requirements of the task. The purpose and importance of the control to the system, the nature of the controlled activity, the type of change to be made, and the extent and direction of change are all important considerations. The task should be analyzed so that speed, range, precision, and force requirements for using the control and the effects of trade-offs among those characteristics are understood. Design should consider such things as the distribution of tasks among limbs, the placement of controls, the type of controls, hazards, overriding controls, and the similarity of controls. The informational needs of the operator are important in terms of locating and identifying the control, determining control position (i.e., setting), and sensing changes in the control position. Important design considerations include the control to display ratio, direction of movement relationships, control resistance, prevention of accidental activation, and control coding. The control to display ratio is the ratio of the distance of movement of the control to that of the moving element of the display (e.g., cursor, pointer). The direction of movement of a control should be related appropriately to the change it inputs in the associated equipment, displays, and system as a whole. Proper direction of movement can improve reaction time, correctness of initial control movement, speed and precision of control adjustment, and operator learning time. For example, try to imagine a new car with a brake that you must pull on instead of pushing with your foot. The amount of resistance provided by the control to the operator is also important and should be based on what best satisfies the particular performance requirements. Controls should be located and designed to prevent accidental activation or movement, and to respond quickly with rate movement in emergencies. Often this can be achieved through such things as recessing, location, locking, sequencing, covering, and designed-in assistance or resistance. Controls can be coded to make them easy to identify. The five most common methods used to code controls are shape, size, mode of operation, labeling, and color.

The workplace and its layout have a significant impact on the ability of the human to support successful mission accomplishment. The layout of controls and displays should be accomplished based on the operator's size, position (seated, standing, prone, etc.), look angle, and the spaces in which he/she can best accomplish control. Controls and displays should be grouped into logical units. An accurate and consistent correspondence should be established between each control and its display. The most important controls should always be located within the optimum dimensions (i.e., the most desirable space for location for a high percentage of the human population). The other controls must be within acceptable dimensions but not necessarily optimum dimensions. These dimensions should be established based on the population of operators.

10.2.1.4 Special attributes of humans.
Humans and machines have different strengths and weaknesses in performing various tasks. More often than not, one is better at a given task than the other. Consequently, in system analysis, it is important to decide which tasks can be best achieved by humans and which can be best achieved by machines.

The human can be viewed and examined as a system. Figure 10-2 depicts a simplified model of the human system. The human sensors detect and encode energies from the physical environment and pass them on for processing/synthesis. The information processing activities manage both the stored and sensed inputs, discriminate signals from noise, assimilate and synthesize information, recognize patterns, make decisions, and initiate the selected responses. It also controls reflexive as well as regulatory processes for adaptation and survival. Memory provides for the storage of both short- and long-term encoded information. Responses are selected, initiated, and carried out in terms of actions such as speech, movement, and other physical and mental adjustments.

Humans have extraordinary sensing and pattern recognition capabilities, ability to learn and modify behavior accordingly, and ability to assimilate information and make decisions. On the other hand, there is considerable variability among humans and they are subject to problems with fatigue and boredom. These differences in ability are also important in designing machines that can integrate with, and be used most effectively by, human operators.

The human sensing system is reliable, consistent, and accurate. It has a much greater mean time between critical failure than most machine counterparts. In many cases when the human operator makes an error, it can be attributed to input information, displays, environment, and so forth. Humans have an extraordinary capacity to recognize patterns and situations when compared to machines. Their ability to recognize terrain features, sensor images, photographic details, and multiple sources of images cannot be achieved by even the most advanced machines.

The ability of humans to adapt and change their behavior based on past experience (i.e., learn) cannot be matched by machines. Humans learn by practicing a skill, trial and error, and transfer of training. All of the learning mechanisms of humans are highly dependent upon timely and accurate feedback regarding success in accomplishing the task. Practicing a skill results in learning by humans when the task is rehearsed over and over again until it becomes automatic to the human system. Trial and error allows operators to attempt various solutions to a task and adopt the one that succeeds most often. Transfer of training occurs when the individuals are able to apply training from a previous task to the task at hand (for example, an individual trained to fly one aircraft being trained to fly another). Individuals can learn faster when they are able to understand and acquire insight into the task rather than depend solely on trial and error and repetitive practice.

Although machines have enormous capacity to rapidly receive, store, manipulate, and display vast amounts of data, generally the human operator must digest

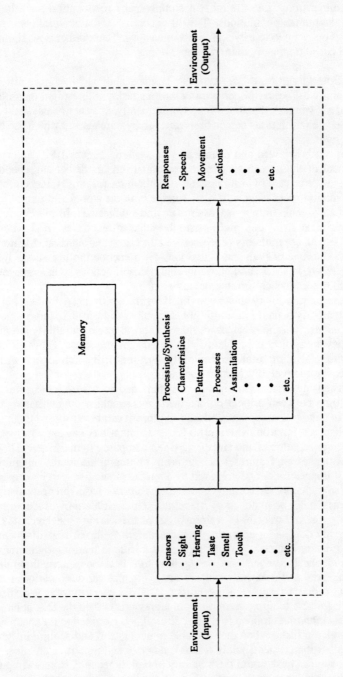

Figure 10-2. Simplified Human System

and understand the important trends and characteristics of the data, fuse and assimilate the important indicators, and make the final choices. Having a machine make such choices requires that we understand precisely what choices are to be made, what information is required to arrive at the choice, and the logic that underlies the decision making process regarding the choice. When this can be achieved, the task can often be delegated to the machine to enhance mission success.

Human abilities, skills, and performance can vary widely and they are subject to problems with fatigue and boredom. Fortunately, the variability in a given individual's performance can be decreased through practice. However, there is a wide range of overall performance capabilities that generally occur when humans address complicated and difficult tasks. Humans get fatigued and bored and their performance degrades with time when they are asked to do the same thing for long periods. Research indicates that an individual has the capacity to continue to perform well but loses motivation (i.e., the will to excel) after an extended period of time. Fatigue and boredom cause individuals to lose their concentration on the task, and this results in errors, mistakes, and accidents. Fatigue and boredom can be lessened by shorter work periods, variation in task assignments, and adjustment in work position (e.g., stand versus sit). Humans become anxious and upset when they lose control of their own safety, are not provided adequate information about what is going on around them, and do not like or trust fellow workers. All of these characteristics can lead to problems or breakdowns within the human system.

10.2.2 System Evaluation

System evaluation should be conducted after system analysis; it provides a measurement of how well a system can be expected to perform under realistic operating conditions or whether one system can be expected to perform better than another. System evaluation is normally achieved through a combination of methods that includes analysis, modeling and simulation, physical testing, and comparative techniques. Humans, unlike machines, vary widely from one to another and from one time to another and require special precautions and attention. In fact, humans have a great tendency to "game" the evaluation—especially once they know the "rules" of what is to be evaluated.

Human factors testing is a subset of system testing and concentrates on answering the following types of questions:

1) Can the human being reliably accomplish the tasks necessary for overall system success?
2) Have the training programs adequately provided the skills necessary for the human (e.g., operators and maintainers) to successfully do their job?
3) Are there discrepancies or deficiencies in the design of human interfaces, procedures, or training?
4) Are human task loading and manpower sufficient?
5) Are essential time and motion factors correct and adequate?
6) Are there any unacceptable safety hazards?
7) Do the operators and maintainers (i.e., users) accept the product?

System evaluation starts with the foundation of good system analysis. Knowing the analytical characteristics, functions, tasks, and expected performance of the operators and machines is an important first step in the evaluation. Modeling and simulation of a system should be performed either by a computer-type simulator, a mock-up, or some hybrid combination of the two. Computer simulations offer the advantages of speed and repetitiveness, and are generally less expensive. Mock-ups provide the equipment (or something like the equipment) and enable the operators to interface as they would in operational applications. Physical testing and comparative techniques involve measurements being made of the actual operator performance on the real system. The principal difference between mock-up testing and the latter is the degree of reality in the systems being used for evaluation.

10.2.2.1 Measuring performance. The effects of performance must be measured and evaluated at several levels. The component or subsystem performance must be measured as well as its impact on the overall system and mission. The evaluator should attempt to understand how task performance impacts function performance and how function performance impacts overall mission performance. Without this insight, the human and machine activities cannot be optimized for overall mission performance. The measures of performance selected are extremely important. Both quantitative and qualitative measures are essential. Success rate, accuracy, timeliness, and numerous other quantitative measures are important.

10.2.2.2 Realistic conditions. System evaluations must occur under realistic conditions if they are to predict meaningful operational performance of the system. In general, realism can be improved by the proper choice of tasks to be employed in the evaluation, the adequate training of test participants, and the psychological conditions existing during the evaluation. Predictions will be good or bad based on how well they deal with the critical operational conditions of the system. Realism has to do with the machine and its behavior, as well as the operators and their behavior. The evaluation scenarios should require operators to work at realistic speeds, provide the same kind of challenges that they would face in combat, and require them to observe the rules of the evaluation. Operators should be chosen and trained to the proper level to achieve realistic results. Operators should be of the same general age, ability, skill level, and experience as those who would employ the system in combat. Operators and evaluators who have some vested interest in the outcome of the evaluation should not be allowed to participate. Their biases and prejudices could influence the results of the evaluation. Often their skill level and knowledge far exceed the norm, and they produce results which cannot be expected from the average combat operators. In extreme cases where it is essential to use highly expert operators in an evaluation, it is also desirable to conduct the evaluation with more typical personnel if the system is likely to be employed over an extended time frame and require a larger numbers of operators.

10.2.2.3 Training operators. Operators must be adequately trained before their performance can be expected to represent realistic system capabilities. This requires that they be given adequate instruction in the tasks, be measured and provided feedback on task performance, and be given sufficient time to reach a representative plateau of performance for the system they are operating. The training provided should take the operators through the various stages of learning until their task performance is within the acceptable bounds of function and system performance.

10.2.3 Questionnaires and Subjective Assessments

Questionnaires and subjective assessments are essential tools for collecting and evaluating the results of system evaluation. The single most valued data point in T&E is the informed opinion of the operator, maintainer, expert, or customer. Consequently, the tester must ensure that methods used to sample and obtain the opinions of the experts are representative and valid. It is extremely important in T&E that the right experts be used to answer the right questions. The tester must seek out, obtain, and present the unbiased opinion of the operational experts and not allow that opinion to be artificially transformed by superficial methods into another domain where it may or may not be properly understood. Consequently, the analytical methods used in evaluation must clearly serve as tools to assist experienced operational experts and operations research analysts in applying their ingenuity, competence, and judgments to the evaluation process.

10.2.3.1 Questionnaires. Questionnaire statements should be made in such a way that they do not bias or load the results. They should be straightforward, simple, clear, relevant, and unbiased (see References 6–9).

10.2.3.2 Subjective assessments. Rating scales are often helpful in obtaining consistent results from subjective assessments.* Rating scales are especially useful when multiple subjects are used to make judgments regarding the effects of human factors. There are four basic types of scales for these purposes (see Figure 10-3).

10.2.3.2.1 Nominal scale. The nominal scale, or classificatory scale, is the most common scale applied and is used to simply place things or items in a given category. For example, one might classify weapon systems in terms of "bomber," "fighter," "helicopter," and "missile" where these categories correspond to roles that the systems are to perform. One important aspect of the nominal scale is equivalence, where equivalence means that the items presented in a given category are the same and the order in which the designations for "bomber," "fighter," "helicopter," and "missile" are presented in the scale is irrelevant. Hence, the order in which the terms are presented on the nominal scale is interchangeable. Nominal scales provide the lowest form of measurement; and, generally, only limited nonparametric statistics can be applied to results obtained from these scales.

10.2.3.2.2 Ordinal scale. An ordinal scale, or ranking scale, has the characteristics of a nominal scale in addition to which it identifies the relative ranking order of the categories. Items appearing higher on the scale have the "greater than" or "preferred to" characteristic in the ranking of utility. The ordinal scale is a higher order form of measurement than the nominal scale since it has an order of terms which is noninterchangeable. Although it depicts the "greater than" characteristic, it does not specify how much greater than. Nonparametric statistics are normally used to analyze the results obtained by ordinal scales.

---

*One of the better-known ratings scales applied in aviation work is the Cooper-Harper Scale (see References 10 and 11).

Nominal	Interval (Cardinal)
Bomber _____	4 Excellent _____
Fighter _____	3 Satisfactory _____
Helicopter _____	2 Marginal _____
Missile _____	1 Unsatisfactory _____

Ordinal	Ratio (Cardinal with Zero Point at Origin)
A Excellent _____	4 Excellent _____
B Satisfactory _____	3 Satisfactory _____
C Marginal _____	2 Marginal _____
D Unsatisfactory _____	1 Unsatisfactory _____
	0 Unusable _____

Figure 10-3. Example Rating Scales

10.2.3.2.3 Interval scale. The interval scale has the characteristics of the ordinal scale plus equal and linear distances between points on the scale. Consequently, interval scales not only possess the "greater than" characteristics but they also specify how much greater than. In an interval scale, the zero point and unit of measurement are arbitrary. The interval scale is a quantitative scale which assigns real numbers to objects in an ordered set. Both nonparametric and parametric statistics can be applied to analyze the results obtained by interval scales.

10.2.3.2.4 Ratio scale. The ratio scale has all of the characteristics of the interval scale and it has a true zero point as its origin. The ratio scale provides the highest form of measurement, and both nonparametric and parametric statistics can be applied to the results obtained from ratio scales.

10.2.3.3 *Justification of rating scales.* It is extremely important that the tester does not incorrectly apply or use the results from rating scales in dealing with human factors. Scales up through the ordinal scale can generally be applied in T&E without problems. Data obtained from the scales can be presented in many ways and provide powerful results on a comparative basis. However, when interval and ratio scales are applied, the tester (much like the statistician who applies a statistical test) should justify the appropriateness of the rating scale being used. That is, if equal and linear distances between points on the scale or a true zero point at the origin are assumed, the tester should justify the assumptions being made. Such justification is of critical importance when parametric statistical methods are to be performed on the numerical results and if they are to lead to valid conclusions and recommendations.

10.2.3.4 *Small sample considerations.* Small sample results in subjective evaluations during T&E are common and derive their major utility from conditions where major discrepancies or benefits are discovered (i.e., where preferences and findings reveal critical, previously unknown information and may overwhelmingly be in favor of or against a specific item). Under conditions where critical information is not revealed and/or results are not overwhelming, the statistical findings associated with the sample become significantly less conclusive as the sample size decreases below samples in the order of 30 to 50. Thus, stated simply, the major analytical contribution made by the small samples associated with subjective evaluations does not usually lie in the power of the statistics. The following paragraphs describe some suggested methods and criteria to accomplish meaningful analyses for small sample evaluations.

10.2.3.4.1 Selection of the sample. Statistical methods normally specify random sampling from the population. However, in some small sample cases, a random sample may not be the most productive. For example, if 10 pilots are selected to test a new G-suit, it probably would be more productive to select pilots that represent a cross section of physical attributes (e.g., tall, short, heavy, and thin) rather than leave it up to chance that a small random sample would be physically representative of the total population.

10.2.3.4.2 Comparisons against an existing item with subjective comparison scale. The following small sample procedure is recommended, assuming that one desires to compare against an existing item by use of the subjective comparison scale of 1, 2, 3, 4, and 5, where 3 is equivalent capability. It is also assumed that one always will carefully evaluate the specific results and comments from the sample, and that the statistical techniques described herein will be used only to augment other parts of the evaluation. The following steps are recommended:
1) Select a measure of success (e.g., individual scores must be 4 or higher to indicate preference).
2) Group sample results in terms of success or failure (e.g., scores of 4 and higher, or 3 and less).
3) Criterion: Require that the majority (or higher number) of the sample prefer the item.

This criterion and method is recommended because it is simple and understandable, and it is statistically reasonable for small sample cases. For example, given a sample of 10 pilots, for an item to be acceptable, 6 of the 10 pilots would have to prefer the item. This translates to a confidence level slightly over 60 percent that the majority of the population would also prefer the item. Conversely, given that a majority of the population truly prefers the item, the probability of accepting the item based on the sample findings is well over .5; thus, there is not an overpowering tendency to reject an acceptable item. Although a specific small sample of 10 was chosen to illustrate the recommended procedure, the above discussion holds in a general sense for all small samples.

10.2.3.4.3 Comparison against an existing item by task evaluation. The following analysis procedure is recommended, assuming that one desires to compare items by use of an absolute scale evaluating task accomplishment:
1) Select a measure of success for task accomplishment (e.g., task score must be 3 or higher).
2) Group scores in terms of successes and failures for old item.
3) Group scores in terms of successes and failure for new item.
4) Perform either Fisher's Exact Probability Test or the Chi Square Contingency Test (2×2 case) to ascertain whether one item is significantly better in task performance from a statistical point of view. Selection of which method to apply

			Percentage in Rating Category
A.	Completely Agree	<u>No Doubt</u> that the product <u>cannot be any better</u>	30
B.	Strongly Agree	Product is <u>very good</u> and <u>very helpful</u>	20
C.	Generally Agree	Product is <u>acceptable</u> and it is <u>helpful</u>	20
D.	Generally Disagree	Product is <u>unacceptable</u>, and <u>minor improvements</u> are required to make it helpful	10
E.	Strongly Disagree	Product is <u>unacceptable</u>, and <u>major improvements</u> are required to make it helpful	10
F.	Completely Disagree	<u>No doubt</u> that the product is <u>unacceptable</u> and must be <u>completely redesigned or rewritten</u>	10

Criterion - All ratings should be A, B, OR C. The operational impact and the ability to resolve problems with ratings of D, E, or F will be investigated in greater detail.

Figure 10-4. Example Ordinal Rating Scale Problem

should be based on sample cell size (see Reference 20, pages 71–74).

5) In all cases, the item should at least meet the minimum required score, or the exception should be explained as to why it is acceptable.

6) Criterion: New item should exceed old item by a value that is operationally significant and at a statistical confidence level of at least 80 percent.

This criterion and method is recommended because it is relatively simple and does not require excessive computations, and it is relatively powerful and straightforward considering the limitations present.

10.2.3.5 Example ordinal rating scale problem. Suppose two different software evaluators rate the acceptability of software documentation in accordance with the ordinal scale depicted in Figure 10-4. The frequency of their 200 ratings is reduced and presented in terms of the percentages shown in the figure. Also described in the figure is the criterion that ratings of A, B, or C are judged as successes while ratings of D, E, or F fail. The data show that 70 percent of the total ratings assigned passed the criterion while 30 percent failed. The analyst would normally carefully examine and categorize the reasons why specific aspects of the software documentation were assigned ratings of D, E, or F. Another question that would be of interest is how well the two raters agreed on the items receiving a pass or fail rating. Suppose the ratings by the two raters were as summarized in the 2 ×

	RATER # 1	RATER # 2	TOTALS
PASS	80	60	140
FAIL	37	23	60
TOTALS	117	83	200

Figure 10-5. Summary of Pass/Fail Ratings by Raters #1 and #2

2 contingency table in Figure 10-5.

The chi square or Fisher's Exact Probability Test could be applied to ascertain whether or not the passing and failing ratings by the two evaluators were statistically different at a given selected level of significance (see References 20 and 21). For example, if we choose a level of significance of .10, we note that the critical value for the chi square statistic with one degree of freedom extracted from a chi square table for a two-sided test is 2.71. The chi square statistic calculated for the values shown above is as follows:

$$x^2 = \frac{200[1(80)(23) - (37)(60)| - 200/2]^2}{(140)(60)(117)(83)} = .192$$

Since .192 is less than 2.71, we accept the null hypothesis that the different ratings by the two raters are not significantly different at the .10 level.

References

[1] MIL-STD-1472, *Human Engineering Design Criteria for Military Systems, Equipment, and Facilities*. Establishes general human engineering criteria for design and development of military systems, equipment, and facilities. Its purpose is to present human engineering design criteria, principles, and practices to be applied in the design of systems, equipment, and facilities.

[2] MIL-H-46855, *Human Engineering Requirements for Military Systems, Equipment, and Facilities*. Establishes and defines the requirements for applying human engineering to the development and acquisition of military systems, equipment, and facilities. These requirements include the work to be accomplished by the contractor or subcontractor in conducting a human engineering effort integrated with the total system engineering and development effort.

[3] Hagin, W. V., Osborne, S. R., Hockenferger, R. L., Smith, J. P., Seville Research Corporation, and Gray, Dr. T. H., Operations Training Division, Williams AFB, Arizona, *Operational Test and Evaluation Handbook for Aircrew Training Devices: Operational Effectiveness Evaluation* (Volume 11) (a handbook), AFHRL-TR-81-44 (II). Brooks AFB, Texas: Air Force Human Resources Laboratory, Air Force Systems Command, February 1982.

[4] Cott, H. P., and Kinkade, R. G. (eds.), *Human Engineering Guide to Equipment Design*. Washington, D.C.: American Institute for Research, 1972.

[5] Morgan, C. T., Chapanis, A., Cook III, J. S., and Lund, M. W. (eds.), *Human Engineering Guide to Equipment Design*. New York: McGraw-Hill Book Company, Inc., 1963.

[6] Dyer, R. F., Matthews, J. J., Wright, C. E., and Yudowitch, K. L., Operations Research Associates, *Questionnaire Construction Manual*. Fort Hood, Texas: U.S. Army Research Institute for the Behavioral and Social Sciences (ARI), July 1976.

[7] Dyer, R. F., Matthews, J. J., Stulac, J. F., Wright, C. E., and Yudowitch, K. L., Operations Research Associates, *Questionnaire Construction Manual Annex*, Literature Survey and Bibliography. Fort Hood, Texas: U.S. Army Research Institute for the Behavioral and Social Sciences (ARI), July 1976.

[8] Gividen, G., *Order of Merit—Descriptive Phrases for Questionnaires* (disposition form). Fort Hood, Texas: Army Research Institute (ARI), Field Unit, 22

February 1973.

[9]Payne, S. L., *The Art of Asking Questions*. Princeton, New Jersey: Princeton Press, 1973.

[10]Cooper, G. E., Ames Research Center, and Harper, Jr., R. P., Cornell Aeronautical Laboratory, *The Use of Pilot Rating in the Evaluation of Aircraft Handling Qualities* (a published technical note), NASA Technical Note D-5153. Washington, D.C.: National Aeronautics and Space Administration (NASA), April 1969.

[11]Martin, L. M., and Mueller, G., *Close-Loop Handling Qualities* (a systems note), Chapter XII. Edwards AFB, California: USAF Test Pilot School, February 1979.

[12]Bennett, E., Degan, J., and Spiegel, J. (eds.), *Human Factors in Technology*. New York: McGraw-Hill Book Company, 1963.

[13]Cakir, A., Hart, D., and Stewart, T., *Visual Display Terminals*. New York: John Wiley and Sons, Inc., 1979.

[14]Chapanis, A., *Research Techniques in Human Engineering*. Baltimore, Maryland: Johns Hopkins University Press, 1959.

[15]Siegel, S., *Nonparametric Statistics for the Behavioral Sciences*. New York: McGraw-Hill Book Company, 1956.

[16]Barley, R. W., Bell Telephone Laboratories, *Human Performance Engineering, A Guide for Systems Designers*. Englewood Cliffs, New Jersey: Prentice-Hall, Inc., 1982.

[17]Singleton, W., Fox, J., and Whitfield, D. (eds.), *Measurement of Man at Work*. London, England: Taylor and Francis, 1971.

[18]Swain, A., and Guttmann, H., *Handbook of Human Reliability Analysis with Emphasis on Nuclear Power Plant Applications*. Washington, D.C.: U.S. Nuclear Regulatory Commission, 1980.

[19]Parsons, H. M., *Man-Machine System Experiments*. Baltimore, Maryland: Johns Hopkins University Press, 1972.

[20]Tate, M. W., and Clelland, R. C., *Nonparametric and Shortcut Statistics*. Danville, Illinois: Interstate Printers and Publishers, Inc., 1957.

[21]Hollander, M., and Wolfe, D. A., *Nonparametric Statistical Methods*. New York: John Wiley and Sons, Inc., 1973.

[22]McFarland, R. A., *Human Factors in Air Transportation: Occupational Health and Safety*. New York: McGraw-Hill Book Company, Inc., 1953.

[23]McNemar, Q., *Psychological Statistics*. New York: John Wiley and Sons, Inc., 1955.

[24]Fletcher, H., *Speech and Hearing in Communication*. New York: D. Van Norstand Company, Inc., 1953.

[25]Fitts, P. M., and Jones, R. H., "Analysis of Factors Contributing to 460 "Pilot-Error" Experiences in Operating Aircraft Controls" in *Selected Papers on Human Factors in the Design and Use of Control Systems* (H. W. Sinaiko, ed.). New York: Dover Publications, Inc., 1961.

[26]Dill, D. B., *Life, Heat and Altitude*. Cambridge, Massachusetts: Harvard University Press, 1938.

[27]Browne, R. C., "Fatigue, Fact or Fiction?" in *Symposium on Fatigue* (W. F. Floyd and A. T. Welford, eds.). London, England: H. K. Lewis & Company, Ltd., 1953.

[28]Beraneks, L. L., *Acoustic Measurements*. New York: John Wiley and Sons, Inc., 1949.

[29] Barnes, R. M., *Motion and Time Study*. New York: John Wiley and Sons, Inc., 1950.

[30] Aero Medical Association, *Aviation Toxicology: An Introduction to the Subject and a Handbook of Data*. New York: McGraw-Hill Book Company, Inc., 1953.

[31] Mowbray, G. H., and Gebhard, J. W., "Man's Senses as Informational Channels" in *Selected Papers on Human Factors in the Design and Use of Control Systems* (H. W. Sinaiko, ed.). New York: Dover Publications, Inc., 1961.

11
Reliability, Maintainability, Logistics Supportability, and Availability

11.1 General (See References 1–3)

The capability of a military system to carry out its intended function can be described in terms of its operational effectiveness and operational suitability. Both operational effectiveness and operational suitability are affected to a large degree by the reliability, maintainability, logistics supportability, and availability (RML&A) of a system. The triangle in Figure 11-1 depicts the interrelationships among reliability, maintainability and, supportability. These characteristics combine to produce operational availability, and in turn, result in system effectiveness. For systems to be effectively applied in combat, they must be available in an operable condition. The systems must be reliable to the extent that they can carry out their intended missions over the time duration required. When systems do experience failure, the appropriate level of personnel, skills, and supplies must be available to carry out the needed repairs. Finally, the systems must be maintainable to allow

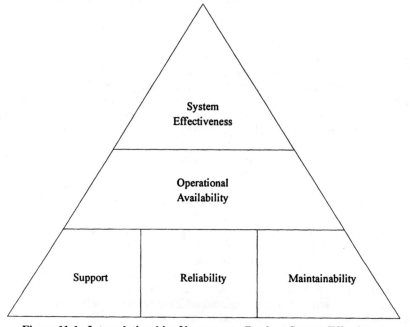

Figure 11-1. Interrelationships Necessary to Produce System Effectiveness

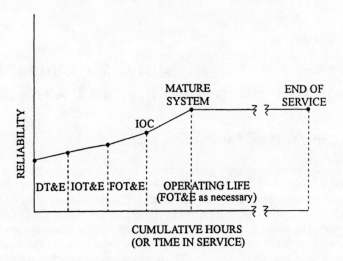

Figure 11-2. Typical Reliability Growth Cycle for a Weapons System

the repairs to be carried out in a timely and efficient manner.

The correct assessment of a system's RML&A during test and evaluation (T&E) is essential to its successful application in combat. The assessment can lead to the timely correction of deficiencies, enhancement of capabilities, and the proven knowledge of how best to support the system in operational application. Because it is desirable to have a system in a state or condition to be successfully operated on demand (or on a continuous basis), most of the common measures of RML&A are expressed in terms of probabilities of success or failure and/or times. When these measures are carefully examined, it can be seen that RML&A goals are best achieved by acquiring systems that break infrequently, are easily and quickly repaired or replaced, reduce or eliminate elements of the maintenance support structure, decrease the mobility requirements per unit, decrease manpower requirements per unit, and decrease overall support costs. Generally, these assessments of RML&A are made in terms of measuring the current capabilities and limitations of the system, as well as trying to predict or project whether it is capable of reaching specified mature values during its life cycle.* Figure 11-2 depicts the typical reliability growth of a system through the various phases of T&E during its life cycle. The system progresses in reliability growth from developmental test and evaluation (DT&E) through initial operational test and evaluation (IOT&E). It further progresses in growth through follow-on operational test and evaluation (FOT&E) to the establishment of initial operational capability (IOC). FOT&E may also occur as necessary during the operating life of the system. This growth often results from the correction of design problems and improvement in manufacturing quality. Although T&E discovers and provides for correction of many of these problems, some are not discovered until the system is fielded and does not perform as expected in its operational environment. Insufficient T&E can result from not having sufficient time, an inadequate supply of test articles, and using a

---

*The concept of reliability growth provides for a system to be improved over time until it reaches specified goals referred to as "mature" values.

Table 11-1. Definitions of Concepts Used in Test and Evaluation of Reliability, Maintainability, Logistics Supportability, and Availability

System Effectiveness is the probability that the system can successfully meet an operational demand within a given time when operated under specified conditions.

System Effectiveness (for a one-shot device such as a missile) is the probability that the system (missile) will operate successfully (kill the target) when called upon to do so under specified conditions.

Reliability is the probability that the system will perform satisfactorily for at least a given period of time when used under stated conditions.

Mission Reliability is the probability that, under stated conditions, the system will operate in the mode for which it was designed (i.e., with no malfunctions) for the duration of a mission, given that it was operating in this mode at the beginning of the mission.

Operational Readiness is the probability that, at any point in time, the system is either operating satisfactorily or ready to be placed in operation on demand when used under stated conditions, including stated allowable warning time. Thus, total calendar time is the basis for computation of operational readiness.

Availability is the probability that the system is operating satisfactorily at any point in time when used under stated conditions, where the total time considered includes operating time, active repair time, administrative time, and logistic time.

Inherent or Intrinsic Availability is the probability that the system is operating satisfactorily at any point in time when used under stated conditions, where all free time, such as standby and delay times, associated with scheduled and preventative maintenance, administrative, and logistics time are excluded.

Design Adequacy is the probability that the system will successfully accomplish its mission, given that the system is operating within design specification.

Maintainability is the probability that, when maintenance action is initiated under stated conditions, a failed system will be restored to operable condition within a specified active repair time.

Serviceability is the degree of ease or difficulty with which a system can be repaired.

---

test environment that is different from the intended operational environment. Also, attempting to conduct T&E and make reliability and maintainability predictions with immature configurations and systems can lead to invalid results. Inadequate design of the equipment can result from failure to properly describe the operational and maintenance requirements and environments, misapplication of parts and materials, software errors, and poor integration of hardware and software. Manufacturing problems can occur as a result of poor parts, workmanship, materials, and processes. When T&E is inadequate, some of these problems do not become apparent until after the systems reach the field environment. Unfortunately, these problems sometimes do not become apparent until attempts are made to apply the systems in the combat environment.

Tables 11-1 and 11-2 provide a list of important definitions for *concepts* and *time categories* used in addressing RML&A.

Most of the definitions presented in Tables 11-1 and 11-2 stress four important characteristics: 1) probability, 2) satisfactory performance, 3) time, and 4) condi-

Table 11-2. Definitions of Time Categories Used in Test and Evaluation of Reliability, Maintainability, Logistics Supportability, and Availability

Operating Time is the time during which the system is operating in a manner acceptable to the operator, although unsatisfactory operation (or failure) is sometimes the result of the judgment of the maintenance technician.

Free time is time during which operational use of the system is not required. This time may or may not be downtime, depending on whether or not the system is in operable condition.

Storage time is time during which the system is presumed to be in operable condition, but is being held for emergency (i.e., as a spare).

Administrative time is the portion of downtime not included under active repair time and logistic time.

Downtime is the total time during which the system is not in acceptable operating condition. Downtime, can, in turn, be subdivided into a number of categories such as active repair time, logistic time, and administrative time.

Active repair time is that portion of downtime during which one or more technicians are working on the system to effect a repair. This time includes preparation time, fault-location time, fault-correction time, and final check-out time for the system, and perhaps other subdivisions as required in special cases.

Logistic time is that portion of downtime during which repair is delayed solely because of the necessity for waiting for a replacement part or other logistics delay.

---

tions of use. Probability is a number between 0 and 1.0 and can be thought of as the proportion of times that a certain event (e.g., nonfailure of a system) will occur if the mission related to the event is repeated an infinite number of times. Satisfactory performance is usually described in terms of criteria that the system must accomplish to successfully complete its mission. Duration of mission (or time) is the period during which one expects the system to perform satisfactorily. Finally, conditions of use describe those environmental and operational conditions under which the system is expected to perform.

11.2 System Cycles (See References 3–18)

11.2.1 System Daily Time Cycle

The RML&A of a system can be examined in terms of its daily and longer-term cycles (i.e., its predicted and/or actual behavior over a period of time). For example, Figure 11-3 depicts the major activities of a system (or some device less than a full system) over the daily status conditions of the system. In general, the system is either in operable condition and is "available" for use, or it is in nonoperable condition and is "not available" for use. When the system is available for use, it might be "off" awaiting use, in some type of "standby" mode operating at less then the full system mode, or "on" and "operating" in the full system mode. If a failure or complaint should occur regarding the operational condition of the system, it would normally be moved into the not available for use category. This would be termed an "unscheduled maintenance" activity. On the other hand, if the system required some form of periodic preventive maintenance, it would be

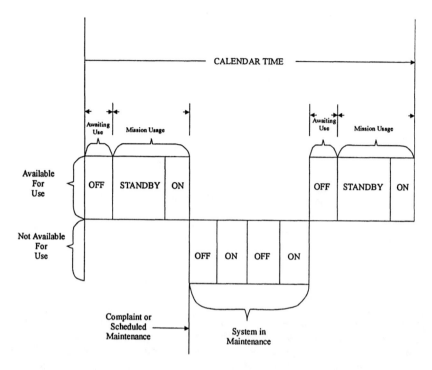

Figure 11-3. Example Daily Time Cycle for System Operation and Repair

moved into the not available for use category for the "scheduled maintenance" activity. When systems are broken and are not available for use, they sometimes must wait for someone to repair them, and/or wait for supply parts to be made available so that they can be repaired. Thus, it is also common to address the time that the system is not available for use in terms of the time lost caused by maintenance and supply.

It should be noted that the system might also be operable and available for use, but not actually in use during the depicted time period. Thus, the utilization rate of a system is important along with all of the other measures in examining the capabilities and limitations of both commercial and military systems.

11.2.2 System Life Cycle

The RML&A behavior of a system over the long term can be examined in terms of its life cycle. Figure 11-4 shows an example of the failure characteristics that a system might depict over its life cycle. Because of its shape, this type of curve is sometimes referred to as a "bathtub" chart. The behavior of the system and its effects on RML&A can be discussed in terms of three important stages: 1) burn-in, 2) useful life, and 3) wear-out.

Many systems experience a higher failure rate in the early hours of their life (see stage 1 on the chart). This higher failure rate, sometimes called "infant mortality," occurs during the "burn-in" period and can result from poor workmanship

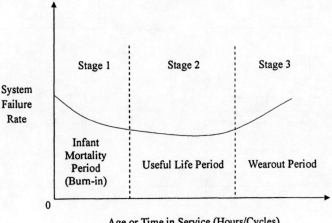

Figure 11-4. System Life Cycle Characteristics

(i.e., manufacturing deviations from the design intent, transportation damage, or assembly and installation errors). The higher failure rate, not uncommon in new equipment, can often be traced to such things as incorrect positioning of parts; improper insulation and protective coatings; contamination and/or impurities in materials; and poor connections, welds, and seals. Often, greater attention to detail in the quality process by those directly responsible for the manufacturing tasks can eliminate many of these problems. System producers sometimes allow a "burn-in" period for their product prior to delivery to help eliminate a high portion of the initial failures and to assist in achieving a higher level of stable operational performance once the system is introduced into the customer's inventory.

The failure rate of a system is at its lowest value and is relatively constant once the system reaches its useful life period. The system may exhibit some signs of wear, and the failure rate slowly increases during its useful life period. Failures during this period are characterized mainly by the occurrence of stress and its effect on system components. For example, such things as excessive temperature, transient and steady state voltage levels, humidity, vibration, shock, altitude, and acceleration forces all can contribute to the failure of system components. Systems designed with higher margins of safety and less variation in the manufacturing process normally fail less during the useful life period. The exponential failure distribution is most commonly accepted as the mathematical model for complex systems to describe and predict failures during this period.

The third stage of a system's life occurs when its failure rate starts to increase because of system wearout. When a system enters this stage, its failure rate generally increases as it increases in age. Wearout failures are due to such things as oxidation and corrosion, shrinking and cracking, friction wear and fatigue, ionic migration of metals in vacuum and on surfaces, and breakdown or leakage of materials. Ideally, it is desirable to overhaul or replace a system before its failure rate becomes unacceptable from wearout. The prediction of when and where to make the overhaul or replacement of the system is a challenging task for both the commercial and military user. Generally, this decision is made based on some form of cost and operational effectiveness analysis as described in Chapter 2.

11.3 System Degradation

The RML&A of many military systems do not reach the potential predicted for them during design. The transition from a theoretical design to actual operations under field/flight conditions introduces degradation factors that often are not accounted for during the design and production or that can be accounted for in laboratory testing. Part of the degradation can be attributed to manufacturing and operating conditions. Poor quality in manufacturing and poor quality in military servicing and maintenance of the equipment can contribute to degradation. Also, sometimes, the military must operate the equipment under conditions that exceed its original design specifications in order to accomplish the mission. This could result either from not knowing what the real requirements for the equipment are or from having situations occur that were not anticipated in the original statement of the specifications.

Modern techniques of total quality management (TQM) are directed at trying to improve the quality processes by shifting emphasis and responsibility to those actually producing the components and systems rather than inspecting and rejecting items after total assembly of the product. Other TQM theories, like reduced variation in components, worker improvement through increased responsibility, and the application of team concepts, are being applied. Both the commercial and military related industries are presently shifting to the TQM concepts of design and production. Only time will reveal whether the TQM concepts can narrow the wide gap between the RML&A predictions for military systems and what can actually be achieved under stressful operational conditions.

11.4 Reliability

The reliability of an item or system is its capability to satisfactorily perform a required function under stated conditions for a specified period of time. This capability can be assessed for both human and machine performance and can be measured in several ways, each of which are relevant to certain types of systems or situations. Two of the more common ways to express system reliability are 1) probability of successful performance and 2) mean time between critical failure (MTBCF).*

Within the Department of Defense, there have been several different points of view applied regarding what constitutes a failure. In the operational world, generally a failure is any condition in which the delivered hardware and/or software does not perform its required functions when operated and maintained by trained personnel. Often in the developmental and contractor world, a failure is not considered to exist until the failed component is precisely isolated and verified. Also, sometimes the "failed" component is replaced through engineering change by another type of component and the assumption is made that the failure will not reoccur. When this type of adjustment is made, the original failure data are removed

---

*A critical failure is defined as one which prevents the mission of the system from being successfully completed.

from the failure data base because the assumption is made that it will not reoccur. This disconnect in interpretation often leads to communications problems regarding the performance and adequacy of systems. For example, when operators act as a result of an observed or otherwise indicated "failed" condition, they make decisions and take actions in order to accomplish the mission or assure safety based on the failed condition. The system is considered to be in a nonoperable state until the observed or indicated condition is evaluated by appropriate experts. Maintenance personnel are required to expend time to evaluate and/or correct the failed condition documented by the operator. If the failed condition is intermittent (or does not exist), the maintenance activity results in a "cannot duplicate" situation. As previously stated, in the developmental and contractor world, the operator failed condition does not exist until the faulty component is precisely isolated and verified. These differences in the definition of what constitutes a "failure" often result in the developers and contractors having a set of numbers for the mean time between failure of a system much higher and more optimistic than the numbers experienced by the operators under field/flight conditions.

11.4.1 Combining Components

When a system contains multiple components (e.g., machine, human), the reliability of the total system will be a function of the reliability of the individual components and how they are combined within the system. Components of a system can be combined either in series or in parallel, or some combination thereof depending upon the basic system design.

11.4.1.1 Components in series. Successful performance of a system composed of individual components in series requires that each and every component perform successfully. If it can be assumed that 1) the individual component failures are independent of each other, and 2) failure of any one component results in the total system failure, then the basic system reliability can be expressed as the product of the individual component reliabilities*:

$$R_{system} = R_1 \times R_2 \times R_3 \times ... R_n$$

where R_{system} = total system reliability, and $R_1, R_2, R_3, ... R_n$ = reliability of individual components.

11.4.1.2 Components in parallel. Successful performance of a system composed of individual components in parallel requires only that at least one of the parallel components perform successfully. The reliability of a system applying this backup or redundancy arrangement can be computed by the following formula when n like functions with m components of equal reliability (R) are operated in parallel**:

$$R_{system} = [1-(1-R)^m]^n$$

---

*When individual component failures are not independent of each other, more complex conditional probability calculations must be employed.

**Probability computations become more complex for arrangements where the functions are not alike and the components have different individual reliabilities.

where

R_{system} = total system reliability
R = reliability of a single component
m = number of components
n = number of like functions

11.4.1.3 Complex systems. The reliability of a complex system can be analyzed in terms of the series/parallel combinations of its individual components. Additionally, it can often be shown that the overall reliability of a highly complex system behaves in accordance with the exponential distribution as follows:

$$R_{system} = e^{-\frac{t}{MTBCF}}$$

where

R_{system} = total system reliability
e = base (2.718) of natural system of logarithms
t = time system to operate
MTBCF = mean time between critical failure

11.4.1.4 Combining subsystems for a highly complex series system. Suppose that a complex system consists of n subsystems in series and the reliability of each of the subsystems has been shown to behave in accordance with the exponential distribution. Then:

$$R_{system} = R_1 \times R_2 \times R_3 \times \cdots R_n$$

$$R_{system} = e^{\frac{-t}{MTBCF_1}} \times e^{\frac{-t}{MTBCF_2}} \times e^{\frac{-t}{MTBCF_3}} \times \cdots e^{\frac{-t}{MTBCF_n}}$$

Therefore, the MTBCF for the total system can be estimated as shown below:

$$MTBCF_s = \frac{1}{\dfrac{1}{MTBCF_1} + \dfrac{1}{MTBCF_2} + \dfrac{1}{MTBCF_3} + \cdots \dfrac{1}{MTBCF_n}}$$

11.4.2 Example Complex System Reliability Problem

Suppose a system consists of four different components (aircraft, control systems, human interfaces, and armament), all of which must perform successfully to complete a 3-hour mission. Further, suppose measurements of MTBCF have been made on the individual components during development testing and their reliability has been shown to behave in accordance with the exponential distribution.* It is desired to provide an estimate of the overall system reliability. Example data and computations are shown below:

---

*One method of verifying that the reliability distributions are indeed exponential would be to perform a goodness of fit statistical test (e.g., chi square, see Reference 30).

$t = 3$ hours

$$\text{MTBCF}_1 = 100 \text{ hours}; \quad R_1 = e^{\frac{-3}{100}} = .97$$

$$\text{MTBCF}_2 = 50 \text{ hours}; \quad R_2 = e^{\frac{-3}{50}} = .94$$

$$\text{MTBCF}_3 = 40 \text{ hours}; \quad R_3 = e^{\frac{-3}{40}} = .93$$

$$\text{MTBCF}_4 = 60 \text{ hours}; \quad R_4 = e^{\frac{-3}{60}} = .95$$

$$\text{MTBCF}_s = \frac{1}{\frac{1}{100} + \frac{1}{50} + \frac{1}{40} + \frac{1}{60} +} = 13.6 \text{ hours}$$

$$R_{\text{system}} = e^{\frac{-t}{\text{MTBCF}_s}} = e^{\frac{-3}{13.6}} = .80$$

$$R_{\text{system}} = R_1 \times R_2 \times R_3 \times R_4 = .97 \times .94 \times .93 \times .95 = .80$$

11.5 Maintainability (See References 19–30)

The maintainability of an item or system is the ability to return or restore it to an operable or committable state. Maintainability can be expressed in terms such as the probability of being able to repair the system within a specified time, the mean downtime (MDT), the mean time to repair (MTTR), and the manpower and skill levels required for repair. Maintainability can also be addressed during T&E in terms of the organizational level where the maintenance must be performed (e.g., organizational, intermediate, and depot), and whether it is scheduled or unscheduled. Scheduled maintenance is maintenance that must be performed on a recurring basis to successfully operate the system (e.g., lubrication, fueling, periodic calibration, and alignment). Unscheduled maintenance is maintenance that must be performed to repair a system to an operable state after it has experienced a random failure or failures.

As discussed earlier, reliability and maintainability are interrelated in their effect on operational availability and system effectiveness. During the design of a system, a trade-off must be made between reliability and maintainability. Figure 11-5 uses an indifference curve to depict the trade-off between system reliability and maintainability to achieve a given level of system availability (see Chapter 2 for a discussion of cost and operational effectiveness optimization techniques). From this curve, it can be noted that if the total budget (represented by the straight line) was spent on reliability, y_1 units of reliability could be built into the system and would achieve the required level of availability. On the other hand, if the total budget was spent on maintainability, x_1 units of maintainability could be built into the system and would achieve the required level of availability. However, the optimum solution and mix of the two occurs at the intersection of the availability level curve, and the straight line budget constraint, the optimum mix being repre-

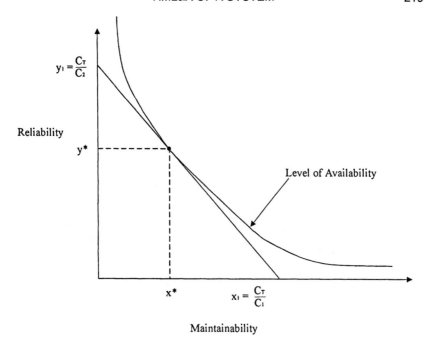

Figure 11-5. Trade-Off of Reliability and Maintainability

sented by a reliability cost of $C_2 y^*$ and a maintainability cost of $C_1 x^*$. This mix provides the highest level of availability that can be achieved for that particular fixed total cost.

Maintainability can also be addressed strictly from the point of view of reducing life-cycle maintenance costs. Figure 11-6 presents curves showing the life-cycle cost of a system versus MTTR. It can be seen that as the investment costs in maintainability are reduced (i.e., MTTR increases), the cost of maintenance increases. Consequently, a trade-off between the investment costs to make the system more maintainable can be made versus the cost of maintenance in terms of MTTR. The point of intersection of the two curves represents the lowest point of the life-cycle costs curve for maintainability investment plus the cost of maintenance.

11.5.1 Maintenance Concept

The maintenance concept for a weapon system specifies how the system will be supported by such things as special support equipment, maintenance organizational levels, and supporting commands. During the planning of T&E, it is important that the test design exercise and assess all aspects of the maintenance concept planned for the weapon system.

The maintenance concept defines the repair philosophies, techniques, and methods that guide support planning and provides a basis for defining support requirements. It should be prepared as an integral part of the system's original design. Thereafter, the concept should serve as an important guide during system develop-

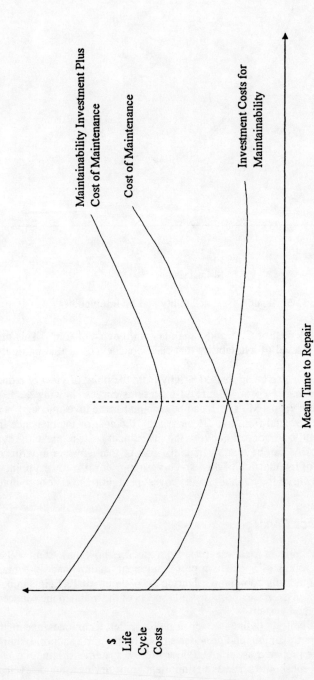

Figure 11-6. Trade-Off of Maintainability and Maintenance Costs

ment. The maintenance concept for a given system must specify the amount and type of support that is planned at all levels of maintenance activity. This specification should be based on the basic design and functional characteristics planned for the system, its complexities, and the conditions under which the system will be used and maintained. The concept should be formulated through a careful analysis of all aspects of the system's structure, function, and use, and be based on the selection of optimum repair policies that support the required system availability and use. If, for example, the system maintenance concept is to replace modular components, then the system design should be to allow modular component replacement. The goal of the support concept should be to provide the level of support essential to meet operational demands, and to do it in the most efficient and economical manner throughout the life cycle of the system.

Many components are repaired on the systems and off the systems in shops, and some are returned to higher level repair activities like depots. The maintenance concept must specifically address these planned activities. If a system is designed to be fully repairable, all of the components should have maximum life expectancy, and should be readily accessible for troubleshooting and removal without the use of special tools. If a system is designed to be maintained by remove and replace, all of the remove and replace items should be readily accessible for removal and replacement only. When systems are highly integrated, the design should provide for keeping the maintainer fully aware of the system state and status, and provide the essential information on actions required to return the system to a full state of operational capability. It would also be very valuable to have an accurate prediction of the reliability value for the next mission, given that no repair has been made to a partially failed fault-tolerant system. The combination of system test and diagnostic capabilities, and the rapid accessibility and repair of the system and its components are all vital parts of the system design and its maintenance concept. Sometimes systems cannot be supported in accordance with the original maintenance concept. Throwaway items may become too expensive, skill levels in the field may not be high enough, and a host of other unplanned circumstances may occur. Under these conditions, the maintenance concept may have to be adjusted and/or revised to deal with the prevailing situation and special circumstances.

11.5.2 Repairability and Serviceability

Repairability is the probability that a failed system can be restored to an operable condition in a specified active repair time. Obviously, this quantitative measure is dependent upon the ease or difficulty with which the failed system can be repaired. The term "serviceability" has been adopted to describe a system's design characteristics in terms of the ease of repair while repairability is a probability involving certain categories of time.

Serviceability to a large extent must be accessed and described in qualitative terms. Often, comparison of different system designs can be made in terms of serviceability and can be rank ordered for performance. Also, a given serviceability scheme for a system can be quantified in an indirect fashion by examining its repairability.

Serviceability is dependent upon many hardware and software characteristics to include engineering design, complexity, the number of accessible test points, the interchangeability of components, and the nature and amount of automatic test and diagnostics capability available for the system. The lack of proper tools and facilities, shortage of work space, shortage of parts, poorly trained personnel and other factors can adversely impact the maintenance and repairability of a system.

11.5.3 Maintenance Personnel and Training

No system can be supported without having a sufficient number of properly trained personnel. The T&E must place special emphasis on addressing the sufficiency of both personnel and training. Personnel must be of the proper skill levels required to accomplish the tasks, and must receive the necessary training prior to the T&E. Their expertise and ability to maintain the test system will form the basis for finalizing plans on how to maintain systems in the field, and for predictions on how well the system maintenance activities can be achieved under wartime conditions.

11.6 Logistics Supportability

Logistics supportability is defined as the degree to which planned logistics and manpower meet system availability and wartime usage requirements. Logistics supportability has a direct impact on the availability of a weapon system and should be assessed during T&E. Elements typically grouped within logistics supportability include the following:

1) Transportability. Transportability is the routine and mobility movement of spare, prime, and support equipment.

2) Manpower Supportability. Manpower supportability addresses the number of people and skill levels required to maintain the system.

3) Training Requirements. Training requirements address the training needs of the persons who maintain the system.

4) Technical Data. Technical data include information necessary for the efficient operation and support of the system as well as the documents used by maintenance personnel to isolate faults and repair a deficiency, and to conduct preventive maintenance and servicing.

5) Supply Support. Supply support consists of petroleum, oil, and lubricants (POL), and the spare parts for supporting a system.

6) Support Equipment (SE). SE includes all equipment required to perform the support function except equipment which is an integral part of the mission equipment. It does not include any of the equipment required to perform mission operations functions. SE includes tools, test equipment, and automatic test equipment (ATE) (i.e., when the ATE is accomplishing a support function); organizational, intermediate, and technical repair center SE; and related computer programs and software.

7) Facilities. Facilities include the system-unique structure (training, test, operations, maintenance, etc.) required for system support.

11.7 Analysis and Predictions

A good estimate of the RML&A that a system can achieve under wartime conditions is sometimes difficult to obtain in a peacetime environment. Nevertheless, these estimates must be made and are essential in the provisioning of systems and supplies for wartime and other emergency situations. The methodology that one might use for such predictions is an important part of the readiness and peacetime planning process.

11.7.1 Mean Time Between Failure

The mean time between failure (MTBF) for a system is the arithmetic average of the time between failures. It can be estimated over a time period by the mathematical relationship shown below.

$$\text{MTBF} = \frac{\text{Total Operating Time}}{\text{Total Number of Failures}}$$

A special class of failures called critical failures was discussed earlier. The MTBCF can be estimated as shown below.

$$\text{MTBCF} = \frac{\text{Total Operating Time}}{\text{Total Number of Critical Failures}}$$

11.7.2 Mean Time To Repair

The mean time to repair (MTTR) of a system or item is the arithmetic average of the time required to repair it when it fails. As such, it is estimated by the summation of all repair times, divided by the total number of failures that occur during the time period under consideration. The mathematical relationship for estimating MTTR is shown below.

$$\text{MTTR} = \frac{\sum_{i=1}^{n} Ri}{n}$$

where
- MTTR = mean time to repair
- Ri = active repair time for the ith repair
- n = the total number of repairs

11.7.3 Mean Downtime

The mean downtime (MDT) of a system is the arithmetic average of the times that a system is down for all reasons. These times include both corrective and preventive maintenance and administrative and logistics downtime during a specified period. All of these items play an important role in achieving the operational availability of systems and, in turn, their end mission effectiveness.

$$\text{MDT} = \sum_{j=1}^{K} \frac{D_j}{K}$$

where
- MDT = mean downtime
- D_j = downtime for the jth reason (includes corrective maintenance time, preventative maintenance time, plus administrative and logistics delay times)
- K = total number of times system was down for any reason

11.7.4 Availability

Availability is defined as the degree to which an item or system is in an operable and committable state at the start of a mission when the mission is called for at a random or unknown point in time. Availability differs from reliability in that reliability depends on an interval of time, whereas availability is taken at an instant of time. A system can be highly available yet experience a considerable number of failures as long as the periods of inoperability are very short. The availability of a system depends not only on how often the system becomes inoperable but how quickly each failure can be repaired. Availability can be expressed in a number of operational terms like mission capable rate (with an associated utilization, or sortie rate when addressing aircraft) or uptime ratio. Actual estimation of the availability of an item or system is accomplished by substituting the appropriate time-based elements into various forms of the availability equation.

11.7.4.1 Mission capable rate.
The availability of a weapon system is usually examined on the basis of a 24-hour period. Mission capable rate is one measure of weapon system availability and is defined as the ratio of time that a weapon system is ready to perform all of its assigned peacetime and wartime missions to the total time that the weapon system is possessed. This rate is usually associated with some specified utilization rate (i.e., sortie rate for aircraft) required for peacetime and/or wartime operations. The sortie rate specifies the number of sorties that can be flown per aircraft per day under specified operational and maintenance concepts. Mission capable rate is often further broken down for complex systems in terms of whether the system is fully mission capable (i.e., capable of performing all aspects of its mission) or partially mission capable (i.e., capable of performing at least one but not all of its assigned wartime missions). Further, the times and reasons for the system's not being mission capable are usually further classified into such categories as maintenance, supply, and administrative. The expression shown below mathematically derives availability in terms of the fully mission capable (FMC) rate:

$$\text{Availability} = A = \frac{\text{Mission Capable Time}}{\text{Total Time}}$$

$$\text{Availability} = \frac{\text{Mission Capable Time}}{\text{Total Time} + \text{Nonmission Capable Time}}$$

11.7.4.2 Uptime ratio.
Availability for communications, electronics, and metrological (CEM) systems is also usually expressed in terms of providing 24-hour service. The state of the system during the 24-hour period is described in

time as being either up (i.e., capable of performing its mission) or down (i.e., not capable of performing its mission). The ratio of uptime to total time (i.e., downtime plus uptime) is defined as the uptime ratio. Downtime can also be examined in terms of the reasons for the downtime (e.g., supply, maintenance, administrative). The basic mathematical expression for CEM system availability is shown below:

$$\text{Availability} = A = \frac{\text{Uptime}}{\text{Total Time}} = \frac{\text{Uptime}}{\text{Uptime} + \text{Downtime}}$$

11.7.4.3 Inherent availability. Inherent availability, also sometimes called intrinsic availability, is a measure of system availability when only operating time and active repair time are considered. It excludes from consideration all free time such as standby and delay times associated with scheduled and preventive maintenance, administrative time, and logistics time. Consequently, it is a best-case measure that provides somewhat of an upper limit on the availability that can possibly be achieved for a given system based upon its internal hardware and software capabilities and limitations. Inherent availability is estimated as follows:

$$A_i = \frac{\text{Successful Operate Time}}{\text{Successful Operate Time} + \text{Active Repair Time}}$$

$$A_i = \frac{\text{MTBF}}{\text{MTBF} + \text{MTTR}}$$

where
A_i = inherent availability
MTBF = mean time between failure
MTTR = mean time to repair

11.7.4.4 Operational availability. Operational availability covers all segments of time during which an item or system is intended to be operational. The same mission capable or up-down relationship exists but is expanded to include operating time plus nonoperating (standby) time (when the equipment is assumed to be operable). Mean downtime includes preventive and corrective maintenance and associated administrative and logistics delay time. Operational availability can be estimated using the expression presented below:

$$A_o = \frac{\text{MTBM}}{\text{MTBM} + \text{MDT}}$$

where
A_o = operational availability
MTBM = mean time between maintenance
MDT = mean downtime

11.7.4.5 Example availability problem. Suppose a system operates, fails, and is repaired in accordance with the data shown in Table 11-3. Provide estimates of its inherent availability (A_i) and its operational availability (A_o).

$$A_i = \frac{\text{MTBF}}{\text{MTBF} + \text{MTTR}} = \frac{240}{240 + 12} = .95$$

$$A_o = \frac{\text{MTBM}}{\text{MTBM} + \text{MDT}} = \frac{120}{120 + 20} = .86$$

11.8 Fault-Tolerant Systems (See References 22–28)

Modern integrated systems can be designed to operate with a high degree of fault-tolerance by prioritization of the critical functions that must be performed and automatic reconfiguration of the system to accomplish those functions. A fault-tolerant system is one that can continue to correctly perform its mission in the presence of hardware failures and software errors. Fault-tolerance techniques have gained wide acceptance and application in computer systems. The advent of very large scale integration (VLSI) in computer design has had a major impact on the design and development of fault-tolerant components and systems.* VLSI has resulted in higher density, decreased power consumption, and decreased costs of integrated circuits. Thus, in turn, it has allowed the implementation of the redundancy approaches used in fault-tolerant computers and weapons systems. Fault-tolerant systems can be designed to continually examine the priority workload to be accomplished and the failure points that have occurred within the system and to automatically reconfigure themselves to accomplish the mission. This type of activity cannot be continued indefinitely because a larger part of the system becomes inoperable with each failure and increasing time. However, reconfiguration can allow a system to perform in a more stable operational state for a given period of time in terms of achieving the high priority activities. Each time a failure occurs that causes the system to drop out of its prioritized operational state, the system rapidly reconfigures and returns to an equivalent operational state. These systems also have the potential to allow the postponement of maintenance until after critical phases of the mission and/or conflict have occurred. A goal of the fault-tolerant design could be to allow maintenance to be postponed until the most convenient and cost-effective time.

It is very important that fault-tolerant systems be easily testable; that is, the capability of the system to perform correctly should be verifiable in a simple, rapid, and straightforward manner. It is extremely important to know the failure status of the system before beginning to use that system. For example, a fault-tolerant system may have the capability to perform a given mission; however, if it has already experienced numerous faults and reconfigured to perform its priority functions, it has lost some capability to tolerate future faults during the next mission. The level of fault-tolerance capability remaining must be understood prior to attempting the next mission so that an informed decision can be made whether to attempt the mission or to repair the system to full status prior to attempting the next mission. These important testability and maintainability concerns should be

---

*An integrated circuit (IC) is one in which more than one component is placed on a single piece of semiconductor material, and all of the components and associated wiring are an integral part of the IC and cannot be physically separated.

Table 11-3. Example Availability Data

Total time period	= 30 days × 24 hours = 720 hours
Total failures	= 3
Total repair time for the 3 failures	= 36 hours
Total maintenance actions	= 6
Total downtime for the 6 actions	= 120
MTBF	= 720/3 = 240 hours
MTTR	= 36/3 = 12 hours
MTBM	= 720/6 = 120 hours
MDT	= 120/6 = 20 hours

considered throughout the design process to ensure that the systems can be effectively maintained and applied in the combat or civilian use environment. They are also of great importance in effectively designing T&E for fault-tolerant systems.

References

[1] DoD Instruction 5000.2, "Defense Acquisition Management Policies and Procedures," Office of the Secretary of Defense, Under Secretary of Defense for Acquisition, 23 February 1991.

[2] DoD Manual 5000.2-M, "Defense Acquisition Management Documentation and Reports," Office of the Secretary of Defense, Under Secretary of Defense for Acquisition, 23 February 1991.

[3] DoD 3235.1-H, *Test and Evaluation of System Reliability, Availability, and Maintainability*, Office of Under Secretary of Defense Research and Engineering, March 1982.

[4] Von Alven, W.H. (ed.), *Reliability Engineering*. Englewood Cliffs, New Jersey: Prentice-Hall, Inc., 1964.

[5] Bazovsky, I., *Reliability Theory and Practice*. Englewood Cliffs, New Jersey: Prentice-Hall, Inc., 1961.

[6] Anderson, R.T., *Reliability Design Handbook*. Chicago, Illinois: ITT Research Institute, 1976.

[7] Michaels, J.V., and Wood, W.P., *Design to Cost*. New York: John Wiley and Sons, Inc., 1989.

[8] Tillman, F.A., and Kuo, C.H.W., *Optimization of Systems Reliability*. New York: Marcel Dekker, Inc., 1980.

[9] Siewiorek, D.P., and Swarz, R.S., *The Theory and Practice of Reliability System Design*. Bedford, Massachusetts: Digital Press, 1982.

[10] Meyers, R., Wong, K., and Gordy, H., *Reliability Engineering for Electronic Systems*. New York: John Wiley and Sons, Inc., 1964.

[11] Military Standardization Handbook 217B, *Reliability Prediction of Electronic Equipment*, September 1974.

[12] Vaccaro, J., and Gorton, H. (RADC and Battelle Memorial Institute), *Reliability Physics Notebook*, RADC-TR-65-33D, AD 624-769, October 1965.

[13] MIL-STD-785A, Reliability Program for Systems and Equipment Development and Production, March 1969.

[14] MIL STD-781B, Reliability Tests: Exponential Distribution, 1967.

[15] MIL-STD-883, Test Methods and Procedures for Microelectronics, May 1968.

[16]Starr, M.K., *Product Design and Decision Theory*. Englewood Cliffs, New Jersey: Prentice-Hall, Inc., 1963.

[17]Jones, E.R. (Wakefield Engineering, Inc.), *A Guide to Component Burn-In Technology*. Wakefield Engineering, Inc., Report 1972.

[18]*Research Study of Radar Reliability and Its Impact on Life Cycle Costs for APQ-113, -114, -120, and -144 Radar Systems*. Utica, New York: General Electric Company, Aerospace Electronic Systems Department, August 1972.

[19]*Engineering Design Handbook, Maintainability, Guide for Design*, Army Material Command, AD 754-202, October 1972.

[20]Blanchard, B.S., and Lowery, E.E., *Maintainability*. New York: McGraw-Hill Book Company, 1969.

[21]Geise, J., and Holler, W.W. (ed.), *Maintainability Engineering*. Martin-Marietta Corporation and Duke University, 1965.

[22]Johnson, B.W., *Design and Analysis of Fault Tolerant Systems*. Reading, Massachusetts: Addison-Wesley Publishing Company, 1989.

[23]Anderson, T., and Lee, P.A., *Fault Tolerance Principles and Practices*. London, England: Hall International, 1981.

[24]Hayes, J.P., "Fault Modeling," IEEE Design and Test, Vol. 2, No. 2, April 1985, pp. 88-95.

[25]Kohari, Z., *Switching and Finite Automata Theory*. New York: McGraw-Hill Book Company, 1978.

[26]Pradham, D.K., *Fault-Tolerant Computing-Theory and Techniques*, Volumes I and II. Englewood Cliffs, New Jersey: Prentice-Hall, Inc., 1986.

[27]Timoc, C., Buehler, M., Griswold, T., Pina, C., Stott, F., and Hess, L., "Logical Models of Physical Failures," *Proceedings of the International Test Conference*, 1983, pp. 546-553.

[28]Tang, D.T., and Chien, R.T., "Coding for Error Control," *IBM System Journal*, Vol. 8, No. 1, January 1969, pp. 48-86.

[29]Chorafas, D.N., *Statistical Processes and Reliability Engineering*. Princeton, New Jersey; Van Nostrand Company, Inc., 1960.

[30]Hoel, P.G., *Introduction to Mathematical Statistics*. New York: John Wiley and Sons, Inc., 1962.

12
Test and Evaluation of Integrated Weapons Systems

12.1 General (See References 1–4)

The effective test and evaluation (T&E) of modern military weapons systems has become increasingly important as the systems have increased in complexity, sophistication, and level of integration. At the same time, this increase in complexity and level of integration has caused T&E to become more difficult.

Each new generation of weapons systems presents new and unprecedented challenges in terms of both range measurement methods and test environments. Present trends indicate that these challenges will have to be addressed in the future under conditions of constrained budgets and overall austere test ranges and facilities. How much physical testing should be done? How and where should it occur? Can analytical modeling and simulation (M&S) be validated to the extent that operators and decision makers are willing to accept its results in lieu of physical testing? There is little doubt that future T&E will require a more cost-efficient approach to both engineering and operational level testing. Cost-efficient T&E as addressed in this chapter is directed at getting the greatest amount of essential test information for the T&E dollars spent.

There are those who maintain that physical open air testing is too expensive, risks compromising the capabilities of systems that emit electronically, and can be effectively replaced by M&S. These advocates of heavy reliance on M&S contend that it has reached a level of maturity that makes it less expensive and equally valid as physical testing.

However, experience has shown that analytical M&S does not have sufficient validity to totally substitute for physical testing of actual weapons systems. Physical testing or actual operational use by the intended combat users has repeatedly revealed serious problems which extensive M&S analyses failed to discover. Consequently, while M&S has a place in T&E, physical testing by the intended combat users remains absolutely vital and is currently the "bottom line" to both effective acquisition and operational T&E of modern complex weapons systems.

There are specific actions which can be taken to ensure that the necessary technology and methods are available to provide more cost-efficient physical testing. These actions require that planning for the overall weapons system's design, operations, maintenance, training, and T&E all be addressed in terms of the system's life cycle and as a total systems problem. The inherent capabilities of modern integrated weapons systems could be capitalized on by "designing in" better T&E monitoring, self-test diagnostics, operational training, and recording capabilities. Greater reliance could be placed on real-world and modified surrogate threat systems that are both movable and transportable. (See Chapter 5 for a discussion of surrogate systems.) Exploiting the capabilities of the global positioning system (GPS) would allow T&E to be conducted at the most cost-efficient locations and should allow a significant reduction in the overall T&E infrastructure.

12.2 Integration Trend (See References 1–15)

The historical evolution of weapons systems can be characterized in terms of segregated, federated, and integrated systems.

12.2.1 Segregated Systems

Early weapons systems were essentially made up of a group of separate subsystems, each independently carrying out important functions of its own, with the primary integrator of the subsystem information being the system operators. The segregated systems contained separate hardware and software within each subsystem that independently carried out the essential functions required to accomplish the mission. Thus, to a large extent, the success or failure of the subsystem in achieving its functions was essentially determined within the individual subsystem and by the effectiveness of its interface to the system operators.

12.2.2 Federated Systems

As military weapons systems became more sophisticated and complex, the increased task loading of system operators, as well as the pursuit of increased benefits from some form of subsystem interfaces, led to federated systems. Federated systems are made up of a group of individual subsystems working together in an attempt to synergistically address the total problems to be solved by the weapons systems. Federated systems, whose subsystems may or may not be effectively interfaced to carry out the total mission, make up the majority of today's major weapons systems.

12.2.3 Integrated Systems

Integrated weapons systems are characterized by the use of shared hardware and software to fuse and synthesize full system information, prioritize essential functions within a given mission, and automatically reconfigure, as necessary, to work around failures to accomplish the mission. Consequently, the subsystem distinctions of integrated systems become blurred if not indistinguishable. Optimally designed integrated weapons systems theoretically should always be more efficient than segregated or federated systems. Efficiency, as discussed here in relation to weapons systems design, is measured primarily in terms of mission achievement at a given level of success with less total system weight and volume.

12.3 System Test and Evaluation

The purpose of T&E is to ensure that only operationally effective and suitable systems are delivered to the operational forces. Consequently, the design of a successful T&E for any system is driven primarily by the operational environment and the military mission that must be accomplished. The dominant measurement of operational utility is how well the system is able to perform its operational mission over time. This includes the necessary training and logistics support systems, as well as the interfaces with other operational systems, and the command, control, communications, and intelligence systems.

Realistic operational test environments for evaluating the capabilities of systems are often difficult to achieve in terms of the density and complexity of a real-world situation. The net result is a difficulty in stressing the systems and their operators to the level they would face in combat. For example, this has become a particularly important issue in electronic warfare as electronic combat (EC) capabilities have evolved into software-intensive designs that must sort, prioritize, and initiate appropriate responses in dense and complex threat environments.

Segregated, federated, and integrated weapons systems have vastly different design concepts and are configured to operate differently; however, the higher level measures of operational utility for all three types of designs are common. These measures must depict how well each system can achieve its overall mission. The mission itself can be broken down into tasks and the tasks into functions that must be successfully completed for mission accomplishment. The methods suggested for this type of mission decomposition are discussed in Chapter 3. The mission completion success of all three types of systems can then be related to the successful completion of the tasks and functions essential to the mission. It is primarily the manner in which the different types of systems internally achieve the essential tasks and functions that differentiates the level of difficulty and approaches that must be taken to conduct T&E of the different types of systems.

12.3.1 Engineering Level Test and Evaluation

Engineering level T&E of systems is defined here as the ability to establish a cause and effect relationship regarding essential functions being carried out by the systems. For example, if an electronic combat jammer is unable to jam a specific threat, then engineering level T&E should provide measurements that establish the basic hardware, software, and human factor causes of this problem.

The engineering level T&E of segregated weapons systems is somewhat straightforward in that each subsystem, as well as its individual components, can be addressed in terms of inputs and outputs for given mission situations. These inputs and outputs are, to a large degree, unique and specific to a constant configuration of subsystem hardware and software.

On the other hand, the engineering level T&E of federated weapons systems is more complex in that it must address the subsystem inputs and outputs as well as any automated system interfaces designed to occur above the subsystem level.

The engineering level T&E of future fully integrated weapons systems will be considerably more difficult than T&E for federated systems. As systems become more highly integrated, directly associating specific hardware and software to the essential functions with the systems that are used as the mission progresses becomes increasingly complex and difficult. Because integrated systems may periodically reconfigure during the mission in response to specific mission situations and failures, it will be vastly more difficult to verify that the appropriate functions are taking place according to system design logic. Also, it will be more difficult to decide what services and repairs should be made to the system prior to the next mission.

Successful weapons systems maintenance will require an unprecedented level of tracking of internal system logic and essential functions. Fortunately, the requirements for successful maintenance are complementary with those for T&E in this regard.

The effective engineering level for T&E of fully integrated weapons systems will require a significant amount of monitoring, measuring, and recording capability not currently available or in use on nonintegrated weapons systems. Ideally, integrated systems would be capable of displaying and recording their status versus time, capability to complete the next mission without repairs, and a listing of the repairs necessary to bring the system back to a 100 percent mission-ready status. This type of information would be highly valuable during the maintenance cycle and preparation for the next mission. It will also be essential for successful T&E. The T&E of integrated weapons systems can benefit greatly from the application of artificial intelligence, expert systems, and/or automated learning applications.

12.3.2 Operational Level Test and Evaluation

The operational level for T&E of segregated, federated, or integrated systems should be conducted on the foundation established by the engineering level T&E, and should be centered primarily on the capability of the system to successfully complete the operational mission.

For segregated systems, it is necessary to evaluate the effective use of individual subsystem information by the system and operators. For federated systems, the capability to effectively interface the subsystem/system information to the system and operators must be verified. Likewise, for the integrated systems, the capability to effectively interface correct fused and synergized information to the system and operators must be verified. The built-in monitoring, measuring, and recording capabilities of integrated systems should be evaluated in physical operational testing to determine their capacities to allow maintenance and support personnel to repair and configure the system.

12.3.3 Testability in Systems

The "testability" of a weapon system can be defined and viewed in many ways. Testability is the ease with which the product or system can be tested to accurately establish its state of functionality. It has also been defined as the capability to probe a system or a product and to assess the functionality of the probed portion. Testability can also be analyzed and viewed from a system functional state and probabilistic viewpoint.

It is generally accepted that there is a high correlation between low cost in production and the ease with which a product can be tested, particularly in the final assembly form. Also, there is great value to the customer (especially the military operators taking a system into combat) to correctly know the functional state of their system and to be able to rapidly locate and repair failures that have occurred. Additionally, it is desirable to be able to probe systems in a manner that can assess where functionality may not yet have failed but is decreasing and failure is likely to occur. In modern integrated weapons systems, this built-in testability can accommodate automatic reconfiguration of the system to provide for higher overall mission reliability. However, as greater testability is built into a system, additional costs are incurred and the issue becomes one of whether the utility gained through the increased testability is worth the additional costs. In the commercial market, this must be addressed through the mechanism of customer satisfaction and profit. In military systems, it is an issue of having systems available for the wartime mission, accomplishing the missions, and incurring the associated costs of those activities. Obviously, the military goal should always be to effectively accomplish the mission and to do it at least cost.

12.3.4 Emphasizing Designed-In Capabilities

More cost-efficient physical testing can be achieved by the increased use of designed-in weapons systems capabilities. Highly integrated systems will require built-in capability for self-test and for recording mission time histories of system failures and reconfiguration actions.

For example, it is within the state of the art to design complex electronic combat systems with self-test features that inject radio frequency (RF) (threat) signals at the front end of the system and automatically evaluate whether the system response is proper. (See Figure 12-1.) Appropriate results from such tests could be immediately communicated to the operator and also recorded for postmission maintenance actions. This information will be essential to repair the system and ensure that it is 100 percent ready for the next mission.

Integrated weapons systems, by their very nature, have significant internal signal processing capability. The design of these systems should take advantage of their internal signal processing capability by including as an integral part of the system the capture, recording, and summarization of essential internal system information as a function of time. At little or no additional expense, these internal operator-alerting and recording capabilities could be expanded during design to meet the system information needs of T&E. This would enhance both mission-to-mission operations and maintenance and would provide the essential information required for engineering level T&E.

If systems are designed with adequate built-in measuring and recording capabilities, both operational and engineering level T&Es could be conducted without the need for extensive postproduction weapons systems modifications for test instrumentation. Instrumenting systems to monitor and record the information required for T&E after the systems are produced is time consuming and expensive—often prohibitively so.

Integrated systems also offer a unique opportunity to achieve greater operational realism in the conduct of operational testing. The built-in self-test features described earlier that would allow electronic combat systems to be stimulated at the RF level could also be used to stimulate electronic combat systems according to specific preselected scenarios. Built-in stimulation capabilities would allow the on-board EW systems to "think" they are operating in a dense electronic signal environment. Since this capability would be programmable, the internally generated threats could be properly synchronized with the actual ground and/or airborne threat (or threats) that are available for physical testing to support more realistic T&E of complex scenarios. (See Figure 12-2.) This type of capability would also greatly enhance the realistic training of system operators.

The examination of the effects of the transmitted electronic waveforms would

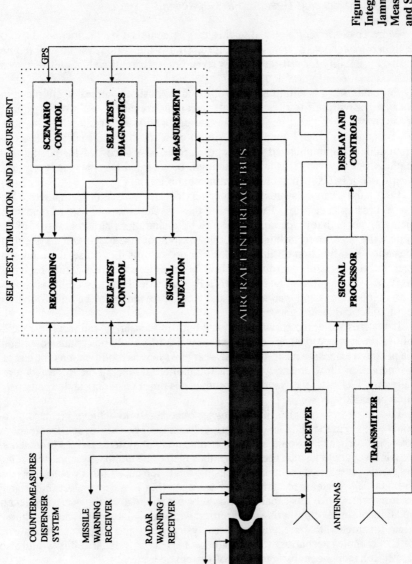

Figure 12-1. Hypothetical Integrated System Showing Jammer and Built-In Measurement, Recording, and Scenario Stimulation

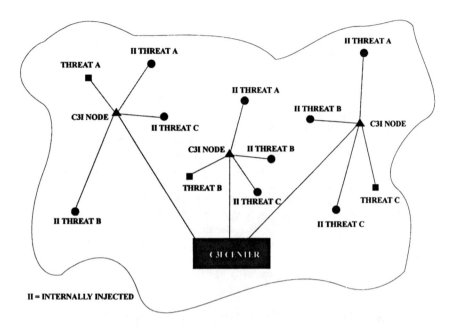

Figure 12-2. Using Built-In Capabilities, Stimulation Can Be Combined with Actual Threats for Greater T&E Effectiveness

center on the actual ground and/or airborne threat (or threats) available; however, the internal stimulation capabilities could be used to provide scenario density. The global positioning system (GPS) provides the capability to maintain an accurate laydown of the threats as well as the capability to accurately determine the location of the weapons system when specific electronic combat responses occur. Such issues as threat prioritization, system response in dense environments, and so forth could be examined, even in limited physical test environments.

12.3.5 Real-World Threats and Surrogates

It should be possible to realize considerable cost efficiencies by greater application of real-world threat systems and appropriate use of modified friendly surrogates. The physical testing problem could be addressed to a large extent by the direct purchase of both Red and Gray real-threat systems on the world market and the increased use of friendly surrogates carefully selected for their capability to be adapted to emulate enemy threat systems. Red systems are those that have a high likelihood of being used against U.S. or friendly forces in conflict and Gray systems are those that belong to allies but, because of weapon system proliferation, possibly could be used against the U.S. or friendly forces.

A major emphasis in the past has been on the development of U.S. "high fidelity" threat simulators. However, with the recent disintegration of the Iron Curtain and the opening up of Eastern Europe, the likelihood increases of realizing considerable cost and time savings by direct purchase of both Red and Gray real-world threat systems. This is not to say that U.S. threat replicas should never be built, but their development should be based on careful technical and management investigations that do not overlook the effective application of less expensive alternatives.

The T&E community has generally taken the attitude that only real threats or "high fidelity" threat simulators can be used for operational physical testing. While no one would argue that these are preferred, a wealth of information can be obtained by the selective use of surrogates. Unbiased engineering evaluations and sound technical judgment should be applied to determine when and where surrogates can be meaningfully used. Surrogates must be selected on the basis of having characteristics that are similar or that can be modified to be made similar to the particular threat and operational techniques of interest. Although they may not satisfy all the testing requirements, surrogates can often provide much needed essential information in a more timely manner and/or at less cost.

Future threat systems used in T&E should be movable and transportable to the maximum extent possible so testing can be accomplished at the most cost-efficient test locations. Not only would this provide additional flexibility in locating tests, it would also provide greater flexibility in tailoring the overall threat to the postulated scenario.

12.3.6 Global Positioning System as a Precision Reference

Another important opportunity for achieving cost efficiency is the extensive use of the GPS as a precision positional reference. This, along with movable and transportable threats, makes it possible to conduct physical testing at practically any operational location worldwide. Historically, testing has been tied to those locations having the capability to provide precision time-space-position information (TSPI). Global TSPI capability, combined with movable and transportable threats, now allows for an expanded test philosophy that seeks out the most cost-efficient T&E locations. This philosophy could provide the potential for conducting much of the one-on-one and other less sophisticated testing at less expensive locations than the existing major test ranges.

12.3.7 Shared Data Bases

The expense of military testing makes it highly desirable that only testing which is essential be conducted, and that duplication in testing be eliminated. This can be achieved to a considerable degree by planning the testing for a system over its entire life cycle (i.e., optimizing the test program over the system's life cycle). Information that is essential to assist in research, development, design, production, operations, training, maintenance, and logistics support could be addressed as an integrated effort. This could allow a single integrated data base to be developed and certified for use by multiple evaluators (i.e., developers, operators, maintainers, independent testers, etc.). Data input to the data base would have to be properly labeled, certified for objectivity, protected for proprietary rights, and controlled for access by only those who have the official need for their use.

12.3.8 Summary

The T&E of modern integrated weapons systems is both difficult and expensive. There are technologies and capabilities that could be applied to significantly reduce the T&E infrastructure and the long-term cost of physical testing. The planning for the overall weapons systems design, operations, maintenance, training, and T&E should be addressed in terms of the system's life cycle and as a total systems problem. Although analytical modeling and simulation are valuable tools in the T&E process, physical testing with actual hardware and software is absolutely essential if T&E is to remain credible. Future T&E should be based primarily on collecting the essential physical testing information at the lowest cost. Additional designed-in weapons systems monitoring and recording capabilities are feasible, desirable, and essential for integrated weapons systems.

Built-in measuring and recording capabilities would allow the integrated weapons systems with standard operational configurations to be operated in known (or measured) threat environments, with TSPI being collected by the GPS. Internally stimulated threat signals could be synchronized with the available external threat environment to provide scenarios of appropriate scope and density. This information could be time-correlated with aircraft position and the weapons system's responses internally recorded by the built-in measuring and recording capabilities.

This built-in system measurement approach, combined with movable and transportable threat systems and GPS, would allow physical testing to be conducted at test locations throughout the world. This would allow the design of each T&E and the selection of the appropriate test locations to be made on the basis of generating the essential system operational and environmental information at the lowest cost.

The combination of designed-in stimulation features with built-in test measuring and recording capabilities, less expensive mobile and transportable ground threat systems, and GPS offers the potential to significantly improve the cost effectiveness of highly essential engineering and operational level testing. These capabilities also offer the opportunity to significantly improve combat training at less cost.

12.4 Automatic Diagnostic Systems (See References 16–33)

The purpose of automatic diagnostic systems (ADSs) is to provide a tool to quickly and accurately ascertain the true operational status of a weapon system and to rapidly fault-isolate those systems indicating failure. ADS is defined as "a system's automatic capability to correctly ascertain the operational status of a system or function, and isolate defective items to a designated ambiguity level. The automatic diagnostic capability may be contained within the system being evaluated, may be a separate piece of equipment, or a combination of both methods." The objective of ADS is to improve the operator and/or maintenance technician's ability to quickly and accurately isolate weapon systems failures by buying systems with carefully "designed-in" diagnostic capability. Past operational experience shows that evolving automated and integrated diagnostic systems have been costly and their attributes are the subject of considerable confusion as they become increasingly complex. When not properly specified and designed, the automatic fault isolation systems contribute little to mission performance and are often less reliable and more difficult to maintain, operate, and understand than the basic systems which they are purchased to support. The probabilistic approach and concepts put forth in this chapter are developed to help identify the important considerations in design, specification, and effectiveness measurements for such systems. With the increasing complexity of modern weapon systems, properly functioning automatic diagnostic capabilities are essential to the achievement of basic system reliability, maintainability, and availability goals. This is especially true of modern integrated avionics systems with designed-in fault and damage tolerance.

This chapter examines the various measures one can employ when addressing the effectiveness of ADSs, and it is designed to be of greatest value to those who have a technical interest in ADSs. The enumeration of some of the various measures identified would be impractical in day-to-day operations at a standard operational unit; that is, collecting data on these functions is not normally a part of the day-to-day unit operational environment. Therefore, it is critical that, prior to operational deployment, sufficient data be collected by dedicated T&E of ADSs. This T&E will require proper system technical expertise, test equipment independent of the ADS, and dedicated data collection. The measures suggested for ADS specification, design, and dedicated T&E should establish that true ADS capability is both worthwhile and essential. Based on special testing of the ADS capabilities for a given weapons system, the operational community should be provided with an assessment of the true capabilities and limitations of the ADS prior to operational deployment. Armed with this information, the operator and maintainer will be in an excellent position to use the ADS on a daily basis to employ and maintain the basic weapons system to its fullest potential.

12.4.1 Purpose of Automatic Diagnostic Systems

There are two operational purposes of ADSs: 1) fault detection to provide a valid operational status of the equipment (i.e., "GO," "NO GO"), and 2) when a system is "NO GO," fault isolation to a given level, such as a line replaceable unit (LRU), module, or component.

An ADS fault detection check will normally be accomplished preparatory to conducting an operational mission; and, if the ADS check indicates that the system is good, the mission will be attempted. It is obvious that it is important for this fault detection check to be highly accurate since an erroneous indication will result in either the mission being flown with a bad system or unnecessary maintenance being performed on a good system. It will be shown later that the probability of conducting a successful mission is the product of ADS reliability and weapon system reliability.

If the preliminary ADS check indicates a system failure or the mission is attempted and is unsuccessful, an ADS fault isolation check will be performed in an effort to determine the cause of the failure so repair action can be initiated. As was the case for the fault detection function, it is important that ADS fault isolation be highly reliable so that unnecessary manpower is not expended on systems which have not failed or bad systems are not placed back into service.

Having an ADS that is not reliable is generally worse than not having an ADS at all. Unless it functions properly and is highly reliable, the ADS will result in lower system effectiveness and wasted manpower. The simplified chart in Table 12-1

Table 12-1. Example ADS Mission Activities

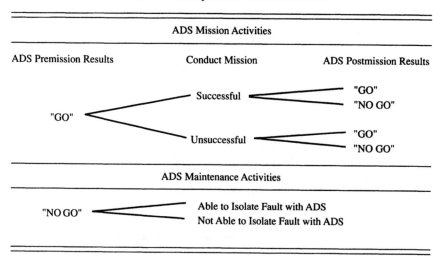

depicts the possible outcomes of the actions of interest during a mission sequence as described above.

12.4.2 Probabilistic Description of Automatic Diagnostic Systems

12.4.2.1 Fault detection. Two sets of probabilities are necessary to describe ADS fault detection. The first set consists of the probabilities of obtaining specific status indications given true system status. For instance, given a good system, what is the probability that the ADS will indicate that the system is in a "GO" status? The second set consists of the probabilities of a system truly being good or bad given a particular ADS status indication. For example, given a "GO" indication, what is the probability that the system is good? Each of these sets of probabilities is useful and necessary in understanding the functioning of ADS fault detection.

The first set of probabilities results from evaluating the four outcomes which may occur when performing an ADS fault detection check to determine the operational status of a system. The ADS may indicate: 1) a good system is good, 2) a good system is bad, 3) a bad system is bad, or 4) a bad system is good. Therefore, there are four conditional probabilities of interest:

$$P_1 = P(\text{"GO" INDICATION/SYSTEM IS GOOD})$$
$$P_2 = 1 - P_1 = P(\text{"NO GO" INDICATION/SYSTEM IS GOOD})$$
$$P_3 = P(\text{"NO GO" INDICATION/SYSTEM IS BAD})$$
$$P_4 = 1 - P_3 = P(\text{"GO" INDICATION/SYSTEM IS BAD})$$

Each of these probabilities is important from an operational standpoint since it provides insight into the utility of the ADS. P_1 gives the probability that the ADS will confirm that a good system is good. Likewise, P_3 is the probability that the

ADS will confirm that a bad system is bad. P_2 and P_4 are the probabilities of the complements of these events, respectively, and, once P_1 and P_3 are known, P_2 and P_4 can be obtained. P_2 is the probability that the ADS will indicate a good system is bad and may result in a mission being delayed or aborted or may result in unnecessary maintenance. P_4 is the probability that the ADS will indicate that a bad system is good and will likely result in an unsuccessful mission.

There are also four outcomes of interest if ADS is viewed from the standpoint of system status given a particular ADS fault detection indication. These are: 1) the system is good when ADS provides a "GO" indication, 2) the system is bad when ADS provides a "GO" indication, 3) the system is bad when ADS provides a "NO GO" indication, and 4) the system is good when ADS provides a "NO GO" indication. Therefore, there are four conditional probabilities of interest:

$$P_5 = P \text{ (SYSTEM IS GOOD/"GO" INDICATION)}$$

$$P_6 = 1 - P_5 = P \text{ (SYSTEM IS BAD/"GO" INDICATION)}$$

$$P_7 = P \text{ (SYSTEM IS BAD/"NO GO" INDICATION)}$$

$$P_8 = 1 - P_7 = P \text{ (SYSTEM IS GOOD/"NO GO" INDICATION)}$$

These probabilities are also important from an operational standpoint because they indicate how reliable the ADS indications are. P_5 gives the probability that the system is good when the ADS indicates it is. Likewise, P_7 gives the probability that the system is bad when ADS indicates it is. P_6 and P_8 are the probabilities of the complements of these events, respectively. P_6 is the probability that the system is bad when ADS indicates it is good, and P_8 is the probability that the system is good when ADS indicates it is bad.

12.4.2.2 Fault isolation.
The second operational purpose of the ADS is to fault-isolate. As was the case with fault detection, two sets of probabilities are necessary to describe ADS fault isolation. The first set consists of the probabilities of obtaining (or not obtaining) correct fault isolation, given true system status. As an example, given that the system is bad, what is the probability that the ADS will provide correct fault isolation? The second set consists of the probabilities of various fault isolation results given that the ADS has or has not fault-isolated. For example, given that the ADS has fault-isolated, what is the probability that the fault isolation is correct? Each of these sets of probabilities is useful and necessary in understanding the ADS fault isolation function.

The first set of fault isolation probabilities arises from looking at the outcomes of the premission ADS checks. As previously discussed, these checks will provide either a "GO" or "NO GO" indication. Assuming that a "NO GO" ADS indication is obtained during the premission ADS check and the ADS fault-isolate function is employed, four outcomes may result. If the system has failed, the ADS may either correctly or incorrectly isolate the fault (or fail to isolate the fault). If the system has not failed, the ADS may verify this, or it may provide an erroneous fault isolation. Therefore, there will be four conditional probabilities associated with the ADS fault isolation function:

$$P_9 = P \text{ (CORRECT FAULT ISOLATION/SYSTEM IS BAD)}$$

$P_{10} = 1 - P_9 = P$ (INCORRECT OR NO FAULT ISOLATION/SYSTEM IS BAD)

$P_{11} = P$ (NO FAULT ISOLATION/SYSTEM IS GOOD)

$P_{12} = 1 - P_{11} = P$ (ERRONEOUS FAULT ISOLATION/SYSTEM IS GOOD)

Each of these probabilities is also important from an operational standpoint. P_9 gives the probability that the ADS will correctly isolate a fault in a failed system and P_{10} is the probability of the complementary event. P_{10} can be written as the sum of two probabilities, P_{10-1} and P_{10-2}, where:

$P_{10-1} = P$ (INCORRECT FAULT ISOLATION/SYSTEM IS BAD)

$P_{10-2} = P$ (NO FAULT ISOLATION/SYSTEM IS BAD)

P_{10-1} is the probability that the wrong LRU, module, etc., will be indicated as bad and may result in erroneous replacement/repair action. P_{10-2} is the probability that the ADS system will not isolate the fault. This condition could result in some method other than the ADS having to be used for fault isolation, or the failure may be reported as a "cannot duplicate." P_{11} is the probability that the system will not erroneously isolate to a nonexistent fault in a good system, and P_{12} is the probability of the complementary event. P_{12} is the probability that a good LRU, module, etc., will be identified as faulty when the system is good. This condition may result in attempts at unnecessary replacement or repair.

Examination of the ADS fault isolation function from the standpoint of whether fault isolation has occurred indicates that there are four outcomes. These are: 1) fault isolation has occurred and it is correct, 2) fault isolation has occurred and it is incorrect (this contains both incorrect isolations when the system is bad and erroneous isolations when the system is good), 3) fault isolation has not occurred and it is correct (the system has not failed), and 4) fault isolation has not occurred and it is incorrect (the system has failed). Therefore, there are four conditional probabilities of interest:

$P_{13} = P$ (CORRECT FAULT ISOLATION/FAULT ISOLATION)

$P_{14} = 1 - P_{13} = P$ (INCORRECT FAULT ISOLATION/FAULT ISOLATION)

$P_{15} = P$ (SYSTEM IS GOOD/NO FAULT ISOLATION)

$P_{16} = 1 - P_{15} = P$ (SYSTEM IS BAD/NO FAULT ISOLATION)

Each of these conditional probabilities is an important operational measure of the fault isolation capability of the ADS. P_{13} gives the probability that fault isolation is correct given that it has occurred; P_{14} is the probability of the complementary event. P_{15} gives the probability that the system has not failed given that fault isolation has not occurred, and P_{16} is the probability of the complementary event. P_{13} is the probability that, once fault isolation occurs, it can be relied on to have correctly isolated the problem, while P_{14} is the probability that reliance on fault isolation will result in unnecessary/unproductive maintenance. P_{15} is the probability that the absence of fault isolation can be correctly relied on to indicate that the system has not failed, and P_{16} is the probability that reliance on the lack of fault isolation will result in a failed system being put back into service.

12.4.2.3 Overall observations.
A close examination of the probabilities specified would indicate that for good ADS design and performance it is highly desirable that the odd numbered probabilities ($P_1, P_3, P_5, P_7, P_9, P_{11}, P_{13}$, and P_{15}) be as close to 1.0 as possible. Conversely, the even numbered probabilities ($P_2, P_4, P_6, P_8, P_{10}, P_{12}, P_{14}$, and P_{16}) should be as close to 0.0 as possible.

12.4.3 Estimating Automatic Diagnostic System Probabilities

To estimate these 16 ADS probabilities, it is necessary to be able to establish an accurate estimate of the true status of the system. Therefore, whether performing laboratory evaluations, conducting development tests and evaluations (DT&Es), or conducting operational tests and evaluations (OT&Es), the evaluations should be designed to produce sufficient data to establish an estimate of the true status of the system for comparison with ADS outcomes. This will require performing the ADS fault detection and fault isolation checks and comparing the results with true system status established by means independent of the ADS. These comparisons should yield data to compute P_1 through P_{16}.

Consideration of the practical aspects of measuring P_1 through P_{16} reveals some problems with obtaining the necessary data during developmental and operational testing. First, from a flight test standpoint, if the premission ADS check indicates that the system is "GO," the mission will ordinarily be flown without verifying the "GO" indication. If the mission is unsuccessful because of a system failure, it may not be apparent whether the "GO" indication was erroneous or the system failed during the mission. On the other hand, if mission results indicate that the system has failed, ADS fault detection and fault isolation checks will ordinarily be accomplished. The net result is a tendency to check the system only when the ADS fault detection results indicate "NO GO" or the mission results indicate the system has failed.

If sufficient data are to be obtained to determine these probabilities, the test methodology must specifically address verification of both "GO" and "NO GO" system status independently of the ADS. Proper ADS testing will require a dedicated effort with the capability to verify ADS indications.

12.4.3.1 Fault detection.
As discussed earlier, determination of P_1 through P_8 depends on designing the test such that an estimate of the true system status can be determined independently of the ADS. If one defines the number of ADS checks performed when the system is later confirmed to be good as N_g and a "GO" indication is provided on N_{gg} of these, then the ratio N_{gg}/N_g is an estimator of P_1. Also, if N_b checks are performed when the system is confirmed to have a failure and N_{bb} of these provide a "NO GO" indication, then the ratio N_{bb}/N_b is an estimator of P_3. Similarly P_2 and P_4 can be estimated by the ratios N_{gb}/N_g and N_{bg}/N_b, respectively, where N_{gb} is the number of ADS checks which indicate the system is bad when it is good and N_{bg} is the number of ADS checks which indicate the system is good when it is bad. This is illustrated in the following chart, which shows ADS fault detection sample size.

$$N = \begin{matrix} N_g = & N_{gg} + N_{gb} \\ + & \\ N_b = & N_{bb} + N_{bg} \end{matrix}$$

where

- N = Number of ADS checks where system status can be confirmed
- N_g = Number of ADS checks when the system is confirmed to be good
- N_b = Number of ADS checks when the system is confirmed to be bad
- N_{gg} = Number of ADS checks which indicate the system is good when it is good
- N_{gb} = Number of ADS checks which indicate the system is bad when it is good
- N_{bb} = Number of ADS checks which indicate the system is bad when it is bad
- N_{bg} = Number of ADS checks which indicate the system is good when it is bad

Determination of P_5 through P_8 can be estimated in a similar manner. P_5 can be estimated by the ratio $N_{gg}/(N_{gg} + N_{bg})$, P_6 by $N_{bg}/(N_{gg} + N_{bg})$, P_7 by $N_{bb}/(N_{bb} + N_{gb})$ and P_8 by $N_{gb}/(N_{bb} + N_{gb})$.

Fault detection probabilities can be summarized as follows:

$$P_1 = P(\text{``GO'' INDICATION/SYSTEM IS GOOD})$$
$$= N_{gg}/N_g$$

$$P_2 = P(\text{``NO GO'' INDICATION/SYSTEM IS GOOD})$$
$$= N_{gb}/N_g$$
$$= 1 - P_1 = 1 - (N_{gg}/N_g) = (N_g - N_{gg})/N_g$$
$$= N_{gb}/N_g$$

$$P_3 = P(\text{``NO GO'' INDICATION/SYSTEM IS BAD})$$
$$= N_{bb}/N_b$$

$$P_4 = P(\text{``GO'' INDICATION/SYSTEM IS BAD})$$
$$= N_{bg}/N_b$$
$$= 1 - P_3 = 1 - (N_{bb}/N_b) = (N_b - N_{bb})/N_b$$
$$= N_{bg}/N_b$$

$$P_5 = P(\text{SYSTEM IS GOOD/``GO'' INDICATION})$$
$$= N_{gg}/(N_{gg} + N_{bg})$$

$$P_6 = P(\text{SYSTEM IS BAD/``GO'' INDICATION})$$
$$= N_{bg}/(N_{gg} + N_{bg})$$
$$= 1 - P_5 = 1 - [N_{gg}/(N_{gg} + N_{bg})]$$
$$= (N_{gg} + N_{bg} - N_{gg})/(N_{gg} + N_{bg})$$
$$= N_{bg}/(N_{gg} + N_{bg})$$

$P_7 = P$ (SYSTEM IS BAD/"NO GO" INDICATION)
$= N_{bb}/(N_{bb} + N_{gb})$

$P_8 = P$ (SYSTEM IS GOOD/"NO GO" INDICATION)
$= N_{gb}/(N_{bb} + N_{gb})$
$= 1 - P_7 = 1 - [N_{bb}/(N_{bb} + N_{gb})]$
$= (N_{bb} + N_{gb} - N_{bb})/(N_{bb} + N_{gb})$
$= N_{gb}/(N_{bb} + N_{gb})$

12.4.3.2 Fault isolation.

Determination of P_9 through P_{16} also depends on designing the test of the ADS such that the weapons system status and fault isolation are verified independently of ADS. In determining these probabilities, the total number of ADS fault isolation checks performed where fault isolation is confirmed by means other than ADS will be denoted as S. If S_b fault isolation ADS checks are performed when the system is later confirmed to have a failure and S_{bc} of these correctly isolate the fault, then the ratio S_{bc}/S_b is an estimator of P_9. Also, if S_g fault isolation ADS checks are performed when the system is later determined to be good and S_{gn} of these correctly indicate that there is no failure, then the ratio S_{gn}/S_g is an estimator of P_{11}. P_{10} and P_{12} can be estimated by the ratios S_{be}/S_b and S_{ge}/S_g, respectively, where S_{be} is the number of fault isolation ADS checks which fail to isolate a fault or incorrectly isolate one and S_{ge} is the number of fault isolation ADS checks which erroneously fault isolate in a good system. This is illustrated in the following chart. S_{be} can be further broken down into S_{bei} fault isolation ADS checks which incorrectly isolate a fault in a bad system and S_{bef} fault isolation ADS checks which fail to isolate a fault in a bad system. $P_{10\text{-}1}$ and $P_{10\text{-}2}$ can be estimated by S_{bei}/S_b and S_{bef}/S_b, respectively. The following chart illustrates ADS fault isolation sample size.

$$S = \begin{matrix} S_b = \dfrac{S_{bc}}{+} \\ S_{be} \\ + \\ S_g = \dfrac{S_{gn}}{+} \\ S_{ge} \end{matrix} \quad S_{be} = \begin{matrix} S_{bei} \\ + \\ S_{bef} \end{matrix}$$

where
S = Number of ADS fault isolation checks where fault isolation is confirmed by means other than ADS
S_b = Number of ADS fault isolation checks when the system is confirmed to be bad
S_g = Number of ADS fault isolation checks when the system is confirmed to be good
S_{bc} = Number of ADS fault isolation checks which correctly isolate a fault in a bad system
S_{be} = Number of ADS fault isolation checks which fail to isolate a fault or incorrectly isolate a fault in a bad system

S_{bei} = Number of ADS fault isolation checks which incorrectly isolate a fault in a bad system

S_{bef} = Number of ADS fault isolation checks which fail to isolate a fault in a bad system

S_{gn} = Number of ADS fault isolation checks which correctly indicate that there is no failure in a good system

S_{ge} = Number of ADS fault isolation checks which erroneously fault-isolate in a good system

P_{13} through P_{16} can be estimated in a manner similar to that used to compute P_9 through P_{12}. P_{13} is estimated by the ratio of $S_{bc}/(S_{bc} + S_{bei} + S_{ge})$, P_{14} by $(S_{bei} + S_{ge})/(S_{bc} + S_{bei} + S_{ge})$, P_{15} by $S_{gn}/(S_{gn} + S_{bef})$, and P_{16} by $S_{bef}/(S_{gn} + S_{bef})$. Each of these probabilities is important from an operational viewpoint because it indicates the confidence which can be placed in ADS fault isolation or lack thereof. P_{14} and P_{16} are the probabilities of the events complementary to those of P_{13} and P_{15}, respectively, and can be estimated once P_{13} and P_{15} are estimated.

Fault isolation probabilities can be summarized as follows:

$P_9 = P$ (CORRECT FAULT ISOLATION/SYSTEM IS BAD)
$= S_{bc}/S_b$

$P_{10} = P$ (INCORRECT OR NO FAULT ISOLATION/SYSTEM IS BAD)
$= S_{be}/S_b$
$= 1 - P_9 = 1 - (S_{bc}/S_b)$
$= (S_b - S_{bc})/S_b$
$= S_{be}/S_b$

$P_{10-1} = P$ (INCORRECT FAULT ISOLATION/SYSTEM IS BAD)
$= S_{bei}/S_b$

$P_{10-2} = P$ (NO FAULT ISOLATION/SYSTEM IS BAD)
$= S_{bef}/S_b$

$P_{11} = P$ (NO FAULT ISOLATION/SYSTEM IS GOOD)
$= S_{gn}/S_g$

$P_{12} = P$ (ERRONEOUS FAULT ISOLATION/SYSTEM IS GOOD)
$= S_{ge}/S_g$
$= 1 - P_{11} = 1 - S_{gn}/S_g$
$= (S_g - S_{gn})/S_g$
$= S_{ge}/S_g$

$P_{13} = P$ (CORRECT FAULT ISOLATION/FAULT ISOLATION)
$= S_{bc}/(S_{bc} + S_{bei} + S_{ge})$

$P_{14} = P$ (INCORRECT FAULT ISOLATION/FAULT ISOLATION)
$= (S_{bei} + S_{ge})/(S_{bc} + S_{bei} + S_{ge})$
$= 1 - P_{13} = 1 - [S_{bc}/(S_{bc} + S_{bei} + S_{ge})]$
$= (S_{bc} + S_{bei} + S_{ge} - S_{bc})/(S_{bc} + S_{bei} + S_{ge})$

$$= (S_{bei} + S_{ge})/(S_{bc} + S_{bei} + S_{ge})$$

$$P_{15} = P \text{ (SYSTEM IS GOOD/NO FAULT ISOLATION)}$$
$$= S_{gn}/(S_{gn} + S_{bef})$$

$$P_{16} = P \text{ (SYSTEM IS BAD/NO FAULT ISOLATION)}$$
$$= S_{bef}/(S_{gn} + S_{bef})$$
$$= 1 - P_{15} = 1 - [S_{gn}/(S_{gn} + S_{bef})]$$
$$= (S_{gn} + S_{bef} - S_{gn})/(S_{gn} + S_{bef})$$
$$= S_{bef}/(S_{gn} + S_{bef})$$

12.4.4 Automatic Diagnostic System Measures of Effectiveness

Section 12.4.2 provides a probabilistic description of ADS. From this description it is evident that there are eight probabilities which are descriptive of the ADS effectiveness. This section examines these probabilistic measures and compares them with some measures currently in common use.

12.4.4.1 Fault detection. The discussion in section 12.4.2 highlights that P_1 through P_8 are each important in evaluating the effectiveness of the ADS. P_1 gives the probability that the ADS will confirm that a good system is good, and P_3 gives the probability that the ADS will confirm that a bad system is bad. The probabilities of the complementary events corresponding to P_1 and P_3, P_2 and P_4, respectively, are also important, and are readily obtained once P_1 and P_3 are known. P_5 is the probability that the system is good when a "GO" ADS indication is obtained, and P_7 is the probability that the system is bad when a "NO GO" ADS indication is obtained. Likewise, P_6 and P_8 are the probabilities corresponding to the complements of the events for P_5 and P_7, respectively, and are readily obtained once P_5 and P_7 are known.

Another measure of interest is the probability that the ADS will provide a valid status indication. Looking first at the percent valid status indications, one sees it given by the relationship:

$$\text{PERCENT VALID STATUS INDICATIONS}$$
$$= \frac{\text{Number of Valid Status Indications}}{\text{Total Number of Status Indications}} \times 100$$
$$= \left[(N_{gg} + N_{bb})/N \right] \times 100$$

From a probabilistic standpoint,

$$P \text{ (VALID STATUS INDICATION)} = P_1 \times P \text{ (SYSTEM IS GOOD)}$$
$$+ P_3 \times P \text{ (SYSTEM IS BAD)}$$
$$= [(N_{gg}/N_g)(N_g/N) + (N_{bb}/N_b)(N_b/N)]$$
$$= [(N_{gg} + N_{bb})/N]$$

Alternatively,

$$P \text{ (VALID STATUS INDICATION)} = P_5 \times P \text{ (``GO'' INDICATION)}$$
$$+ P_7 \times P \text{ (``NO GO'' INDICATION)}$$
$$= [N_{gg}/(N_{gg} + N_{bg})][(N_{gg} + N_{bg})/N]$$
$$+ [N_{bb}/N_{bb} + N_{gb})][(N_{gb} + N_{bb})/N]$$
$$= [(N_{gg} + N_{bb})/N]$$

P (SYSTEM IS GOOD) is the probability of the system's being good at some random point in time. The test design must make provisions for determining system status at random points in time if a sufficiently representative sample size is to be obtained to estimate this probability. This can be accomplished if system status is verified independently of the ADS. As indicated in Section 12.4.2, this is also necessary to obtain estimates for P_1 through P_8. P (SYSTEM IS GOOD) can be estimated by N_g/N. Likewise, P (SYSTEM IS BAD) = N_b/N.

Generally, in the operational employment of the system, if the premission ADS check is "GO" and the mission is conducted and is unsuccessful, there will be no way to know whether the initial ADS check provided erroneous status or the system failed during the mission. An appropriate model to estimate the probability that the mission will be successful given that the ADS has provided a "GO" indication can be written in terms of the weapon system reliability (WSR) as:

$$P \text{ (SUCCESSFUL MISSION/``GO'' INDICATION)}$$
$$= P_5 \times \text{WSR}$$
$$= [N_{gg}/(N_{gg} + N_{bg})] \times \text{WSR}$$

WSR is the probability that a system will complete a specified mission, given that the system was initially capable of performing that mission. WSR is given by the relationship:

$$\text{WSR} = e^{-t/\text{MTBCF}}$$

where

e	=	the base of the natural logarithms
t	=	the mission duration (hours)
MTBCF	=	mean time between critical failure (hours)

A critical failure is a failure which prevents the successful completion of the mission, and MTBCF is given by the relationship:

$$\text{MTBCF} = \frac{\text{Total Operating Hours}}{\text{Number of Critical Failure}}$$

P (SUCCESSFUL MISSION/"GO" INDICATION) is important from an operational viewpoint because it gives the probability of a successful mission if the ADS is relied on to decide whether a mission should be conducted. It is obvious from this relationship that, if the ADS indication is relied on, the probability of a successful mission is the product of weapon system reliability and ADS reliability. Thus, if weapon system reliability is 1.0 and ADS reliability is only .9, the probability of a successful mission will only be .9.

12.4.4.2 Fault isolation.

P_9 through P_{16} are the probabilistic measures of ADS fault isolation. P_9 gives the probability that the ADS will correctly isolate a fault in a bad system, and P_{11} gives the probability that the ADS will not erroneously isolate a fault in a good system. The probabilities of the events complementary to P_9 and P_{11}, P_{10} and P_{12}, respectively, are also important and are readily available once P_9 and P_{11} are known. Likewise, P_{13} gives the probability that, if fault isolation occurs, it will be correct, and P_{15} gives the probability that, if there is no fault isolation, the system has not failed. P_{14} and P_{16} are the probabilities for the complementary events for P_{13} and P_{15} and are easily obtained once P_{13} and P_{15} are determined.

12.4.4.3 Additional measures.

Measures of built-in-test (BIT) fault detection effectiveness which are often used include percent BIT fault detection (FD), percent BIT cannot duplicate (CND), and percent BIT false alarm (FA).

Percent BIT fault detection is defined as:

$$\text{Percent BIT FD} = \frac{\text{Number of Confirmed Failures Detected in BIT}}{\text{Number of Confirmed Failures Detected by All Methods}} \times 100$$

$$= N_{bb} / N_b \times 100$$

From a probabilistic standpoint:

$$\text{Percent BIT FD} = P_3 \times 100$$

This measure is P_3 expressed as a percentage. To obtain this measure, an audit trail must be maintained; that is, for each confirmed failure of the system, a record is maintained documenting whether the failure was detected by the ADS fault detection function.

Percent BIT CND is defined as:

$$\text{Percent CND} = \frac{\text{Number of BIT CNDs}}{\text{Total Number of BIT Failure Indications}} \times 100$$

$$\text{Percent CND} = [N_{gb}/(N_{gb} + N_{bb})] \times 100$$

$$= P_8 \times 100$$

A BIT CND is defined as "an on-equipment, BIT indication of a malfunction that cannot be confirmed by subsequent troubleshooting by maintenance personnel." It should be noted that, in Mil Standard 1309C, BIT CND is defined as a BIT fault indication that cannot be confirmed by the next higher level of maintenance. This is an important measure in that it is the probability expressed as a percent that a failure indication will be erroneous.

T&E OF INTEGRATED WEAPONS SYSTEMS

Percent BIT FA is given by the relationship:

$$\text{Percent FA} = \frac{\text{Number of Erroneous BIT Indications}}{\text{Total Number of BIT Indications}} \times 100$$

From a probabilistic standpoint,

$$\begin{aligned}\text{Percent FA} &= [P_2 \times P \text{ (SYSTEM IS GOOD)} \\ &\quad + P_4 \times P \text{ (SYSTEM IS BAD)}] \times 100 \\ &= [(N_{gb}/N_g) \times (N_g/N) + (N_{bg}/N_b) \times (N_b/N)] \times 100 \\ &= [(N_{gb} + N_{bg})/N] \times 100\end{aligned}$$

Alternatively,

$$\begin{aligned}\text{Percent FA} &= [P_6 \times P \text{ ("GO" INDICATION)} \\ &\quad + P_8 \times P \text{ ("NO GO" INDICATION)}] \times 100 \\ &= [\{N_{bg}/(N_{gg} + N_{bg})\}\{(N_{gg} + N_{bg})/N\} \\ &\quad + \{N_{gb}/(N_{bb} + N_{gb})\}\{(N_{bb} + N_{gb})/N\}] \times 100 \\ &= [(N_{bg} + N_{gb})/N] \times 100\end{aligned}$$

Measures of BIT fault isolation capability include percent fault isolation (FI) and percent retest OK (RTOK).

Percent fault isolation is defined by the relationship:

$$\text{Percent FI} = \frac{\begin{array}{c}\text{Number of Fault Isolations in Which}\\ \text{BIT Effectively Isolated the Fault}\end{array}}{\begin{array}{c}\text{Number of Confirmed Failures}\\ \text{Detected by All Methods}\end{array}} \times 100$$

$$= (S_{bc} / S_b) \times 100$$

From a probabilistic standpoint,

$$\text{Percent FI} = P_9 \times 100$$

Percent FI is equivalent to P_9 expressed as a percentage if the numerator of the above expression is taken to mean faults which are unambiguously isolated to a single item node (e.g., driver, receiver, connector, wire) or to a specified maximum number of items (i.e., an ambiguity group of x items) and if, for each confirmed failure, there has been a corresponding attempt to fault-isolate with ADS. Also, the numerator of this expression must exclude all fault isolations which result in RTOK as described below.

Percent RTOK is defined by the relationship:

$$\text{Percent RTOK} = \frac{\begin{array}{c}\text{Number of Units (LRU, SRU) That}\\ \text{RTOK at a Higher Maintenance Level}\end{array}}{\begin{array}{c}\text{Number of Units Removed}\\ \text{as a Result of BIT}\end{array}} \times 100$$

$$= \left[(S_{bei} + S_{ge}) / (S_{bc} + S_{bei} + S_{ge})\right] \times 100$$

From a probabilistic standpoint,

$$\text{Percent RTOK} = P_{14} \times 100$$

This measure is the probability, expressed as a percentage, that a good unit will be removed because of erroneous BIT fault isolation. It includes those instances where a system is bad but the wrong unit is isolated and those instances where the system is good but an erroneous fault isolation is made. This is P_{14} expressed as a percentage.

12.4.5 Summary

The previous discussion has highlighted the fact that there are a number of probabilities which are descriptive measures of the ADS capability to accomplish the purpose for which it was designed. These measures collectively describe the capability of integrated diagnostic systems and should be used in stating requirements, writing specifications, designing systems, and performing T&E. The test design for ADS must be such that sufficient data are obtained to determine estimates for these MOEs. This will require that efforts be directed specifically at determining estimates of the true status of the system and verifying ADS fault isolations by means other than the ADS. Obtaining this "system truth" is essential to gathering sufficient information to fully describe the capabilities and limitations of the ADS. Determination of system truth, however, should not and need not detract from conducting the mission in an operationally realistic manner.

The capabilities of ADS can be rigidly specified in terms of a series of probabilistic statements and measures as developed in this chapter. Dedicated testing to establish the true ADS capability of a system is both worthwhile and essential in the verification of the design. (This will require the proper technical expertise, test equipment independent of the ADS, and dedicated data collection.) More comprehensive specification and measurement of ADS capabilities provide the potential to help achieve basic system reliability, maintainability, and availability goals. The measures specified should be used in the design and procurement of automatic diagnostic systems. Dedicated testing as discussed should be conducted to verify designs and to establish true ADS capabilities.

Once the ADS capabilities of a system are "designed in," deployed, and verified by test measurements of the ADS, some of the MOEs identified may be more useful at the operational wing level than others. At the operational wing level, the emphasis on interpretation of the measures will shift from ADS effectiveness to the tracking and maintenance of basic weapon system effectiveness. This, of course, assumes that the ADS has been adequately designed, functions properly, and is effective in ascertaining the true operational status of the system and is correctly performing fault isolation.

References

[1] DoD Directive 5000.1, "Defense Acquisition," Office of the Secretary of Defense, Under Secretary of Defense for Acquisition, 23 February 1991.

[2] DoD Instruction 5000.2, "Defense Acquisition Management Policies and Procedures," Office of the Secretary of Defense, Under Secretary of Defense for Ac-

quisition, 23 February 1991.

[3]DoD Manual 5000.2-M, "Defense Acquisition Management Documentation and Reports," Office of the Secretary of Defense, Under Secretary of Defense for Acquisition, 23 February 1991.

[4]DoDD 5000.3, Test and Evaluation. Establishes policy and guidance for the conduct of test and evaluation in the Department of Defense, provides guidance for the preparation and submission of a Test and Evaluation Master Plan and test reports, and outlines the respective responsibilities of the Director, Operational Test and Evaluation (DOT&E), and Deputy Director, Defense Research and Engineering (Test and Evaluation (DDDR&E (T&E)).

[5]Johnson, B. W., *Design and Analyses of Fault-Tolerant Digital Systems*. Reading, Massachusetts: Addison-Wesley Publishing Company, Inc., 1989.

[6]Agrawal, V. D., *Test Generation for VLSI Chips*. Washington, D.C.: IEEE Computer Society Press, 1988.

[7]Penny, W. M., and Lau, L. (eds.), *Integrated Circuits*. New York: Van Nostrand Reinhold Co., 1972.

[8]Barna, A., *Integrated Circuits*. New York: John Wiley and Sons, Inc., 1973.

[9]Nelson, W., *Accelerated Testing*. New York: John Wiley and Sons, Inc., 1990.

[10]Hayes-Roth, F., Waterman, D. A., and Lenat, D. B. (eds.), *Building Expert Systems*. Reading, Massachusetts: Addison-Wesley Publishing Company, Inc., 1983.

[11]Booth, T. L., *Sequential Machines and Automata Theory*. New York: John Wiley and Sons, Inc., 1967.

[12]Kohavi, Z., *Switching and Finite Automata Theory*. New York: McGraw-Hill Book Company, 1978.

[13]Vishwani, D., and Seth, S. C., *Test Generation for VLSI Chips*. Washington, D.C.: Computer Society Press, 1988.

[14]Farrell, J. L., *Integrated Aircraft Navigation*. New York: Academic Press, 1966.

[15]Eichblatt, E. J., Jr. (ed.), *Test and Evaluation of the Tactical Missile*. Progress in Astronautics and Aeronautics, Vol. 119. Washington, D. C.: American Institute of Aeronautics and Astronautics, 1989.

[16]Chen, T. H., and Breuer, M. A., "Automatic Design for Testability Via Testability Measures," *IEEE Transactions in Computer-Aided Design*, Vol. CAD-4, January 1985, pp. 3-11.

[17]Aboulhamid, E. M., and Cerny, E., "A Class of Test Generators for Built-In-Testing," *IEEE Transactions in Computers*, Vol. C-32, October 1983, pp. 957-959.

[18]Daniels, R. G., and Bruce, W. C., "Built-In Self-Test Trends in Motorola Microprocessors," *IEEE Design and Test of Computers*, Vol. 2, No. 2, April 1985, pp. 64-71.

[19]Mead, C., and Conway, L., *Introduction to VLSI Systems*. Reading, Massachusetts: Addison-Wesley Publishing Company, 1980.

[20]Hayes, J. P., "Fault Modeling," *IEEE Design and Test of Computers*, Vol. 2, No. 2, April 1985, pp. 85-95.

[21]Johnson, B. W., "Fault-Tolerant Microprocessor-Based Systems," *IEEE Microprocessors*, Vol. 4, No. 6, December 1984, pp.6-21.

[22]Vesely, W. E., *Fault Tree Handbook*. Washington, D.C.: U.S. Nuclear Regulatory Commission, 1981.

[23]Thorng, H.C., *Introduction to the Logical Design of Switching Systems*. Reading, Massachusetts: Addison-Wesley Publishing Company, Inc., 1964.

[24]Lee, S.C., *Digital Circuits and Logic Design*. Englewood Cliffs, New Jersey: Prentice-Hall, 1976.

[25]Brzozowski, J.A., and Yoeli, M., *Digital Networks*. Englewood Cliffs, New Jersey: Prentice-Hall, 1973.

[26]Sheng, C.L., *Introduction to Switching Logic*. New York: Intext Educational Publishers, 1972.

[27]Ryne, V.T., *Fundamentals of Digital Systems Design*. Englewod Cliffs, New Jersey: Prentice-Hall, 1973.

[28]Hill, J.H., and Peterson, G.R., *Introduction to Switching Theory and Logical Design*. New York: John Wiley and Sons, Inc., 1968.

[29]Miller, R.E., *Switching Theory, Vol II*. New York: John Wiley and Sons, Inc., 1965.

[30]Wood, Jr., P.E., *Switching Theory*. New York: McGraw-Hill Book Company, Inc., 1968.

[31]Booth, T.L., *Digital Networks and Computer Systems*. New York: John Wiley and Sons, Inc., 1971.

[32]Friedman, A.D., and Menon, P.R., *Fault Detection in Digital Circuits*. Englewood Cliffs, New Jersey, Prentice-Hall, 1970.

[33]Chang, H.Y., Maming, E.G., and Metze, G., *Fault Diagnosis Digital Systems*. New York: John Wiley and Sons, Inc., 1970.

13
Measures of Effectiveness and Measures of Performance

13.1 General (See References 1–10)

A measure, in an elementary sense, is a device designed to convey information about the phenomena being addressed. Ideally, in test and evaluation (T&E), these measures will be made in a quantitative manner with an accuracy and precision sufficient to serve their purpose. (See Chapter 3 for additional discussion on measurement). On the other hand, there are many complex T&E situations where qualitative judgment of an expert is also essential to capturing and conveying the important information. Consequently, both quantitative and qualitative measures are essential and play a vital role in T&E.

The measure itself is different from the criteria for success. For example, the measure is a quantity or quality being examined to help determine whether a system does what it is designed to do and the level of military value or utility it achieves when it is applied in a combat mission. An example measure could be the percentage of correct detection and display of a given type of threat signal within 4 seconds. The 4-second time specified might be selected on the basis of how quickly the threat system can react and fire at an aircraft entering its defense zone. The decision criteria for success related to this measure could be that 85 percent of threat type A must be detected and correctly displayed within 4 seconds of threat activation. From this, it can be seen that the measure itself and the criteria for success are distinctly different things.

When measuring military test items and systems, there are two basic questions of interest. The first is the very fundamental one "Does the test item or system do what it is supposed to do?" That is, does the test item or system meet the engineering and design specifications for which it was purchased? This measurement activity is focused on the device itself, and these types of measurements are commonly referred to as measures of performance (MOPs). Engineering and technical performance of the test item or system is the major item of concern when making these measurements. The second question of interest is "What is the military worth or value of the test item or system?" This is a more difficult but very fundamental question that should be addressed and answered at least for the first time early in the process when dealing with concepts and systems. It is certainly not inconceivable that the test item or system could work perfectly (i.e., do exactly what it was engineered and designed to do) yet have very little military value in winning an engagement or conflict. In fact, several cases have occurred during the last 10 years where over a billion dollars was spent on developing systems that met their engineering design specifications but were unable to satisfactorily perform their operational mission. T&E measures addressing the military worth or value of the test item or system are commonly referred to as measures of effectiveness

(MOEs). As discussed in earlier chapters, the effectiveness of a test item or system is described in terms of its capability to successfully carry out a military task or mission when called upon. Thus, MOEs are used to describe how well these operational activities are achieved. Engineering and technical performance, also discussed in earlier chapters, deals primarily with the issue of the test item or system meeting its design specifications. Although we have separated performance and effectiveness to aid in the presentation and discussion, the two are not necessarily independent or mutually exclusive. Indeed, there is a great dependence and some overlap between the two.

Success in the overall acquisition process starts with accurate information on what the requirement is, and proceeds through specification, design, production, and procurement of what is essential to meet the requirement. A close linkage of these activities is mandatory for success. Thus, it is important that the early analyses examining test items and weapons systems accurately establish their potential military worth and value, that is, the military worth or value that they will bring to the mission if they perform as predicted. This type of analysis is sometimes referred to as mission area analysis or mission area assessment, and is used to help identify areas where potential improvements in military capability are required. Later analyses performed in support of the cost and operational effectiveness analysis (COEA) also contribute to the process of establishing the military worth of potential improvements.

Successful test items and systems must have military worth or value, and they must meet their engineering design and performance specifications. Ideally, the MOPs specified to determine whether a test item or system meets its design specifications should be closely linked to the MOEs. That is, if T&E demonstrates that a system meets its MOPs, it should have sufficient capability to carry out its task or mission and meet its MOEs. If not, the MOPs are not sufficiently linked to the MOEs, or they have been underspecified for the task or mission. Conversely, when a test item or system meets its MOEs, it should have sufficient inherent capability to meet its MOPs. Again, if not, the linkage is not sufficient, or the MOPs have been overspecified.

Generally, a single MOE or MOP is not sufficient to fully address the capability (or lack of capability) for a test item or system. Some MOEs and MOPs may be more relevant, descriptive, and important than others for the specific situation being addressed. Consequently, multiple MOEs and MOPs are generally necessary to characterize the value of the test item or system, and some MOEs and MOPs must be given greater consideration in the process of establishing the degree to which the test item or system has achieved its military value.

The development of MOPs and MOEs should be addressed by use of both a "systems engineering approach" and an "operational tasks/mission approach." The systems engineering approach examines the test item or system from the perspective of meeting engineering design specifications and requirements. The operational tasks/mission approach examines the test item or system in terms of its military value or utility in contributing to the accomplishment of the combat mission. The development of MOPs and MOEs can also be viewed as both a "top-down" and a "bottom-up" problem. When addressing the T&E problem from the viewpoint of top-down, we are interested in examining and understanding what the system does as a total entity, and then decomposing it down into the individual parts and determining how well they contribute to the total system. On the other hand, when the T&E problem is viewed as a bottom-up problem, the approach becomes one of examining T&E of the individual parts, and then progressively moving up to a full system or higher level examination.

This concept of addressing MOPs and MOEs for a system and its component parts is depicted in Figure 13-1. The full system and its component parts are shown at the top of the chart, and MOPs for the full system as well as its parts are depicted at the center of the chart. Conversely, the operational tasks and mission are shown at the bottom of the chart, and the MOEs associated with the individual tasks as well as the full mission are depicted at the center of the chart. Ideally, the functional relationship between the MOPs for the various parts of the system and the MOPs for the total system will be such that when the individual parts meet their specified values, the overall system will meet its specified values. It is also desired that a similar functional relationship exist between the MOEs for the various operational tasks and full mission accomplishment (i.e., when MOEs are met for individual parts, the MOEs for the full system will be met). Finally, successful achievement of the MOPs for both the parts and system should result in successful achievement of the MOEs for the individual tasks as well as the overall mission.

13.2 Systems Engineering Approach (See References 11–22)

The nature of the test item or system for which T&E is being conducted will have a significant impact on the specifics of the T&E measures. For example, a communications device requires a different type of testing and measurement than a rifle, and T&E measures for a tank are different from those for a gas mask. Furthermore, a gas mask may be perfectly suitable for use by a ground soldier but totally unusable by a pilot.

Every test item or system can normally be thought of as part of a hierarchy of parts that make up a larger system. This concept is illustrated in Figure 13-2. The test item or system can also normally be broken down into smaller parts or systems (e.g., components, modules, and subsystems) that can each be addressed by MOPs at a very fundamental level of T&E. When we address these parts at the engineering level, MOPs are generally the appropriate T&E measures being applied. For example, we will likely be at the level of measuring voltage, current, power output, receiver sensitivity, velocity, impact, weight, etc., addressing these measures. As parts are aggregated to higher levels which can be related to the accomplishment of operational tasks or missions, MOEs must become the measures of greater concern. (That is, the main reason for having MOPs is to eventually aggregate up to a level of acceptable performance that results in the test item or system being able to meet its MOEs, and successfully accomplish the operational tasks or mission that achieves its military worth.) Most systems are generally examined from a systems engineering approach in terms of how they "stack up" or relate to their engineering design specifications and to the performance of other friendly systems as well as enemy systems. For example, the MOPs for an electronic warfare jammer must be highly related to the threat systems which it is designed to counter. These MOPs treat performance parameters such as output power, frequencies, and waveforms, that are specifically designed to counter the enemy systems based on the engineering design features of the enemy systems. However, eventually the military worth or value of the friendly jamming system must be established based on its contribution to winning and surviving in combat. For example, how does

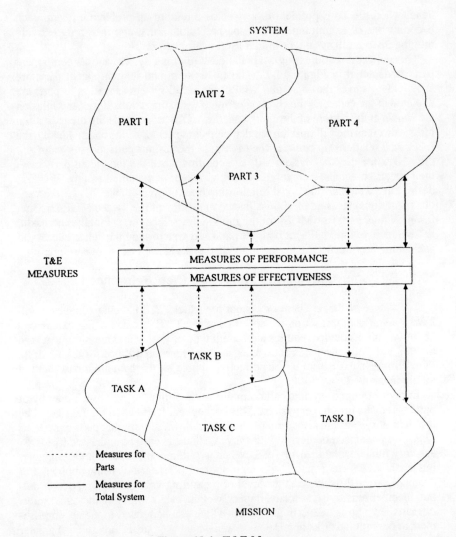

Figure 13-1. T&E Measures

the jammer contribute to destroying enemy targets over time and how does it contribute to the survival of friendly aircraft over time? The MOPs are necessary to design and build the system, and the MOEs are necessary to establish its military worth and value. Each of the individual parts of a system should be subjected to various levels of T&E and thus each requires MOPs and sometimes MOEs. MOEs are always essential at some aggregate level when the military value of what is being achieved is examined. Therefore, one important consideration in the scoping problem of any T&E activity is where we are in the system development hierarchy and to what extent T&E must examine the various parts in terms of MOPs for components, modules, subsystems, systems, etc. Similarly, it is also important to address the development activity for MOEs in terms of the operational tasks and

MEASURES OF EFFECTIVENESS AND PERFORMANCE 257

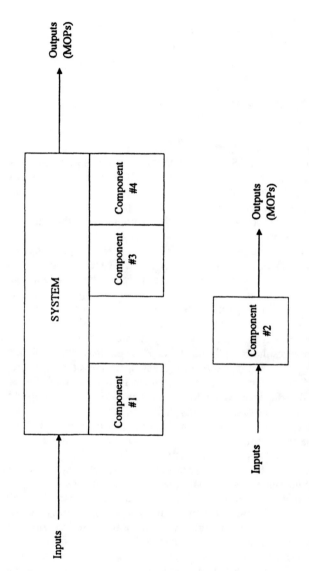

Figure 13-2. Illustration of Systems Engineering Approach to T&E

missions that must be performed by the appropriate parts and systems.

When the systems engineering approach is applied, the T&E of interest could be one examining a single component, a module, a subsystem, a single system, a few systems, or many systems working in concert to accomplish a task, mission, or higher level military goal. That is, we are interested in T&E of the full system as well as the individual parts, and we are interested in making sure that successful accomplishment of the MOPs for the system will result in successful accomplishment of MOEs for the operational tasks and mission. Therefore, engineering T&E carried out on specific parts must accurately test for the harshness of the operating environment plus whatever safety margin must be designed into the equipment.

13.3 Operational Tasks/Mission Approach (See References 23–37)

At the highest level, military capability is essential to the national security and defense of the United States. The national military strategy leads to the identification of military roles and missions for which a capability must be attained and maintained for the well-being of the country. Table 13-1 provides a very simplified depiction of the roles and typical missions of aerospace power. These example missions form the basis for the examination of the military worth or value of aerospace systems.

Aerospace systems perform four basic roles: aerospace control, force application, force enhancement, and force support. These roles and missions distinguish what the Air Force specializes in versus other services within the Department of Defense. Roles define the broad purposes or functions of the aerospace forces. Missions define the specific tasks that must be accomplished, not the capabilities or organizations within the Air Force that accomplish them. Thus, roles and missions are defined by objectives, not by the platforms or weapons used. Aerospace control assures the friendly force use of the environment while denying its use to the enemy. Force application brings aerospace power to bear directly against enemy surface targets. Force enhancement increases the ability of friendly aerospace and surface forces to perform their missions. Force support helps attain and sustain operations of the friendly aerospace forces.

Most aerospace forces are capable of performing multiple roles and missions, and some can perform multiple roles and missions in unique ways. There is also a relationship and dependency among the roles. For example, force application is dependent upon force support and could not be achieved without successfully accomplishing the mission of base operability and defense.

All aerospace systems are designed and applied to contribute to carrying out these specific missions. System engineering designs and system operational characteristics must address and focus on one or more of these missions.

In turn, the military missions can be broken down into operational tasks that must be performed to successfully carry out each military mission. (See Chapter 3 for additional discussion on the breakdown of military missions, tasks, etc.) Once a determination is made, through the higher level analysis, that a mission and its associated tasks are essential to the strategy being pursued, the T&E problem can be scoped to a lower level. It then becomes one of examining the test item or system in terms of successfully accomplishing the specific tasks and missions for which it is being acquired. Thus, the MOEs and MOPs can then be focused on

MEASURES OF EFFECTIVENESS AND PERFORMANCE

Table 13-1. Roles and Typical Missions of Aerospace Power

Roles	Typical Missions
Aerospace Control	Counterair
	Counterspace
Force Application	Strategic Attack
	Interdiction
	Close Air Support
Force Enhancement	Airlift
	Air Refueling
	Spacelift
	Electronic Combat
	Surveillance and Reconnaissance
	Special Operations
Force Support	Base Operability and Defense
	Logistics
	Combat Support
	On-Orbit Support

those specific tasks and missions.

Figure 13-3 depicts a simplified operational mission environment. This environment consists of all of the elements (favorable as well as unfavorable) that affect being able to accomplish the mission. The mission activities are shown at the top of the chart and are broken down in terms of premission, mission, and postmission. All of these activities must be performed in an operational environment that is affected by both nature and the enemy. The test item or system must operate within the friendly Command, Control, Communications, and Intelligence (C3I) system, and it must interface and be compatible with other systems and other services, and sometimes other countries. The activities necessary to prepare for, support, carry out, recover, as well as prepare for the next mission can be described in terms of tasks that must be successfully accomplished in the operational flight/field environment. Factors such as chemical and nuclear warfare significantly increase the complexity of these activities. Each of these important activities and tasks require MOEs that are capable of reflecting the contribution (in terms of military worth or value) of the test item or system to the mission. Ideally, these MOEs can be clearly related from the bottom up and top down with regard to accomplishment of the overall mission. Certain MOEs will be more descriptive at the aggregate level for the mission, and others will be more descriptive at the lower levels of investigation. It is extremely important that the level of examination of any T&E be specific and thorough enough to allow the military processes to be well understood, and to allow the areas of excessively high risks to be removed or controlled.

Premission activities shown at the upper left of the diagram must be capable of preparing and maintaining the system in a state that is ready for use in its operational mission. For example, in the case of an attack fighter aircraft, this would consist of all of the support activities necessary to get the aircraft ready for the

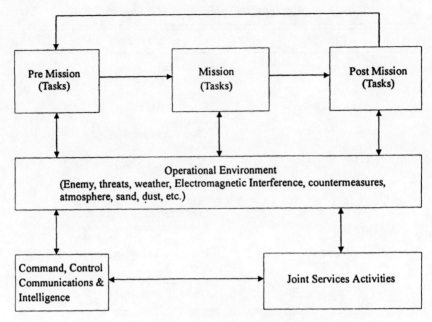

Figure 13-3. Simplified Operational Mission Environment

mission to include training activities as well as the premission planning effort that is essential in preparing for the mission. Premission tasks include a large number of activities that are related to both operational suitability and operational effectiveness. Each activity and task should have MOEs identified for it and cover areas such as aircraft servicing, armament loading, etc. Aircrew planning for the mission is also an important task performed during premission activities.

The mission itself can be depicted in terms of the operational tasks that must be achieved for success. Interfaces and interactions with other systems (to include the enemy) are important. Many mission tasks must be performed in accordance with exact timing and planning in order to complement higher level military planning and actions. Other tasks may require execution that was not necessarily a desired part of the original mission plan. Often external effects such as weather; battlefield smoke, dust, and clutter; and enemy reactions may require different tasks or emphasis during the mission. Planning and training for these contingencies is difficult but essential. MOEs must be capable of capturing and communicating the effects of these complex situations. Sometimes, it is difficult if not impossible to capture the sum total of these multiple situations in a quantitative value. Consequently, the consideration of multiple MOEs, along with the judgment of operators, operations analysts, intelligence personnel, maintainers, and other test team experts, is essential to making these measurements and accomplishing effective T&E.

Postmission activities include the tasks necessary to complete the previous mission and to provide information necessary for the next premission cycle. For example, in the case of the attack fighter aircraft, the aircrew must be debriefed and the aircraft repaired as required for the next mission. Information regarding mis-

MEASURES OF EFFECTIVENESS AND PERFORMANCE

Table 13-2. Example Roll-Up/Roll-Down Examination of MOPs

Level I MOPs	Level II MOPs	Level III MOPs
Input Power	Detection Range	Prevention of Enemy
Voltages	Identification &	Missile Launches
Currents	Classification Time	Effects on Enemy Missile
Temperature Tolerances	Jamming Range	Tracking Loops
Weight	System Reaction Time	Decrease in Enemy P_k
Volume		(probability of kill)
Receiver Sensitivity		
Power Output		
Frequencies & Tolerances		
Bandwidth		

sion accomplishment and battle damage assessment against the enemy as well as failure of or damage to friendly aircraft systems is important. On the operational suitability side, both the aircrew and the aircraft must be recycled and prepared for future missions.

For most systems, there is a special category of characteristics referred to as critical system characteristics which historically drive the design, cost, and risk associated with various types of systems. These characteristics include such things as energy efficiency, transportability, electronic counter-countermeasures, etc. These characteristics must be identified early in the system acquisition process to allow cost performance trade-offs. Adequate test and evaluation of MOPs and MOEs for these characteristics are essential to allow trade-offs and for the tracking of progress throughout the acquisition process.

Table 13-3. Example Aircraft System Requirements Matrix

System Capabilities and Characteristics	Thresholds	Objectives
Supersonic Cruise (Mach)	2.1M	2.5M
Sustained Speed at Sea Level	1.3M	1.5M
Dash	2.5M	2.7M
System Availability	90%	95%
Mission Capable Rate	90%	95%
Not to Exceed Break Rate	10%	5%
Fix Rate Within 4 Hours	95%	97%
Radar (Multiple Track)	20 tracks	30 tracks
Search (No. of Targets)	40	50
Track (No. of Targets)	6	10
Detection Range (NM)	150	200
Track Range (NM)	80	100
Terrain Following	Yes	Yes
Minimum Altitude (ft)	200 (All Weather)	100 (All Weather)

13.4 Example Measures of Performance

MOPs are highly dependent upon the mission, task, and level of T&E activity being performed. They are also highly dependent upon the nature of the equipment being examined. Table 13-2 depicts an example "roll-up/roll-down" examination of MOPs for an electronic combat system designed to protect an aircraft. The more fundamental measures are at Level I and address such things as power, volume, weight, etc. These measures can be rolled up to Level II measures addressing ranges and times that the system is capable of achieving against specific threat requirements. The Level III MOPs depicted here are aggregated up to a level that might cause some of them also to be considered as MOEs. If one chose to start first with the Level III MOPs, they could have been rolled down to Level II and Level I MOPs. An extremely important part of this process is that the specification values (i.e., criteria) that the MOPs must achieve are determined by the capabilities of the enemy systems that must be countered. Consequently, the MOPs must be properly specified against the correct threats and the correct operational environment if the system is to be successful.

Requirements and specifications for systems are sometimes presented in matrices that identify MOPs and the associated threshold and objective values that are to be achieved. Table 13-3 presents a simplified example matrix showing MOPs and their associated threshold/objective values for an aircraft system. Most real systems will need complex multidimensional matrices to adequately address MOPs. System capabilities and characteristics cite MOPs that should be realistic, meaningful, and germane to the mission need, including mission planning requirements. Capabilities and characteristics address areas such as performance, support, range, capacity, survivability, reliability, maintainability, availability, operational effectiveness, and operational suitability. The MOPs cited in these matrices should be those necessary to result in a test item or system that can successfully accomplish the operational mission. It is also important that the units of measure and data collection cited for the MOPs be consistent with the instrumentation and data collection capabilities possible and available. As the program matures, these values reflect system-specific performance and support that form the basis for the contractual specifications. The "objective" value represents the design goal and the "threshold" value represents the value below which the utility of the system becomes questionable. Sometimes ranges of values are more appropriate and should be used instead of point values. The values presented in these types of matrices are generally changed over time through various system trade-offs and as the system matures. Early in the acquisition process, the number of capabilities and characteristics listed are normally few in number and are operationally oriented. As the system matures and becomes better defined, subelement or new characteristics and capabilities are added as necessary to fully define the desired system performance. T&E must focus on measurement of the accomplishment of these factors at all appropriate levels.

13.5 Example Measures of Effectiveness

MOEs are also highly dependent upon the mission and tasks being performed and the level of the T&E activity being performed. For example, Table 13-4 illustrates the principle of roll-up and roll-down of MOEs. The highest level measure

MEASURES OF EFFECTIVENESS AND PERFORMANCE

Table 13-4. Example Roll-Up/Roll-Down Examination of MOEs

Level I MOEs	Level II MOEs	Level III MOEs
Planned mission strategy vs execution over time		
Effects of deviation of planned mission execution		
Force summaries for scheduled and achieved missions		
Battle damage assessment results vs time	Enemy targets destroyed vs time	Movement of the forward edge of the battle area (FEBA) vs time
Number and types of weapons delivered by both sides		
Suppression of enemy and friendly defenses	Friendly forces surviving vs time	
Kill assessments against enemy and friendly forces		
Number of missed attack opportunities and reasons vs time		
Number of blue forces engaged by type and threat vs time		
Types and results of counter and counter-countermeasures vs time		

depicted in the chart is shown on the right side of the table. This measure describes who controls the ground area over time (similar measures could also be used to address the air and space situations). This example MOE addresses movement of the forward edge of the battle area (FEBA) versus time. It could be rolled down to lesser measures at Level II that address enemy targets destroyed and friendly forces surviving versus time. Level II measures could be rolled down to MOEs of even greater specificity at Level I. Obviously, one could start the process at the Level I depicted (or lower) and roll up the MOEs to Level III, etc. The specifics of which measures are the most appropriate for a given test item or system depend upon how the system fits into the combat environment and which measures are most appropriate to describe its military value to success in the task or mission.

Table 13-5 presents a mission characteristics matrix that identifies example areas where specific MOEs and associated criteria could be established. For example, if the system being examined was a transport aircraft, the MOE for payload could be expressed in terms of ton-miles versus time. Threshold and objective values are also important but are not shown in the illustration to save space. The areas presented are simple examples but are specific enough to allow examination at a detailed level. They also could be aggregated up and expressed in terms of higher level functions and tasks for which higher level MOEs could be established as previously discussed.

13.6 Example Measures of Effectiveness for Electronic Combat Systems

Electronic Combat (EC) can be described in terms of three important tasks: 1) Electronic Warfare (EW), 2) Command, Control, and Communications Counter-

Table 13-5. Example Mission Characteristics Matrix

Mission Function	Operator & Passenger Factors
Payload	Seating Area
Interfaces	Climate
Travel	Work Space
Speed	Controls/Display
Distances	Lighting
Acceleration	Visibility
Operational Environment	Storage
Road Surfaces	Communications
Paved/Nonpaved	Driver to Passenger
Snow/Ice	With Pedestrians
Load	Radio
Width	Safety
Road Crown	Egress/Ingress
Weather Conditions	Fire Extinguishers
Base Facilities	Other Hazards
Emissions	
Vehicle in Commission Rate	Readiness
Interoperability	
	Response Time
Factors Related to Flight/ Field Operations	Warm-Up
	Windshield Clearing
Concurrent Operations	Repair Time
Stability	Deployment
Road Grade	Mode (on/off, height/length/width/
Winds	axles)
Loads	Time to Prepare for Shipment
Speed	Time to Restore Function at
Electromagnetic Interference	Operating Location
	Number of Days Operating Away
Survivability	from Home Station
	Hours per Day Required to Be Available
Emergency Standby Operations	Hours per Day Engine Runs
Fuels	Off/on Cycle Per Day (maximum)
Acceleration	Decontamination
Exposure	
Storage	
Nuclear/Biological/Chemical Threats	
Electromagnetic Threats	

(Continued on next page.)

MEASURES OF EFFECTIVENESS AND PERFORMANCE

Table 13-5. (continued) Example Mission Characteristics Matrix

Support Factors	
Organization	Technical Data
Home Station	Scheduled Inspections/Maintenance
Deployed	Intervals
Refuel	Manpower
Response Time	Commercial Compatibility
Concurrent	Subsystem Reliability &
Truck	Subsystem Break Rate
Hydrant	Manpower
Operator Service	Spaces
Supporting Equipment	Military Specialty Code
Towing	Supply Response Time
Jacks	
Maintenance Equipment	
Diagnostics	
Test	
Repair	

measures (C^3CM), and 3) Suppression of Enemy Air Defenses (SEAD). EW is military action using electromagnetic energy to determine, exploit, reduce, or prevent hostile use of the electromagnetic spectrum. It contributes significantly to the counterair mission by allowing friendly aircraft to operate with a greater degree of freedom and less attrition in environments with enemy threats present. It also includes actions designed to retain the friendly use of the electromagnetic spectrum. C^3CM is military action involving defensive and offensive operations in a strategy that is designed to deny information to the enemy, to protect friendly C^3, to influence enemy actions, and to degrade or destroy enemy C^3 capabilities. C^3CM, supported by intelligence operations, integrates the use of operations security, military deception, jamming, and physical destruction. SEAD is directed at gaining freedom of action to perform friendly force missions by neutralizing, destroying, or temporarily degrading enemy air defense systems. All EC systems are designed to perform, or help perform, one or more of these important force enhancement tasks.

MOEs for EC systems should be selected on the basis of what measures best describe the value or utility of the EC system when it is applied in its operational combat mission. Operational expertise and judgment combined with scenario/threat knowledge and mission area analyses are all important tools that help the operator in the MOE selection process. Because of the complexity and the fact that no single MOE or test environment is sufficient to address all aspects of an EC system, it is necessary to conduct T&E over a range of conditions and in multiple test environments. Many of the testing situations provide comparisons of the effects (i.e., MOEs) with and without the new EC capability present. Test environments include such activities as measurement facilities, integration laboratories, hardware-in-the-loop test facilities, installed test facilities, and actual flight/field testing on open air ranges.

Table 13-6 provides an example partial listing of the various types of EC systems applied in combat, the desired mission benefits from the systems, and sug-

Table 13-6. Example MOEs for Electronic Combat Systems

System	Desired Benefit	Measures of Effectiveness
All Electronic Combat Systems	They open up opportunities for more effective and less restrictive tactics and procedures	1) More enemy targets destroyed per mission 2) Less friendly aircraft lost per mission
Self-Protection Jamming Systems	1) Prevent weapons from being fired 2) Cause weapons to miss given that they are fired 3) Cause delay, disruption, and confusion of operations at threat weapons systems	1) Reduction in weapons fired (wet vs dry) 2) Reduction in weapons that hit (wet vs dry) 3) Increased delay, disruption, and confusion of operations at threat weapons systems (wet vs dry)
Standoff Jamming Systems (EW/GCI/ACQ Radars)	1) Prevent friendly aircraft from being acquired, tracked, and fired upon 2) Cause weapons to miss given that they are fired 3) Cause delay, disruption, and confusion of operations at EW/GCI/ACQ radars	1) Reduction in friendly aircraft being acquired, tracked, and fired upon (wet vs dry) 2) Reduction in weapons that hit (wet vs dry) 3) Increase in delay, disruption, and confusion of operations at EW/CGI/ACQ radars (wet vs dry) 4) Decrease in capability of enemy command and control system to assess the size and intent of strike force
Radar Warning Receivers	1) Provide timely and accurate threat warning (range and bearing) 2) Provide timely and accurate launch warning (range and bearing)	1) Percent of timely and accurate threat warning (range and bearing) 2) Percent of timely and accurate launch warning (range and bearing) 3) False alarm rate
End Game Warning Systems	1) Provide timely and accurate end game warning 2) Cue and initiate timely and accurate self-protection actions (chaff, flares, maneuvers, etc.)	1) Percent of timely & accurate end game warning 2) Percent of timely and accurate end game self-protection actions 3) False alarm rate

(Continued on next page.)

MEASURES OF EFFECTIVENESS AND PERFORMANCE

Table 13-6. (continued) Example MOEs for Electronic Combat Systems

System	Desired Benefit	Measurements of Effectiveness
Chaff, Flares, and Maneuvers	Prevent weapons kill during end game	Decrease in percent of weapons kill during end game (wet vs dry)
Standoff Jamming Systems (Communications C^3CM)	1) Prevent friendly aircraft from being acquired, tracked, and fired upon 2) Cause delay, disruption, and confusion of enemy operations 3) Deny/degrade enemy command and control capability	1) Reduction in friendly aircraft being acquired, tracked, and fired upon (wet vs dry) 2) Increase in delay, disruption, and confusion of operations at EW/GCI/ACQ and threat tracking radar (wet vs dry) 3) Decrease in number of communications handled by command and control system
Wild Weasel/HARM (SEAD)	1) Prevent friendly aircraft from being acquired, tracked, and fired upon 2) Cause delay, disruption, and confusion of enemy operations 3) Kill enemy threat systems	1) Reduction in friendly aircraft being acquired, tracked, and fired upon (wet vs dry) 2) Increased delay, disruption, and confusion of operations at EW/GCI/ACQ and threat tracking radars (wet vs dry) 3) Number of enemy threat systems killed

Wet is jamming present.
Dry is no jamming present.
EW is electronic warfare.
GCI is ground control intercept.
ACQ is acquisition.
C^3CM is command, control, and communications countermeasures.
HARM is high speed antiradiation missile.
SEAD is suppression of enemy air defenses.

gested MOEs that might be evaluated. The table addresses all EC systems to include self-protection and standoff jamming systems; radar warning receivers; end game warning systems; expendable countermeasures such as chaff, flares, and maneuvers; and wild weasel systems. End game systems are those designed to alert and save friendly aircraft from an inbound missile that will likely be in the vicinity of the aircraft. Wild weasel aircraft are aircraft designed to seek out and destroy enemy threat systems located on the ground. Ideally, the ground threat systems will be jammed or destroyed before they can shoot down friendly aircraft. All of these MOEs are directed at establishing the required contribution of the test item or system to the accomplishment of the combat mission.

13.7 Example Measures of Effectiveness for Operational Applications, Concepts, and Tactics Development

Most T&E discussed thus far has been directed at the acquisition of a weapons system. However, once a decision is made to acquire and produce systems, the major developmental decisions have been made; the production, money, and configurations have been committed; and the remaining T&E activities are clearly military activities primarily of the operational combat user. These activities are designed to optimize the operational application of the equipment, to develop suitable tactics and procedures, to train appropriate personnel, and to carry out the logistics supportability process for the incoming weapons systems. Many of the MOEs previously applied are still appropriate, but they may require some "fine-tuning" to the combat application. A major portion of the logistics supportability effort by the operational user has to do with establishing and maintaining operational data bases, software changes and operational flight program (OFP) updates, and the initiation and maintenance of the logistics supply system. These military operational activities, much like the establishment of operational requirements, require operational combat user expertise and are primarily the domains of the operational major commands. Ideally, these nonacquisition functions are disassociated to the degree possible from the political and procurement activities of acquisition test and evaluation and are grouped into major command endeavors that address operational applications, concepts, and tactics developments. These activities include operational concept demonstrations (OCDs), tactics developments and evaluations (TD&Es), OFP updates, and supportability assessments (SAs) and are best accomplished by the major commands that operate the systems in combat.

References

[1] Air Force Manual 1-1, "Basic Aerospace Doctrine of the United States Air Force," Volumes I and II. Washington, D.C.: Headquarters U.S. Air Force, March 1992.

[2] DoD Directive 5000.1, "Defense Acquisition," Office of the Secretary of Defense, Under Secretary of Defense for Acquisition, 23 February 1991.

[3] DoD Instruction 5000.2, "Defense Acquisition Management Policies and Procedures," Office of the Secretary of Defense, Under Secretary of Defense for Acquisition, 23 February 1991.

[4] DoD Manual 5000.2-M, "Defense Acquisition Management Documentation and Reports," Office of the Secretary of Defense, Under Secretary of Defense for

Acquisition, 23 February 1991.

[5]DoDD 5000.3, Test and Evaluation. Establishes policy and guidance for the conduct of test and evaluation in the Department of Defense, provides guidance for the preparation and submission of a Test and Evaluation Master Plan and test reports, and outlines the respective responsibilities of the Director, Operational Test and Evaluation (DOT&E), and Deputy Director, Defense Research and Engineering (Test and Evaluation (DDDR&E (T&E)).

[6]Przemieniecki, J. S. (ed.), *Acquisition of Defense Systems*. AIAA Education Series. Washington, D.C.: American Institute of Aeronautics and Astronautics, Inc., 1993.

[7]Duckworth, W. E., Gear, A. E., and Lockett, A. E., *A Guide to Operational Research*. New York: John Wiley and Sons, Inc., 1977.

[8]Brewer, G., and Shubik, M., *The War Game, A Critique of Military Problem Solving*. Cambridge, Massachusetts: Harvard University Press, 1979.

[9]Dupuy, T., *Understanding War*. New York: Paragon House, 1987.

[10]Dalkey, N. C., "The Delphi Method: An Experimental Study of Group Opinion," RM-5888-PR. Santa Monica, California: 1969.

[11]Davidson, D., Suppes, P., and Siegel, S., *Decision Making, An Experimental Approach*. Stanford, California: Stanford University Press, 1957.

[12]Dyer, J. S., Farrel, W., and Bradley, P., "Utility Functions for the Test Performance," *Management Science* 20, 1973, pp. 507-519.

[13]Feller, W., *An Introduction to Probability Theory and Its Applications*, Volume 2. New York: John Wiley and Sons, Inc., 1966.

[14]Checkland, P. B., *Systems Thinking, Systems Practice*. New York: John Wiley and Sons, Inc., 1981.

[15]McKean, R. N., *Efficiency in Government Through Systems Analysts*. New York: John Wiley and Sons, Inc., 1958.

[16]Miller, J. R., III, *Professional Decision Making*. New York: Praeger Publishers, 1970.

[17]Nash, J. F., "The Bargaining Problem," *Econometrica*, 18, 1950, pp. 155-162.

[18]Pratt, J. W., Raiffa, H., and Schlaifer, R. O., *Introduction to Statistical Decision Theory*. New York: McGraw-Hill Book Company, 1965.

[19]Savage, L. J., *The Foundations of Statistics*. New York: John Wiley and Sons, Inc., 1954.

[20]Schlaifer, R. O., *Analysis of Decisions Under Uncertainty*. New York: McGraw-Hill Book Company, 1969.

[21]Miles, L. D., *Techniques of Value Analysis and Engineering*. New York: McGraw-Hill Book Company, 1972.

[22]Defense Systems Management College, *Systems Engineering Management Guide*, 1982.

[23]Fishburn, P. C., *Decision and Value Theory*. New York: John Wiley and Sons, Inc., 1964.

[24]Debreu, G., "Topological Methods in Cardinal Utility Theory," *Mathematical Methods in the Social Sciences*. Stanford, California: Stanford University Press, 1959.

[25]Fishburn, P. C., "Utility Theory," *Management Science*, 14, 1968, pp. 335-378.

[26]Fishburn, P. C., *Utility Theory for Decision Making*. New York: John Wiley and Sons, Inc., 1970.

[27] Fishburn, P. C., *The Theory of Social Choice*. Princeton, New Jersey: Princeton University Press, 1973.

[28] Fried, C., *An Anatomy of Values*. Cambridge, Massachusetts: Harvard University Press, 1970.

[29] Luce, R. D., and Raiffa, H., *Games and Decisions*. New York: John Wiley and Sons, Inc., 1957.

[30] Raiffa, H., *Decision Analysis*. Reading, Massachusetts: Addison-Wesley, 1968.

[31] Rapoport, A. (ed.), *Game Theory as a Theory of Conflict Resolution*. Dordrecht, Holland: Reidel Publishing Company, 1974.

[32] Keeney, R. L., and Raiffa, H., *Decisions with Multiple Objectives*. New York: John Wiley and Sons, Inc., 1976.

[33] Arrow, K. J., *Social Choice and Individual Values* (2nd ed.). New York: John Wiley and Sons, Inc., 1963.

[34] Bauer, R. A., *Social Indicators*. Cambridge, Massachusetts: M.I.T. Press, 1966.

[35] Braithwaite, R. B., *Theory of Games as a Tool for Moral Philosophers*. Cambridge, England: Cambridge University Press, 1955.

[36] Cochrane, J. L., and Zeleny (eds.), *Multiple Criteria Decision Making*. Columbia, South Carolina: University of South Carolina Press, 1973.

[37] Miser, H. J., and Quade, E. S., L. Elsevier. *Handbook of Systems Analysis: Craft Issues and Procedural Choices*. New York: Elsevier Science Publishing Co., 1988.

14
Measurement of Training

14.1 General (See References 1–10)

Training is the process of helping one gain knowledge or understanding of, or skill in, by study, instruction or experience.* "Transfer of training" is a very special concept used in training to denote the influence of past learning on present learning. Training applies control and manipulation outside the trainee to facilitate learning. It normally involves some form of instruction in the tasks to be performed, after which the trainee practices to accomplish the tasks under the appropriate degree of realism.

"Realism" is the extent to which the training equipment is identical to regular on-the-job equipment. The amount of realism that training equipment must have regarding any particular feature depends upon how much learning can occur with and without the feature and how much it costs (realism is also sometimes referred to as "fidelity"); that is, training equipment is designed to train, not necessarily to be totally realistic. For example, a training device may be perfectly suitable for the training of an operator in a given task without including certain expensive features intended simply to add realism. In fact, sometimes a well-designed training device may intentionally deviate from realism in order to better support learning. However, when they are found to be essential to the learning process, the conditions relevant to performance by the trainee must be realistic. Identifying which conditions are relevant and establishing the amount of realism that must be present are situation-dependent, requiring definition and study of the training problem by experts, proper design of the training system to maximize the benefit of positive training, and the implementation of cost trade-offs regarding benefits. Training always requires measurement of the trainee's performance and feedback to the trainee on performance success or failure. Ideally, the measurement of the trainee's performance will be objective. This is usually achieved through some form of scoring or quantitative measurement of the trainee's performance in the completion of tasks. Generally, the training process is continued until the trainee reaches a plateau of performance that is acceptable or, in some cases, until further improvement in performance is negligible. These highly desirable training features will be illustrated later in this chapter in the description of an example complex training system called the Air Combat Maneuvering Instrumentation (ACMI) system and in the discussion of training features and measures.

---

*To train is to form by instruction, discipline, or skill; to teach so as to make fit, qualified, or proficient; to make prepared for a test of skill.

14.2 Learning by Humans (See References 11–16)

The ability of human beings to learn from what they do far exceeds what can be duplicated by machines. Machines can store, sort, and compare large amounts of information fed into them, but human beings are far more capable of adapting their behavior and changing what they do (i.e., learning) on the basis of past experience (see Chapter 10 for additional discussion of human factor evaluations). Generally, human beings are much better at adapting and reacting to changing situations than machines. Human beings can learn by practicing a skill, such as playing a musical instrument, over and over until it becomes ingrained. Humans also learn through trial and error, in which they attempt various approaches to a problem and adopt what is successful (e.g., playing checkers or solving a puzzle). Finally, humans are able to transfer training from previous experience (e.g., with some explanation, a person who can fly one aircraft can generally be taught to fly another aircraft more easily than someone who hasn't flown at all).

The human process of learning optimizes performance by the recognition and use of familiar characteristics in a present situation based on past experience. Any person or system that learns must be capable of recognizing familiar characteristics in a situation, generalizing from one situation to another, and then selecting responses that optimize performance. These acts require storage (or retention) of past characteristics and, when required, the ability to recall and relate them to the situation at hand.

14.3 Transfer of Training (See References 17–23)

The principle or basis of transfer of training is the idea that there is continuity to the behavior of human beings. How people react as they experience new situations is highly dependent upon the way they have reacted in similar situations in the past, and how well satisfied they were with the results of those actions. Thus, new trainees come to each new training situation with their own data base of experiences. The trainee's data base of experience influences how he/she will perform during training. Additionally, what is experienced during training is likely to have some influence on the trainee's success. When the experiences a trainee has in one stage of training carry over to benefit the trainee during later stages of training, positive transfer of training is said to have occurred.

Generally, for positive transfer of training to occur, both the situation and the actions involved in the present learning must be similar to those of past learning, and the rate of learning is accelerated. For example, when an aircrew receives training in a flight simulator, it is desirable that positive transfer takes place and what he/she learns in the simulator carries over to the experience with the aircraft. A well-designed training program seeks to control the training environment so that maximum positive transfer of training always takes place.

On the other hand, negative transfer of training and habit interference can occur. When the present situation is similar to that of past learning but the actions required for success are different, and especially when they are the opposite of actions required in past learning, the rate of learning is retarded. In fact, this situation is extremely difficult because the trainee must "unlearn" what was previously the proper thing to do to achieve success. This condition is highly undesirable and is called negative transfer of training or habit interference. Proper design of both the operating systems and the training systems can do much to minimize these undesirable conditions.

14.4 Example Complex Training System (See References 24–32)

There are many complex training systems in use throughout the Department of Defense. These systems are used to provide training to military personnel on the operational tasks and missions for such vehicles as helicopters, tanks, aircraft, ships, submarines, and spacecraft. The ACMI system is one such system and illustrates some of the important Test and Evaluation (T&E) principles and measures associated with these types of devices. T&E is important to these systems because they must be subjected to it, and they, in turn, are sometimes used in the conduct of T&E on other training systems and weapon systems.

The original ACMI system was designed by Cubic Corporation of San Diego, California, and provides a means of improving the training for aircrews whose aircraft are equipped with the system and who fly within the coverage zone of the system. The original version of the system was developed for the U.S. Navy in the late 1960s. The system has been deployed widely throughout the world in a variety of configurations and under various names. The example system depicted in Figure 14-1 uses a combination of distance-measuring equipment, airborne sensors, and inertia-measuring equipment to provide real time measurement of position, velocity, acceleration, and attitude of the aircraft flying within its coverage. It provides a capability for training aircrews in air combat maneuvering, tactics, and procedures. Different versions of these types of systems also provide a capability to train aircrews in the recognition of air-to-air, air-to-ground, ground-to-air, and ground-to-ground weapons envelopes and weapons delivery. The training system consists of instrumentation that allows radio communications and presents real time graphic and alphanumeric displays of aircraft and weapon status data. Computer simulations to score missile firings and weapons releases can be used for training in achieving weapons firing envelopes, and for evaluations of multiple air and ground player activities. The ACMI presentations can also be recorded and reviewed later for mission task evaluation and debriefing. Data can be summarized in a variety of formats for both real time and postmission display and evaluation.

The ACMI system can be described in terms of four subsystem activities: 1) the tracking instrumentation subsystem (TIS), 2) the aircraft instrumentation subsystem (AIS), 3) the control and computational subsystem (CCS), and 4) the display and debriefing subsystem (DDS).

14.4.1 Tracking Instrumentation Subsystem

The TIS consists of an appropriate deployment of ground stations to provide coverage of a specified volume of air and ground space and number of aircraft. The ground stations are located to maintain line-of-sight on the aircraft and to allow for a multilateral solution of the player locations and parameters. Aircraft can be tracked as high activity or low activity players. The designation of an aircraft as high activity allows for better accuracy and precision during maneuvering and performance of high interest tasks.

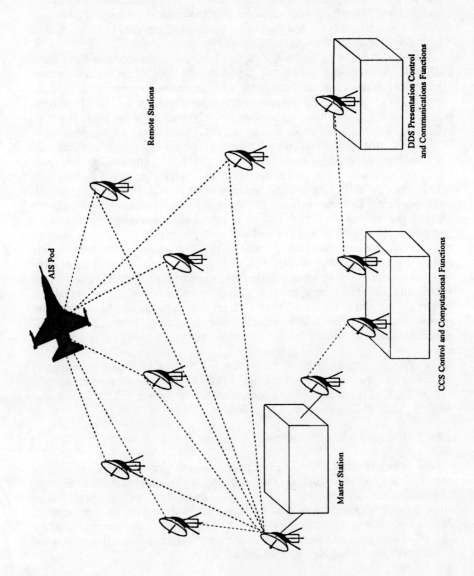

Figure 14-1. Example ACMI System

14.4.2 Aircraft Instrumentation Subsystem

The AIS is generally packaged in a pod configuration to allow it to be easily adapted to various aircraft. It operates from standard power available at the launchers of the aircraft. It is often packaged in a missile shell which is compatible with the aircraft and contains such components as a transponder, a digital interface unit, an inertial reference unit, an air data sensor unit, and a digital data link receiver and transmitter. These units measure flight data and events which are transmitted to the TIS for computation of space positioning and other essential information on all AIS equipped players. In addition to transmitting flight parameters, the AIS can monitor and transmit to the ACMI important electronic and fire control information such as switchology, gun and missile firings, and weapons release signals.

Each ground-to-air and air-to-ground transmission consists of a digital data message and ranging tones. The AIS pod receives an uplink ranging and data message from one of the ground stations designated as the interrogator and returns a downlink ranging and data message which all ground stations may receive. The downlink data message contains such parameters as aircraft attitude, velocity, pressure barometric, and missile firing data. The uplink data message contains such information as attitude and velocity corrections for updating the inertial reference unit data processor and pod identification.

14.4.3 Control and Computational Subsystem

The CCS consists of a digital computer system which computes and smoothes the position and orientation data of each aircraft, simulates missile launches and weapons releases, computes hits/misses, and serves as the computational executive for the overall ACMI system.

Attitude and velocity corrections are derived in the CCS real time filter based on the tracking data from the TIS combined with attitude, velocity, and barometric data downlinked from the AIS pod. These corrections are used to update the inertial reference unit platform state, forming a closed loop between AIS, TIS, and CCS.

14.4.4 Display and Debriefing Subsystem

The DDS consists of a variety of cathode ray tubes and large screen three-dimensional and alphanumeric displays. The DDS is used to monitor and control missions, and to replay previously recorded missions for review, analysis, and debriefing.

The ACMI system allows the performance of critical tasks to be recorded and carefully reviewed by each trainee as well as supervisors and instructors. It serves as a highly quantitative measurement system as well as an excellent feedback tool.

14.4.5 ACMI System Training Attributes

The quantitative measurements made by the ACMI are "fake-proof" and allow a trainee to progress in performance from simple basic maneuvers to highly complex two-sided engagements. The system allows for participants to be given adequate instruction in the tasks to be performed, to be objectively scored and recorded in their performance, and to repeat/review the tasks until sufficient proficiency is obtained. The interactive and feedback capabilities provided by the ACMI for participants, instructors, and supervisors are outstanding. These activities can occur at the one-on-one, few-on-few, many-on-many, and/or other appropriate training levels, and can be replayed over and over with multiple aspects for multiple training sessions.

14.5 Example Measures for Training Activities (See References 33–40)

The data collected, measurements made, and potential information fed back to the trainees, instructors, and supervisors are extremely important. Example data products and measures that have potential benefit in a variety of aircrew training situations are presented in Tables 14-1 through 14-13. Tables 14-1 through 14-10 address data/information and measures for Blue forces (i.e., friendly forces). Tables 14-11 through 14-13 present similar example data/information for enemy or opposing forces. (In this chapter only, all tables are located at the end of the chapter, after the references.) Various performance activities are identified within the mission areas and are suggested for measurement.

For discussion purposes, the data products for Blue air operations are divided into the following mission areas: air-to-surface; air-to-air; reconnaissance; electronic combat (EC); command, control, communications, and intelligence (C^3I); airlift; and a general category which includes those operations which do not fit into the other identified categories. For each mission, measures useful to each aircrew/flight commander and to each package commander are identified.* Summaries of this information are then used to provide information to the Blue Mission Commander and to the Blue Force Commander.

Data products for enemy air operations are divided into surface-to-air and air-to-air mission areas. For each mission, the measures which are useful for the aircrew, flight commander, ground defenses, and package commander are identified. Finally, useful data products for the enemy mission commander and enemy force commander are also presented. Pertinent C^3I data for the enemy forces are embedded in each area.

All of these types of data products and measures are available within the instrumentation at various testing and training ranges. One example location is the Red Flag Measurement and Debriefing System (RFMDS) located at Nellis AFB, Nevada. It is desirable that training activities have the data capture and processing ability to rapidly summarize, display, and provide hard copy results through the instrumentation systems on significant events of high interest play by individual aircraft (i.e., aircrew), flight, threat, etc. For example, the instrumentation system should be capable of rapidly providing an aircrew with a summary list of key events that occurred during his/her flight profile through the range complex, and/or list key events for an individual flight of aircraft. This same type of information should be available for each threat as well as all other players operating within the testing or training exercise. It is also important that the data/information process-

---

*A package commander is a higher level official in charge of the application of a mix of aircraft and weapons systems carrying out an attack or multiple aircraft/weapons systems operations against enemy forces.

ing system have the capability to rapidly summarize pertinent information across aircraft packages and across missions. Flexibility to sort and examine unforeseen situations using this same type of summary information is of considerable value in attempting to train personnel and to optimize the air warfare process.

When these measures are addressed, gun, missile, and other weapon system probability of kill (P_k) estimates should be appropriately degraded as a result of the presence and application of EC. It is important in the training process that kill calculations appropriately reflect countermeasures (CM), counter-countermeasures (CCM), flares, maneuvers, and other P_k degrading factors.

The addition of appropriate aircraft and threat interfaces through electronic jamming pods and/or instrumentation packages is important to reflect the critical air and ground crew switch actions and corresponding results. Interfaces are also often required to stimulate radar warning receivers and other on-board equipment for realistic training. The installation of smoke generators and/or strobe lights on pods and threat systems to signify target kills and enable more meaningful play of real time kill and removal can significantly improve the benefit of training. Automated selection/deselection of high activity instrumented players throughout a mission can lead to more accurate and improved data products. For example, the instrumentation system should be capable of automatically switching players from low to high activity and vice versa based on measured parameters and logic algorithms associated with the location and intensity of the flight activity. Minimum reliance upon a labor-intensive console operator function is highly desirable for these types of training systems. The instrumentation playback should provide for rapid identification and display of the most significant portion of each aircrew's flight profile where high activity play is of greatest interest during debriefings. Finally, it is often desirable to merge data from both notional and noninstrumented aircraft or ground players into the overall instrumented training scenario. Certain instrumentation and playback systems are capable of these actions which lead to more realistic examination of penetration and combat in heavily defended areas.

References

[1] Chapanis, A., *Research Techniques in Human Engineering*. Baltimore, Maryland: The Johns Hopkins Press, 1959.

[2] Meister, D., and Rabideau, G.F., *Human Factors Evaluation in System Development*. New York: John Wiley and Sons, Inc., 1965.

[3] Krumm, R.L., and Kirchner, W.K., "Human Factors Checklist for Test Equipment, Visual Displays and Ground Support Equipment," Report No. AFSWC-TN-56-12. Kirtland AFB, New Mexico: Air Force Special Weapons Center, 1956.

[4] Lindahl, L.G., "Movement Analysis as an Industrial Training Method," *Journal of Applied Psychology*, 29, pp. 420-436, 1945.

[5] Hatze, H., "Biomechanical Aspects of a Successful Motion Optimization," Kami P.V. (ed.), *Biomechanics V-B*. Baltimore, Maryland: University Park Press, 1976.

[6] Bekey, G.A., and Gerlough, D.L., "Simulation," in Mahol, R.E. (ed.), *Systems Engineering Handbook*. New York: McGraw-Hill, Inc., 1965.

[7] Campbell, D.T., and Stanley, J.C., *Experimental and Quasi-Experimental Designs for Research*. Chicago, Illinois: Rand McNally, 1963.

[8] Woodson, W.E., and Conover, D.W., *Human Engineering Guide for Equipment Designers* (2nd ed.). Berkeley, California: University of California Press,

1966.

[9]Cochran, W.G., and Cox, G.M., *Experimental Designs*. New York: John Wiley and Sons, Inc., 1950.

[10]Edwards, A.L., *Experimental Design in Psychological Research* (3rd ed.). New York: Holt, Rinehart, and Winston, 1968.

[11]Schmidt, R.A., *Motor Control and Learning*. Champaign, Illinois: Human Kinetics Publishers, 1982.

[12]Schendel, J.D., Shields, J.L., and Katz, M.S., *Retention of Motor Skills: Review*, (Technical Paper 313). Alexandria, Virginia: U.S. Army Research Institute for the Behavioral and Social Sciences, 1978.

[13]Barnes, R.M., *Motion and Time Study*. New York: John Wiley and Sons, Inc., 1950.

[14]Seminara, J.L., and Gerrie, J.K., "Effective Mockup Utilization by the Industrial Design-Human Factors Team," *Human Factors* 8(4), 1966, pp. 347-359.

[15]Obermayer, R.W., "Simulation, Models, and Games: Sources of Measurement," *Human Factors*, 6(6), 1964, pp. 607-619.

[16]Hurlock, R.E., and Montague, W.E., *Skill Retention and Its Implications for Navy Tasks: An Analytical Review*, (NPRDC Special Report 82-21). San Diego, California: Navy Personnel Research and Development Center, 1982.

[17]Holland, J.G., and Henson, J.B., "Transfer of Training Between Quickened and Unquickened Tracking Systems," Report No. 4703. Washington, D.C.: Naval Research Laboratory, 1956.

[18] Davis, D.R., *Pilot Error: Some Laboratory Experiments*. Cambridge, England: Applied Psychology Research Unit, Medical Research Council, 1948.

[19]Hughes, R., Brooks, R., Graham, D., Sheen, R., and Dickens, T., "Tactical Ground Attack: On the Transfer of Training from Flight Simulator to Operational Red Flag Range Exercise," *Proceedings of The Human Factors Society 26th Annual Meeting*. Santa Monica, California: Human Factors Society, 1982.

[20]Knowles, W.B., Jr., "Aerospace Simulation and Human Performance Research," *Human Factors*, 9(2), 1967, pp. 149-159.

[21]Sitterley, T.E., Zaitzeff, L.P., and Berge, W.A., *Degradation of Learned Skills—Effectiveness of Practice Methods on Visual Approach and Landing Skill Retention* (D180-15082-1). Seattle, Washington: The Boeing Aerospace Company, 1972.

[22]Chapanis, A., "Men, Machines, and Models," *American Psychologist*, 16(3), 1961, pp. 113-131.

[23]Flagle, C.D., "Simulation Techniques," Flagle, C.D., Huggins,. W.H., and Roy, R.H., (eds.), *Operations Research and Systems Engineering*. Baltimore, Maryland: The John Hopkins Press, 1960.

[24]Gum, D.R., Albery, W.B., and Basinger, J.D., *Advanced Simulation in Undergraduate Pilot Training: An Overview* (AFHRL-TR-72-59(1), AD-A030 224). Wright-Patterson AFB, Ohio: Advanced Systems Division, Air Force Human Resources Laboratory, 1975.

[25]Baker, C.A., and Grether, W.F., "Visual Presentation of Information," WADC-TR-54-160. Wright-Patterson AFB, Ohio: Aeronautical Medical Laboratory, Wright Air Development Center, 1954.

[26]White, W.J., *Acceleration and Vision*, WADC-TR-58-333. Wright-Patterson AFB, Ohio: Wright Air Development Center, 1958.

[27]Voss, J.F., "Effect of Target Brightness and Target Speed Upon Tracking Proficiency," *Journal of Experimental Psychology*, 49, 1955, 237 (222).

[28]Steedman, W.C., and Baker, C.A., "Target Size and Visual Recognition," *Human Factors*, 2(3), 1960, pp. 120.

[29]U.S. Navy Electronics Laboratory, "Suggestion for Designs of Electronic Equipment" (Rev. ed.). San Diego, California: 1960.

[30]Winer, B.J., Statistical Principles in Experimental Design. New York: McGraw-Hill, Inc., 1962.

[31]Meyers, J.L., *Fundamentals of Experimental Design*. Boston, Massachusetts: Allyn and Bacon, 1966.

[32]Natrella, M.G., *Experimental Statistics*, National Bureau of Standards Handbook 91. Washington, D.C.: U.S. Government Printing Office, 1966.

[33]Bartlett, F.C., "The Measurement of Human Skill," *Occupational Psychology*, 22, 1948, pp. 83-91.

[34]Killion, T.H., *Electronic Combat Range Training Effectiveness*. Williams AFB, Arizona: Operations Training Division, Air Force Human Resources Laboratory, 1986.

[35]Stevens, S.S., "Mathematics, Measurement, and Psychophysics," Stevens, S.S. (ed.), *Handbook of Experimental Psychology*. New York: John Wiley and Sons, Inc., 1951.

[36]Kempthorne, O., *The Design and Analysis of Experiments*. New York: John Wiley and Sons, Inc., 1952.

[37]Lindquist, E.F., *Design and Analysis of Experiments in Psychology and Education*. Boston, Massachusetts: Houghton Mifflin Co. 1953.

[38]Training Analysis and Evaluation Group (TAEG), *Effectiveness of Simulation for Air-to-Air and Air-to-Surface Weapons Delivery Training*, (Technical Note 6-83). Orlando, Florida: Training Analysis and Evaluation Group, 1983.

[39]Sitterley, T.E., and Berge, W.A., *Degradation of Learned Skills—Effectiveness of Practice Methods on Simulated Space Flight Skill Retention* (D180-15081-1). Seattle, Washington: The Boeing Aerospace Company, 1972.

[40]Churchman, C.W., Ackoff, R.L., and Arnoff, E.L., *Introduction to Operations Research*. New York: John Wiley and Sons, Inc., 1957.

Table 14-1. Example Training Measurements and Feedback Information for a Blue Force Air-to-Surface Package

Data/Information for Each Aircrew and Flight Commander	Data/Information for the Package Commander

Objectives
  Planned profile with significant events
  Actual profile with significant events
  Deviations from planned to actual profile and impact on overall mission effectiveness
Aircraft identification and IFF squawk
Time-over-target
Weapons delivered by type and BDA results for primary, secondary, and defensive air-to-air
Early warning radar information
  Target detections by source, time, and range
  Target identifications by time, range, and aircraft
  Command and Control information including time enemy threat sites were passed information
EC support aircraft impacts
Engagement summary data (surface-to-air and air-to-air)
  Threat control status and mode (e.g., autonomous, radar, optics)
  Types of threats engaged including site number
  Times of engagements
  Times engagements were complete
	Objectives
	  Planned mission strategy (i.e., timing and orchestration of force packages, etc.)
	  Actual mission execution (i.e., timing and orchestration of force package, etc.)
	  Deviations from planned to actual mission and impact on overall mission effectiveness
	Sortie summary for scheduled and actual flown
	BDA results
	  Number of weapons delivered by type
	  Air-to-surface results for primary and secondary targets including Blue losses
	  Air-to-air results including enemy and Blue losses
	Number of missed attack opportunities and reasons
	Summary of surface-to-air engagements involving air-to-surface aircraft
	  Number of Blue aircraft engaged, by threat type
	  Number of engagements where countermeasures were used
	  Types and results of countermeasures used
	  Number of countermeasure opportunities not used
	  Results of not using countermeasures
	  Results of engagements
	Number of missed engagement opportunities and reasons

(Continued on next page.)

Table 14-1. (continued) Example Training Measurements and Feedback Information for a Blue Force Air-to-Surface Package

Data/Information for Each Aircrew and Flight Commander	Data/Information for Package Commander
P_k, quality of engagement (indexed by P_k value) Results Effects of countermeasures (e.g., ECM, flares, maneuvers) employed Missed engagement opportunities including cause	Summary of air-to-air engagements involving air-to-surface aircraft Number of enemy aircraft flown Number of Blue air-to-surface aircraft engaged by enemy air and results Number of enemy air-to-air aircraft engaged by Blue air-to-surface aircraft and result Effects of countermeasures on air-to-air engagements Number of missed engagement opportunities

IFF is identification friend or foe
BDA is bomb damage assessment
P_k is probability of kill
ECM is electronic countermeasures

Table 14-2. Example Training Measurements and Feedback Information for a Blue Force Air-to-Air Package

Data/Information for Each Aircrew and Flight Commander	Data/Information for the Package Commander

Objectives
  Planned profile with significant events
  Actual profile with significant events
  Deviations from planned profile to actual profile and impact on overall mission effectiveness
Aircraft identification and IFF squawk
Type of control and source
Vectoring information if received
Position, speed, and direction of target formations
Early warning radar information
Target detections by source, time, and range
Target identifications by time, range, and aircraft
EC support aircraft impacts
RWR indication/identification
Did engagement result in advantage for Blue or enemy?
Engagement summary data
Range and relative aircraft positions at firings
Were firings observed by attacked aircraft?
Types and amounts of ordnance used
	Objectives
	  Planned mission strategy (i.e., time and orchestration of force package, etc.)
	  Actual mission execution (i.e., timing and orchestration of force package, etc.
	  Deviations from planned to actual mission and impact on overall mission effectiveness
	Sortie summary for scheduled and actual flown
	Summary of surface-to-air engagements involving Blue air-to-air aircraft
	Number of Blue air-to-air aircraft engaged, by threat type
	Number of engagements where countermeasures used
	Results of countermeasures used
	Number of countermeasure opportunities not used
	Results of not using countermeasures
	Results of engagements
	Number of missed engagement opportunities and reasons
	Summary of air-to-air engagements involving Blue air-to-air aircraft

(Continued on next page.)

MEASUREMENT OF TRAINING 283

Table 14-2. (continued) Example Training Measurements and Feedback Information for a Blue Force Air-to-Air Package

Data/Information for Each Aircrew and Flight Commander	Data/Information for the Package Commander
Were ECM/evasive actions taken by aircraft (type, time of initiation, duration)? If not, why not? Were air-to-air aircraft engaged by surface-to-air threats? If so, what were parameters of engagement? Results of engagements (with and without EC support) Missed engagement opportunities and reasons	Number of enemy aircraft flown Enemy tactics: timing, formations, corridors, targets Number and types of countermeasures used Results of firings and countermeasures Number of enemy aircraft engaged by Blue air-to-air Number and types of weapons fired Number of firings not observed by attacked aircraft Number and types of countermeasures used Results of firings and countermeasures Number of missed engagement opportunities and reasons

EC is electronic combat
RWR is radar warning receiver
ECM is electronic countermeasures

Table 14-3. Example Training Measurements and Feedback Information for a Blue Force Reconnaissance Package

Data/Information for Each Aircrew Flight Commander	Data/Information for the Package Commander

Objectives
  Planned profile with significant events
  Actual profile with significant events
  Deviations from planned to actual profile and impact on overall mission effectiveness
Aircraft identification and IFF squawk
Time-over-target
Early warning radar information
Target detections by source, time, and range
Target identifications by time, range, and aircraft
Command and Control information including times enemy threat sites were passed information
EC support aircraft impacts
Engagement summary data
  Threat control status and mode (e.g., autonomous, radar, optics)
  Types of threats engaged including site number
  Times of engagements
  Times threats fired including shot parameters
  Times engagements were complete
  P_k, quality of engagement (indexed by P_k value)
  Results
  Effects of countermeasures (e.g., ECM, flares, maneuvers) employed
Missed engagement opportunities including cause
	Objectives
	  Planned mission strategy (i.e., timing and orchestration of force package, etc.)
	  Actual mission execution (i.e., timing and orchestration of force package, etc.
	  Deviations from planned to actual mission and impact on overall mission effectiveness
	Sortie summary for scheduled and actual flown
	Summary of surface-to-air engagements involving reconnaissance aircraft
	  Number of engagements where countermeasures were used
	  Results of engagements
	  Results of countermeasures used
	  Number of engagements where countermeasures were not used
	  Results of engagements where countermeasures were not used
	  Results of not using countermeasures
	  Number of missed engagement opportunities
	Summary of air-to-air engagements involving reconnaissance aircraft
	  Number of enemy aircraft flown
	  Number of RECCE aircraft engaged by enemy air
	  Effects of countermeasures on air-to-air engagements
	  Results of engagements and countermeasures used
	  Missed countermeasure opportunities and results

(Continued on next page.)

Table 14-3. (continued) Example Training Measurements and Feedback Information for a Blue Force Reconnaissance Package

Data/Information for Each Aircrew Flight Commander	Data/Information for the Package Commander

RECCE target acquisition
  Photo RECCE
    Target positions on film for forward, low pan, and vertical cameras
    Percent of EEI taken from photo
    Number of targets per mission
  TEREC
    Number of emitters detected
    Accuracy of emitter locations
    Percent of data passed by datalink
	RECCE target acquisition
	  Photo RECCE
	    Percent of targets on film
	    Percent of EEI taken from photo
	    Number of targets per mission
	  TEREC
	    Number of emitters detected
	    Accuracy of emitter locations
	    Percent of data passed by datalink

IFF is identification friend or foe
EC is electronic combat
Pk is probability of kill
EEI is essential elements of information
RECCE is reconnaissance
TEREC is Tactical Electronic Reconnaissance

Table 14-4. Example Training Measurements and Feedback Information for a Blue Force Electronic Combat Package

Data/Information for Each Aircrew and Flight Commander

C³CM: Compass Call, Quick Fix, etc.
  Objectives
    Planned profile with significant events
    Actual profile with significant events
    Deviations from planned to actual profile and impact on overall mission effectiveness
  Aircraft identification and IFF squawk
  Time on station including start and stop times
  Jamming types and times including starting and stop times
  Early warning radar information
    Percent of messages passed successfully during nonjam periods
    Percent of messages passed successfully during jam periods
  Engagement summary data
    Total number of intercepts
    Percent autonomous operation of IADS during nonjam periods
    Percent autonomous operations of IADS during jam periods
    Number of GCI intercepts during nonjam periods
    Number of GCI intercepts during jam periods
    Time of DEF CAP during nonjam periods
    Time of DEF CAP during jam periods
    Number of autonomous attacks by enemy air during nonjam periods
    Number of autonomous attacks by enemy air during jam periods
    Number of attacks by enemy air on Compass Call and results
    Number of missed engagement opportunities caused by C³CM and reasons

Electronic Warfare Functions (EF-111 and other jammers)
  Objectives
    Planned profile with significant events
    Actual profile with significant events
    Deviations from planned to actual profile and impact on overall mission effectiveness
  Aircraft identification and IFF squawk
  Time on station including start and stop times
  Jamming type and times including start and stop times
  Early warning radar information
    Target detections (all Blue aircraft) by source, time, and range for jam and nonjam periods
    Target identifications by time, range, and aircraft for jam and nonjam periods
    Command and Control information including times passed and jamming status
    Detections prior to TOT during jam and nonjam periods
    Detection ranges for jamming periods using standoff jamming versus close in jamming
  Engagement summary data (all Blue aircraft)
    Types of threats engaged including site number
    Times of engagements
    Times threats fired including shot parameters
    Times engagements were complete

(Continued on next page.)

Table 14-4. (continued) Example Training Measurements and Feedback Information for a Blue Force Electronic Combat Package

---

Data/Information for Each Aircrew and Flight Commander

P_k, quality of engagement (indexed by P_k value)
Engagement results
Jamming types, timing, and effects
Number of attacks by enemy air on EF-111 and result
Number of missed engagement opportunities caused by EW including reasons

Electronic Warfare Functions (Self-Protection)
Objectives
Self-protection employed (e.g., jammer type and status)
Engagement summary data
    Types of threats engaged including site number
    Times of engagements
    Times threats fired including shot parameters
    Times engagements were complete
    RWR indications/identifications
    P_k, quality of engagement (indexed by P_k value)
    Engagement results
    Degradation in engagement results from self-protection tactics
        Terrain masking
        Chaff/flare
        ECM pod
        Maneuvers
    Number of missed engagement opportunities caused by self-protection jamming and reasons

SEAD Including Wild Weasel Effect
Objectives
    Planned profile with significant events
    Actual profile with significant events
    Deviations from planned to actual profile and impact on overall mission effectiveness
Aircraft identification and IFF squawk
Time on station including start and stop times
Weapons delivered by type and BDA results
    Number of radars killed by F-4G
    Number of missed opportunities by F-4G and causes
    Number of radars killed by non-Weasels with F-4G assistance
    Number of radars killed by non-Weasels without F-4G assistance
    Air-to-air BDA
Early warning radar information
    Target detections by source, time, and range
    Target identifications by time, range, and aircraft
    Command and Control information including times enemy threat sites were passed information
    EC support aircraft impacts
Engagement Summary
    Threat control status and mode (e.g., autonomous, radar, optics)

(Continued on next page.)

Table 14-4. (continued) Example Training Measurements and Feedback Information for a Blue Force Electronic Combat Package

Data/Information for Each Aircrew and Flight Commander
Types of threats engaged including site numbers
Times of engagements
Times threats fired including shot parameters
Times engagements were complete
P_k, quality of engagement (indexed by P_k value)
Results
Number of Weasels killed by IADS
Number of non-Weasels killed by IADS
Number of Weasels killed while attacking IADS
Number of nonengagements resulting from Weasel presence
Effects of countermeasures employed
Total time radar shut down because of Weasels
Missed engagement opportunities including cause

C^3CM is Command, Control, and Communications countermeasures
IFF is identification friend or foe
IADS is integrated air defense system
GCI is ground control intercept
DEF CAP is defensive captive air patrol
TOT is time-over-target
P_k is probability of kill
EW is electronic warfare
RWR is radar warning receiver
ECM is electronic countermeasures
SEAD is suppression of enemy air defenses
BDA is bomb damage assessment
EC is electronic combat

MEASUREMENT OF TRAINING

Table 14-5. Example Training Measurements and Feedback Information for a Blue Force Electronic Combat Package

Data/Information for the Electronic Combat Package Commander

Objectives
    Planned mission strategy (i.e., time and orchestration of force package, etc.)
    Acutal mission execution (i.e., timing and orchestration of force package, etc.)
    Deviations from planned to actual mission and impact on overall mission effectiveness
Sortie summary for scheduled and actual flown
Package time line including start and stop times
Jamming types and times including start and stop times
BDA results of SEAD aircraft
    Number of weapons delivered by type
    Number of radars killed by F-4G
    Number of radars killed by non-Weasels with F-4G assistance
    Number of radars killed by non-Weasels without F-4G assitance
Air-to-air results including enemy and Blue losses
Summary of surface-to-air engagements
    Number of Blue aircraft engaged, by threat type
    Number of engagements where countermeasures were used
    Results of countermeasures used
    Number of countermeasure opportunities not used
    Results of not using countermeasures
    Total time radar shut down because of Weasels
Summary of air-to-air engagements involving EC SEAD aircraft
    Number of enemy aircraft flown
    Number of Blue EC SEAD aircraft engaged by enemy air and results
    Countermeasures used and results
    Missed countermeasure opportunities and results
    Compass Call/Quick Fix, etc.
        Percent of messages passed successfully during nonjam periods
        Percent of messages passed successfully during jam periods
    EF-111 and other jammers
        Target detections (all Blue aircraft) by source, time, and range for jam and nonjam periods
        Target identifications by time, range, and aircraft
            Command and Control information including times passed and jamming status
        Detection prior to TOT during jam and nonjam periods
        Detection ranges for jamming periods using standoof jamming and close in jamming
Engagement summmary data
    Compass Call/Quick Fix, etc.
        Types of threats engaged including site number
        Times of engagements
        Times threats fired including shot parameters
        Times engagements were complete
        P_k, quality of engagements (indexed by P_k value)
        Results
        Jamming effects
        Missed opportunities due to jamming

(Continued on next page.)

Table 14-5. (continued) Example Training Measurements and Feedback Information for a Blue Force Electronic Combat Package

Data/Information for the Electronic Combat Package Commander

EF-111 and other jammers
    Number of attacks by enemy air on EF-111 and results
    Percent autonomous operations of IADS during nonjam periods
    Percent autonomous operations of IADS during jam periods
    Number of GCI intercepts during nonjam periods
    Number of GCI intercepts during jam periods
    Time in DEF CAP during nonjam periods
    Number of autonomous attacks by enemy air during nonjam periods
        Number of autonomous attacks by enemy air during jam periods
        Number of attacks by enemy air on Compass Call and results

BDA is bomb damage assessment SEAD is suppression of enemy air defenses EC is electronic combat TOT is time-over-target P_k is probability of kill IADS is integrated air defense system GCI is ground control intercept DEF CAP is defensive captive air patrol

Table 14-6. Example Training Measurements and Feedback Information for a Blue Force C³I Package (e.g., Rivet Joint, AWACS, ABCCC)

Data/Information for Each Aircrew/Flight Commander	Data/Information for the C³I Package Commander

Objectives
  Planned profile with significant events
  Actual profile with significant events
  Deviations from planned to actual profile and impact on overall mission effectiveness
Aircraft identification and IFF squawk
Time on station
Early warning radar information
Target detections by source, time, and range
Target identifications by time, range, and aircraft
EC support aircraft impacts
Engagement summary data
  Types of threats engaged including site number
  Times of engagements
  Countermeasures employed
  Results
  Missed engagement opportunities and reasons
C³I available and employed by Blue force
Impact of enemy EC on Blue C³I
Impact of enemy EW on Blue C³I
Impact of Blue EC on Blue C³I
Impact of Blue EW on Blue C³I
Number of threat calls by Rivet Joint
	Objectives
	  Planned mission strategy (i.e., timing and orchestration of force package, etc.)
	  Actual mission execution (i.e., timing and orchestration of force package, etc.)
	  Deviations from planned to actual mission and impact on overall mission effectiveness
	Sortie summary for scheduled and actual flown
	Engagement summary data
	  Number of engagements
	  Number of engagements impacted by C³I
	  Result of C³I impact
	  Number of missed engagement opportunities and reason
	C³I available, and employed by Blue force
	Impact of enemy EC on Blue C³I
	Impact of enemy EW on Blue C³I
	Impact of Blue EC on Blue C³I
	Impact of Blue EW on Blue C³I
	C³I available, and employed by enemy force
	Impact of Blue EC on enemy C³I
	Impact of Blue EW on enemy C³I

C³I is command, control, communication, and intelligence
AWACS is airborne warning and control system
ABCCC is airborne command, control, and communications

IFF is identification friend or foe
EC is electronic combat
EW is electronic warfare

Table 14-7. Example Training Measurements and Feedback Information for a Blue Force Airlift Package

Data/Information for Each Aircrew/Flight Commander	Data/Information for Airlift Package Commander
Objectives	Objectives
Planned profile with significant events	Planned mission strategy (i.e., timing and orchestration of force package, etc.)
Actual profile with significant events	Actual mission execution (i.e, timing and orchestration of force package, etc.)
Deviations from planned to actual profile and impact on overall mission effectiveness	Deviations from planned to actual mission and impact on overall mission effectiveness
Aircraft identification and IFF squawk	Sortie summary for schedule and actual flown
Computed air release point	Drop results
Mode of delivery including delivery error	Number of drops
Early warning radar information	Mode of delivery
Airlift aircraft detections by source, time, and range	Delivery accuracy (e.g., circular errors)
Airlift aircraft identifications by time, range, and aircraft	Summary of surface-to-air engagements
Command and control information including time enemy threat site was passed information	Number of Blue aircraft engaged, by threat type
Engagement summary data (surface-to-air and air-to-air)	Number of engagements where countermeasures were used
Types of threats engaged including site number	Results of countermeasures used
Times of engagements	Number of countermeasure opportunities not used
Times threats fired including shot parameters	Results of not using countermeasures
Times engagements complete	Number of missed engagement opportunities and reason
P_k, quality of engagement (indexed by P_k value)	Summary of air-to-air engagements involving airlift aircraft
Results of engagements	Number of enemy aircraft flown
Effects of countermeasures employed	Number of airlift aircraft engaged by enemy aircraft and result
Missed opportunities including cause	Effects of countermeasures
	Missed opportunities and results

IFF is identification friend or foe
P_k is probability of kill

MEASUREMENT OF TRAINING

Table 14-8. Example Training Measurements and Feedback Information for Package of Other Blue Force Mission Elements (e.g., Bomber Aircraft, Tanker Aircraft)

Data/Information for Each Aircrew/Flight Commander	Data/Information for Commander of Other Mission Elements (e.g., Bomber Aircraft, Tanker Aircraft)

Objectives
  Planned profile with significant events
  Actual profile with significant events
  Deviations from planned to actual profile and impact on overall mission effectiveness
Aircraft identification and IFF squawk
Time-over-target
Weapons delievered by type and BDA results for primary, secondary, and defensive air-to-air
Early warning radar information
Target detections by source, time, and range
Target identifications by time, range, and aircraft
Command and control information including time enemy threat sites were passed information
EC support aircraft impacts
Engagement summary data
  Types of threats engaged including site number
  Times of engagements
  Times threats fired including shot parameters
  Time engagements complete
  P_k, quality of engagement (indexed by P_k value)
Results
  Effects of countermeasures employed
  Missed opportunities including cause
	Objectives
	  Planned mission strategy (i.e., timing and orchestration of force package, etc.)
	  Actual mission execution (i.e, timing and orchestration of force package, etc.)
	  Deviations from planned to actual mission and impact on overall mission effectiveness
	Sortie summary for scheduled and actual flown
	BDA results
	  Number of weapons delivered by type
	  Air-to-surface results for primary and secondary targets
	  Air-to-air results including enemy and Blue losses
	Summary of surface-to-air engagements
	  Number of Blue aircraft engaged, by threat type
	  Number of engagements where countermeasures were used
	  Results of countermeasures used
	  Number of countermeasure opportunities not used
	  Results of not using countermeasures
	  Number of missed engagement opportunities and reasons
	Summary of air-to-air engagements by threat and aircraft type
	  Number of enemy aircraft flown
	  Number and type of aircraft engaged by enemy air and results
	  Effects of countermeasures on air-to-air engagements
	  Missed countermeasure opportunities and results

IFF is identification friend or foe
BDA is bomb damage assessment
EC is electronic combat
P_k is probability of kill

Table 14-9. Example Training Measurements and Feedback Information for a Blue Force Mission Commander

Objectives
    Planned mission strategy (i.e., time and orchestration of all force packages, etc.)
    Acutal mission execution (i.e., timing and orchestration of all force packages, etc.)
    Deviations from planned to actual mission and impact on overall mission effectiveness
Sortie summary for scheduled and actual flown
BDA results by mission package
    Number of weapons delivered by type
    Number of enemy surface-to-air assets killed by type
    Number of enemy air-to-air assets killed by type
Time-over-target summary
    Number of aircraft ± min TOT
    Percent all aircraft on time by package
    Impact of mission planning on TOT results
Summary of surface-to-air engagements by package
    Number of Blue aircraft engaged, by threat type and results
    Number of engagements where countermeasures were used
    Results of countermeasures used
    Number of countermeasure opportunities not used
    Results of not using countermeasures
    Number of missed engagement opportunities and reasons
Summary of air-to-air engagements by package
    Number of enemy aircraft flown
    Number of Blue aircraft engaged by enemy and results
    Number of enemy aircraft engaged by Blue and results
    Effects of countermeasures on air-to-air engagements
    Number of missed countermeasure opportunities and effect
    Number of missed opportunities and reason
Early warning radar information
    Number of aircraft detected by package
    Enemy C^3I information passed including Blue EC impact
Blue EC impact
    EC support aircraft impacts
    Compass Call/Quick Fix, etc., effects
    Percent of messages passed successfully during nonjam periods
    Percent of messages passed successfully during jam periods
    EF-111 and other jammers
        Percent of detections during nonjam periods
        Percent of detections during jam periods
        Detections prior to TOT during jam and nonjam periods
        Detection ranges for jam period using standoff jamming versus close in jamming
Blue C^3I effects
    Impact of enemy EC on Blue C^3I
    Number of C^3I calls passed to Blue aircraft
Airlift results
    Number of drops
    Mode of delivery
    Accuracy measurements (e.g., circular error probable)

BDA is bomb damage assessment
TOT is time-over-target
C^3I is command, control, communications, and intelligence
EC is electronic combat

Table 14-10. Example Training Measurements and Feedback Information for Blue Force Commander

Objectives
    Planned campaign/exercise strategy (i.e., timing and orchestration of all force packages, etc.)
    Actual campaign/exercise execution (i.e., timing and orchestration of all force packages, etc.)
    Deviations from planned to actual campaign/exercise and impact on overall campaign/exercise effectiveness
Sortie summary for scheduled and actual flown
BDA results by mission package for campaign and exercise
    Number of weapons delivered by type
    Number of enemy surface-to-air assets killed by type
    Number of enemy air-to-air assets killed by type
Time-over-target summary for campaign and exercise
    Number of aircraft ± min TOT
    Percent all aircraft on time by package
    Impact of mission planning on TOT results
Summary of surface-to-air engagements for campaign/exercise
    Number of Blue aircraft engaged, by threat type and results
    Number of engagements where countermeasures were used
    Results of countermeasures used
    Number of countermeasure opportunities not used
    Results of not using countermeasures
    Number of missed engagement opportunities and reason
Summary of air-to-air engagements for campaign/exercise
    Number of enemy aircraft flown
    Number of Blue aircraft engaged by enemy and results
    Number of enemy aircraft engaged by Blue and results
    Effects of countermeasures on air-to-air engagements
    Number of missed countermeasure opportunities and effect
    Number of missed engagement opportunities and reasons
Early warning radar information for campaign/exercise
    Number of aircraft detected by package
    Enemy C^3I information passed including Blue EC impact
    EC support aircraft impact
        Compass Call/Quick Fix, etc., effects
            Percent of messages passed successfully during nonjam periods
            Percent of messages passed successfully during jam periods
        EF-111 and other jammers
            Percent of detections during nonjam periods
            Percent of detections during jam periods
            Detections prior to TOT during jam and nonjam periods
            Detection ranges for jam period using standoff jamming versus close in jamming
Blue C^3I effects
    Impact of enemy EC on Blue C^3I
    Number of C^3I calls passed to Blue aircraft
Airlift results for campaign/exercise
    Number of drops
    Mode of delivery
        Accuracy measurements (e.g., circular error probable)

BDA is bomb damage assessment; TOT is time-over-target
C^3I is command, control, communications, and intelligence; EC is electronic combat

Table 14-11. Example Training Measurements and Feedback Information for Surface-to-Air Enemy Forces

Data/Information for Regimental Commander

Objectives
    Planned mission strategy (i.e., timing and orchestration of surface-to-air defense systems)
    Actual mission execution (i.e., timing and orchestration of surface-to-air defense systems)
    Deviations from planned to actual mission and impact on overall mission effectiveness
In commission summary for threat sites scheduled and actual operation
    Number of threat sites attrited by type (show time each site killed)
    Number of Blue aircraft attrited by type (show time each A/C killed)
Early warning radar information
    Total targets detected by early warning radar sites
    Total targets identified by early warning radar sites and aircraft type
    Command and control information summary including number of messages successfully sent and number unsuccessful
    Effects of Blue EC support aircraft on IADS net
Engagement summary (surface-to-air)
    Number of Blue aircraft engaged by site and type aircraft
    Number of weapons fired by threat type at type aircraft
    Number of engagements where Blue countermeasures were employed against enemy
    Results with and without countermeasures
    Number of engagements countermeasures were employed by enemy against Blue
    Results
    Number of countermeasure opportunities not employed and cause
    Number of missed engagement opportunities and reasons
Engagement summary data (SEAD aircraft) Wild Weasels
    Number of sites engaged by SEAD aircraft by type aircraft and weapon
    Number of sites attrited by SEAD aircraft (time, type threat, type aircraft, and type weapon)
    Total all number of weapons fired at sites by type
    Number of weapons fired by site at SEAD aircraft
    Number of SEAD aircraft attrited by site, type weapons, and time attrition occurred
    Number of engagements where countermeasures were employed against enemy
    Results
    Number of countermeasures employed by enemy against Blue
    Results
    Number of countermeasure opportunities not employed and cause
    Missed engagements and cause
Capability to provide a summary presentation of significant events from the above information in terms of pictorial profiles and individual threat chronological event listings

(Continued on next page.)

Table 14-11. (continued) **Example Training Measurements and Feedback Information for Surface-to-Air Enemy Forces**

Data/Information for Each Threat/Battery Commander

Objectives
    Planned operating procedures with significant events
    Actual operating procedures with significant events
    Deviations from planned procedures and impact on overall mission effectiveness
    Total time site was operational during the mission
Early warning radar information
    Target detections by source, time, and range
    Target identifications by type A/C, IFF squawk, time, and range
    Command and control information including time enemy warning information was sent and times received at threat sites
    Blue EC support A/C effects on message transmission
Engagement summary data (surface-to-air)
    Types of A/C engaged including IFF squawk
    Times of engagements
    Times threats fired including shot parameters (AZ, EL, range)
    Aircraft parameters at times of firings (speed, direction, altitude, AGL, range)
    Times engagements completed
    P_k, quality of engagement (indexed by P_k value)
    Results with and without countermeasures
    Effects of countermeasures employed by target to include times and types
    Effects of counter-countermeasures employed by threat site to include times, types
    Missed engagement opportunities including cause
Engagement summary data (SEAD aircraft) Wild Weasels
    Types of SEAD A/C engaged including IFF squawk
    Times of engagements
    Times SEAD A/C fired at sites including shot parameters (speed, directions, altitude, range) by type weapon fired
    P_k, quality of engagement (indexed by P_k value)
    Results with and without countermeasures
    Effects of countermeasures employed by threat sites to include times and types
    Missed engagement opportunities including cause
Capability to provide a summary presentation of significant events from the above information in terms of pictorial profiles and individual threat chronological event listings

---

A/C is aircraft
EC is electronic combat
IADS is integrated air defense system
SEAD is suppression of enemy air defense
IFF is identification friend or foe
AZ is azimuth
EL is elevation
AGL is above ground level
P_k is probability of kill

Table 14-12. Example Training Measurements and Feedback Information for Air-to-Air Enemy Forces

Data/Information for Mission Commander

Objectives
    Planned mission strategy (i.e., timing and orchestration of all force packages)
    Actual mission execution (i.e., timing and orchestration of all force packages)
    Deviations from planned to actual mission and impact on overall mission effectiveness
Sortie summary for scheduled and actual flown
In commission summary for threat sites scheduled and actual operational
BDA results surface-to-air
    Number of Blue aircraft killed by each threat site and type of Blue aircraft and mission
    Number of threat sites killed by Blue by time and site number
BDA results air-to-air
    Number of Blue aircraft killed by enemy air-to-air
    Number of enemy aircraft killed by Blue air-to-air
Summary of surface-to-air engagements by package
    Number of Blue aircraft engaged by threat type
    Number of engagements where countermeasures were used
    Results of countermeasures used
    Number of countermeasure opportunities not used
    Results of not using countermeasures
    Number of missed engagement opportunities and reasons
Summary of air-to-air engagements by package
    Number of enemy aircraft flown
    Number of Blue aircraft engaged by enemy and results
    Number of enemy aircraft engaged by Blue and results
    Effects of countermeasures on air-to-air engagements
    Number of missed countermeasure opportunities and effects
    Number of missed engagement opportunities and cause
Early warning radar information
    Number of aircraft detected by package
    Enemy C^3I information passed including Blue EC impact
EC support aircraft impacts
    Compass Call/Quick Fix, etc., effects
        Percent of messages passed successfully during nonjam period
        Percent of messages passed successfully during jam period
    EF-111 and other jammers
        Percent of detections during nonjam period
        Percent of detections during jam period
        Detections prior to TOT during jam and nonjam period
        Detection ranges for jam period using standoff jamming versus close in jamming
C^3I effects
    Impact of enemy EC on Blue C^3I
    Number of C^3I calls passed to Blue aircraft

(Continued on next page.)

MEASUREMENT OF TRAINING

Table 14-12. (continued) Example Training Measurements and Feedback Information for Air-to-Air Enemy Forces

Data/Information for Each Aircrew/Flight Commander

Objectives
    Planned profile with significant events
    Actual profile with significant events
    Deviations from planned profile to actual profile and impact on overall mission effectiveness
Aircraft identification and IFF squawk
Type of control and source
Vectoring information, if received
Position, speed, and direction of target formations
Early warning radar information
    Target detections by source, time, and range
    Target identifications by time, range, and aircraft
    EC support aircraft impacts
RWR indication/identification
Did engagements result in advantage for Blue or enemy?
Engagement summary data
    Range and relative aircraft positions at firings
    Were firings observed by attacked aircraft?
    Types of ordnance used
    Were ECM/evasive actions taken by aircraft (type, time of initiation, duration)? If not, why not?
    Were air-to-air aircraft engaged by surface-to-air threat? If so, what where engagement parameters?
    Results of engagements (with and without EC support)
    Missed engagement opportunities and reasons

BDA is bomb damage assessment
C^3I is command, control, communications, and intelligence
EC is electronic combat
IFF is identification, friend or foe
RWR is radar warning receiver
ECM is electronic countermeasures

Table 14-13. Example Training Measurements and Feedback Information for the Enemy Force Commander

Objectives
  Planned campaign/exercise strategy (i.e., timing and orchestration of all force packages, etc.)
  Actual mission execution (i.e., timing and orchestration of all force packages, etc.)
  Deviations from planned to actual mission and impact on overall mission effectiveness
Sortie summary for scheduled and actual flown
In commission summary for threat sites scheduled and actual operational by mission for campaign/exercise
BDA results surface-to-air by mission for campaign/exercise
Number of Blue aircraft killed by each threat site and type of Blue aircraft and mission
Number of threat sites killed by Blue, by time and site number
BDA results air-to-air by mission for campaign/exercise
Number of Blue aircraft killed by enemy air-to-air
Number of enemy aircraft killed by Blue air-to-air
Summary of surface-to-air engagements by mission for campaign/exercise
Number of Blue aircraft engaged by threat type
Number of engagements where countermeasures were used
Results of countermeasures used
Number of countermeasure opportunities not used
Results of not using countermeasures
Number of missed engagement opportunities and cause
Summary of air-to-air engagements by mission for campaign/exercise
  Number of enemy aircraft flown
  Number of Blue aircraft engaged by Red and result
  Number of enemy aircraft engaged by Blue and result
  Effects of countermeasures on air-to-air engagements
  Number of missed countermeasures opportunities and effect
  Number of missed engagement opportunities and cause
Early warning radar information for campaign/exercise
  Number of aircraft detected by package
  Red C^3I information passed including Blue EC impact
EC support aircraft impacts
  Compass Call/Quick Fix, etc., effects
    Percent of messages passed successfully during nonjam periods
    Percent of messages passed successfully during jam periods
  EF-111 and other jammers
    Percent of detections during nonjam periods
    Percent of detections during jam periods
    Detections prior to TOT during jam and nonjam periods
    Detection ranges for jam period using standoff jamming versus close in jamming
Blue C^3I effects for campaign/exercise
  Impact of enemy EC on Blue C^3I
  Number of C^3I calls passed to Blue aircraft

BDA is bomb damage assessment
C^3I is command, control, communications, and intelligence
EC is electronic combat
TOT is time-over-target

15
Joint Test and Evaluation

15.1 General

The U.S. Military Services are organized and conduct T&E in a highly specialized and functional manner. Each individual Service brings, based on its unique mission and contribution to defense, a distinct and powerful capability to the warfighting effort. Although each Service has the capability to carry out broad and impressive individual warfare, the capabilities of all of them are complementary, and they are designed and equipped to operate and fight jointly. When the U.S. military forces are being scaled down to a smaller overall force, it becomes increasingly important that the forces are operated and tested in the joint environment within which they will be employed in combat.

In 1970, the Blue Ribbon Defense Panel Study chartered by the Secretary of Defense recommended that joint testing by the Services be expanded and, since that time, joint testing has been significantly strengthened and increased. Early emphasis in joint test and evaluation (JT&E) was on operational T&E, but evolving multi-Service needs and issues led to expansion of the effort to include joint development and operational T&E. The Blue Ribbon Defense Panel Report also recommended that continuing responsibility for JT&E be vested in the Office of the Secretary of Defense (OSD), Director of Development Test and Evaluation (DDT&E).

15.2 Joint Test and Evaluation Program
(See References 1-9)

The JT&E program brings two or more Services together to formally address joint matters and resolve joint issues that cannot be as effectively addressed or resolved through independent Service actions. JT&E may also include other Department of Defense (DoD) agencies or the Joint Chiefs of Staff (JCS). The program involves JT&E nomination, selection, funding of appropriate joint costs, review of progress, providing overall guidance and direction, and incorporating the results of joint Services T&E into the proper DoD procedures and operations.

Several important agencies, groups, and committees contribute to the JT&E program. These include the OSD Director of DDT&E, the Director of Operational Test and Evaluation (DOT&E); the Joint Staff; the JT&E Planning Committee, the Senior Advisory Council (SAC); the Technical Advisory Board (TAB); the Technical Advisory Group (TAG); and the staffs of each of the Services.

The requirements for JT&E can be initiated by Congress, DoD, the unified and specified commands, or DoD components (i.e., the individual Services). JT&E is normally not directed at any particular acquisition program or to the acquisition cycle itself. JT&E is designed to: 1) assess the interoperability between Service systems in joint operations and develop solutions to identified problems; 2) evaluate and provide recommendations for improvements in joint technical and operational concepts; 3) develop and validate system development and testing methodologies having multi-Service application; and 4) evaluate technical or operational performance of interrelated/interacting systems under realistic conditions.

JT&E is conducted under conditions that are considered most appropriate and affordable for the task. To the degree possible, testing is conducted in the most realistic operational environment feasible. Specific scenario tailoring and downsizing, modeling and simulation, laboratory tests, and other forms of T&E are used as the situation dictates.

OSD delegates management responsibility for a specific JT&E program to one of the military Services or, through the JCS, to a specified or unified command. This organization (i.e., lead component or lead Service) nominates a Joint Test Director who manages the program. The lead Service and other DoD components furnish resources to support the JT&E as necessary.

15.2.1 Joint Test and Evaluation Nomination and Joint Feasibility Study Approval

The formal structure of the JT&E nomination and joint feasibility study (JFS) approval process is depicted in Figure 15-1. This process describes the key events that take place up through the chartering of the Joint Test Force and the test. Key steps include nomination, review and approval of the nomination for feasibility study, and review and approval of the feasibility study which generally leads to chartering of the JT&E.

Nominations for JT&E are requested by OSD and are received by approximately mid-April of each year. Nominations may originate within any one of the Services, or the DoD. Each of the Services has its own process for originating and nominating joint tests. In May, the nominations are reviewed by the JT&E Planning Committee to assess their jointness, criticality in terms of defense capabilities and posture, and potential for resolution/development of a solution through a JT&E effort. These nominations are also reviewed from a technical perspective by the TAB. The nominations are rank ordered by the Planning Committee and the more significant issues are forwarded to the SAC in early June.

The SAC is composed of senior general officers or civilian equivalents from the Services, OSD, and the Joint Staff, and is co-chaired by the DDT&E and DOT&E. It reviews the nominations from the Planning Committee, considers the technical inputs from the TAB, assigns priorities to the nominations, and recommends to DDT&E and DOT&E those which should proceed to a JFS. The SAC's recommendations serve as a commitment by the Services to provide the necessary points of contact, resources, and support to conduct the JFS.

The DDT&E and DOT&E screen all JT&E nominations, consider the recommendations of the JT&E PC and the SAC, and select those to be carried forward to feasibility study. The SAC's recommendations are reviewed by DDT&E and DOT&E which, in turn, direct the Services to provide a plan of action and milestones (POA&M), including personnel and funding requirements for the conduct of the selected JFSs. After review of each POA&M, the DDT&E and DOT&E approve selected JFSs to be conducted; provide appropriate funding; and, based

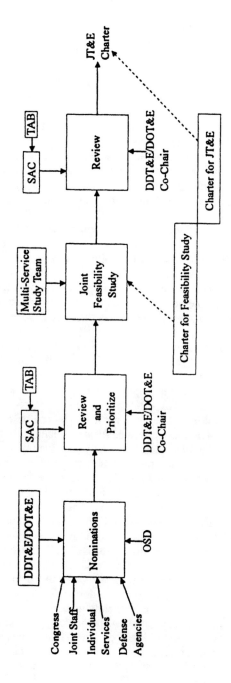

Figure 15-1. JT&E Nomination and JFS Approval Process

on the recommendations of the SAC, designate the lead and participating Services.

The lead Service nominates a JFS Director, and each participating Service nominates a Deputy Study Director. These individuals must be approved by OSD. If OSD disapproves a given Service nomination, another candidate must be submitted for OSD approval.

The JFS normally takes approximately one year to complete and is a major step leading to the chartering of a JT&E. It initially provides a preliminary program test design and an assessment of feasibility. After approval by the DDT&E of the preliminary study effort, the JFS team prepares a full-scale Program Test Design that contains details regarding the tasks to be accomplished and a proposed methodology. The methodology includes the objectives and measures of effectiveness that must be satisfied to fully address each task, the recommended analytical approach, and testing that must be accomplished. The proposed JT&E schedule, budget, and Joint Test Force organization are also included in the JFS products.

The JFS efforts are reviewed periodically by both DoD management and the TAB to increase the chances of producing a successful study that leads to the chartering of the JT&E. When necessary, the Joint Test Force can also request the assistance of a TAG for in-depth scientific and technical input. The initial review of the JFS effort usually concentrates on the objectives to be achieved, the measures to be applied, and the methodology for getting the JT&E accomplished. Once these important initial concepts are formulated, the more detailed development of the study can proceed.

Generally, the problems addressed in a JFS are broad issues that require a multidisciplinary approach from a Joint Service perspective. Many of these issues are technical in nature and require a high level of scientific and technical expertise. Thus, the JFS team normally requires operational, engineering, analytical, logistical, financial, and other areas of expertise.

Because of their complexity and costs, most JT&Es can best be addressed by some combination of physical testing and analytical modeling and simulation. Some success in cost reduction has also been achieved by conducting the JT&E as a part of or immediately after large-scale training exercises conducted by the Services.

Although a JFS conducted by a Joint Test Force is the preferred method for establishing the need and feasibility of a JT&E, special circumstances sometimes dictate the need for alternative methods of producing the feasibility study. When the alternative methods are employed, the affected Services must be involved in the study and the results must be reviewed and approved by the SAC and the DDT&E. These alternative methods include the DDT&E electing to find an existing joint activity or committee to conduct the feasibility study, employing a Federally Funded Research and Development Center or an independent contractor to conduct the study and produce the program test design document for the Joint Test Force to execute, and chartering a JT&E directly without first conducting a feasibility study. When any of the alternative methods are applied, it is essential that the Services required to carry out the T&E be intimately involved in the process to ensure that the study and plan produced will have maximum value to the Services.

Sometimes, a General Officer Steering Group is set up to provide operational input to the JT&E and to ensure rapid feedback of the JT&E findings into the appropriate organizations within each service and the Joint Staff. When this is done, the General Officer Steering Committee is chaired by an officer from the lead Service with other participating Services represented on the General Officer Group.

15.2.2 Joint Feasibility Study Review and Chartering of the Joint Test Force

Once a feasibility study is completed, the results, recommendations, and program test design are coordinated within the involved Services. During the second week in May, the study results and the program test design are briefed to a combined meeting of the Planning Committee and the TAB. The Planning Committee and the TAB then make separate recommendations to the SAC on appropriate issues and which JT&Es should be chartered for testing.

The SAC takes into consideration the recommendations of the Planning Committee and the TAB and makes its own recommendations to the DDT&E and DOT&E on which JT&Es should be chartered for testing. The DDT&E and DOT&E, in turn, select the JT&Es to be chartered and appoint the lead Service for each selected Joint Test Force. The chartering of new JT&Es normally occurs in October when the fiscal year funding is authorized by Congress.

15.3 Activating and Managing the Joint Test Force (See References 10–17)

The Joint Test Force resources required to effectively conduct a JT&E are made up of two major parts: 1) the Joint Test Force staff, and 2) the test support force from the individual services required to support and execute the joint tasks and missions. The Joint Test Force staff carries out the overall planning, management, analysis, evaluation, and reporting activities of the test. The test support force is made up of individuals, units, and equipment from the Services needed to conduct the test. The test support force is not assigned to the Joint Test Director staff, but provides the support as tasked and agreed on in the planning documents for the test.

Ideally, the feasibility study is conducted by a Joint Test Force, which continues on with the important efforts after the JT&E has been chartered. When this is the case, many of the highly experienced personnel who worked on the feasibility study are available to help finalize the design, develop the test plan, execute the test, analyze the results, and complete the final reports and briefings.

One of the first and important actions after a JT&E has been chartered is to select appropriate highly qualified personnel. The Joint Test Director from the lead Service and the Deputy Test Directors from the participating Services are nominated by the Services and approved by OSD.

It is extremely important that these individuals bring the proper amount and mix of experience to the leadership of the JT&E. These individuals should be selected on the basis of Joint Service experience, operational experience in the area under test, functional experience, and past performance and accomplishments in their specialized field of endeavor.

It is also important to obtain personnel for the Joint Test Force who are experienced in planning and directing operational test and evaluation and who are knowledgeable about important procedures in carrying out T&E. Personnel who have served as test directors and managers in Service operational test organizations are often ideal candidates to serve on the Joint Test Force.

The Joint Test Force must contain a sufficient number of operational and support experts who are knowledgeable about the employment doctrine and operational procedures of the systems and equipment involved in the JT&E. These individuals should be experts in the missions and tasks being performed and are required to obtain operational realism in the planning and execution of the JT&E.

Intelligence experts who are knowledgeable about potential enemy operations and systems also serve as key members of the Joint Test Force. Realistic T&E requires that both the threats and operational environments be representative of the combat situation.

Systems analysts and operations research analysts are an important part of the JT&E team. These individuals have the education and background to address broad problems as well as highly specialized JT&E issues. They should be used to help define the problems and issues to address, formalize the T&E design, analyze the data, and draw conclusions from the test results. Operations research analysts also know how to develop and use analytical models and simulations in problem solutions. Their objectivity in the JT&E effort is extremely important.

Engineers, technicians, and other scientific experts provide a valuable addition to the Joint Service test teams. The required number and types of these experts are situation-dependent. Many complex JT&Es are extremely technical in nature and must have sufficient scientific and technical support. Experts in weapon systems and in instrumentation systems are highly valuable members of the team.

Other types of key staff members on the Joint Test Force include resource management and security specialists. These individuals are essential to ensure the appropriate handling of such things as supply/logistics, personnel, administration, budget and fiscal matters, physical and operational security, and general administrative matters.

15.3.1 Command Relationship

Joint Test Forces have been organized in various structures under the DDT&E, DOT&E, and the Services to accomplish their charter. One example organization that has worked extremely well is depicted in Figure 15-2. The Joint Test Director reports directly to DDT&E, and the Deputy Test Directors report to the Joint Test Director. A General Officer Steering Group as discussed earlier is formed by the lead Service with members from the participating services. Although not directly in the Joint Test Force chain of command, this Group plays a highly valuable role in the test. It makes sure that the Joint Test Force is operating in accordance with Service doctrine and procedures, that the appropriate test units are made available by the Services, and that all important findings of the test are immediately implemented by the participating Services. In short, the General Officer Steering Group is responsible for making sure the maximum benefit is derived for the Services from the resources expended in the test.

Administratively, the JT&E is conducted according to the procedures of the lead Service. The Joint Test Director is responsible to OSD for achieving the objectives on schedule and exercises the day-to-day management over the test resources. Personnel rating arrangements and policies are dealt with on an individual basis, unless covered by standing memorandums of agreement. Normally,

JOINT TEST AND EVALUATION

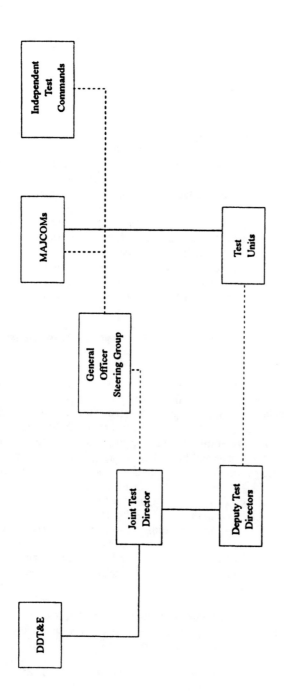

DDT&E is Director Development Test and Evaluation

MAJCOMs is Major Commands

Figure 15-2. Example Command Relationship for a Joint Test and Evaluation

the Deputy Test Directors from the various Services serve as the point of contact for the test support that must be provided by the individual Services. When a test is set up in a different way, the relationship between the Deputy Test Directors from the Services and the support forces is normally spelled out in planning documents before the testing commences.

15.3.2 Example Joint Test Force Organization

Past Joint Test Forces have been organized in a variety of ways. Some have been organized along the lines of staff organizations within their Services. Others have been organized along the lines of major functions that the organization must accomplish. Figure 15-3 is a simple example organizational structure found in many past Joint Test Forces. The operations organization is designed to have the expertise on leading and carrying out the planning and accomplishment of the test. Generally, projects and tasks within the JT&E are addressed and achieved by a team consisting of appropriate individuals from throughout the Joint Test Force. The analysis and reports organization normally is assigned the operations analysts and other key individuals necessary to collect the data, process it, analyze the results, and help complete the reports and briefings on the JT&E findings. The support organization normally provides the engineering and technical support, and the administrative support necessary for the organization to carry out the JT&E.

15.4 Developing the Joint Test and Evaluation Resource Requirements

Estimates of the JT&E resource requirements are made very early in programs (i.e., during the feasibility study) and are refined throughout the planning period. The lead Service documents the resources required for all the Services in the JT&E consolidated resource estimate. Each Major Command (MAJCOM) and other organizations that support a joint test must take the appropriate action to program the resources identified in the Command budget exercises. These requirements are reviewed by each participating Service to determine and document their impact on the mission of the MAJCOMs and other organizations.

The funding requirements for JT&E are shared among OSD, the Services, and other JCS or DoD agencies involved in the test. The funds provided by OSD are primarily those associated with the joint and unique costs of the test. These costs include such unique activities as the feasibility study; test design and planning support; the development, procurement, installation, and operation of special instrumentation; transportation, travel, and per diem for the Joint Test Director and staff; modification of test articles; transporting test support equipment; contractor support for the Joint Test Force staff; test unique training; cost for use of unique test facilities and ranges; and data collection, reduction analysis, and reporting. Other costs not specifically delineated in existing guidance, including facility related costs, are negotiated between OSD and the Services.

15.5 Preparing the Test Design (See References 18–26)

Once a feasibility study has been coordinated within the Services and approved

JOINT TEST AND EVALUATION

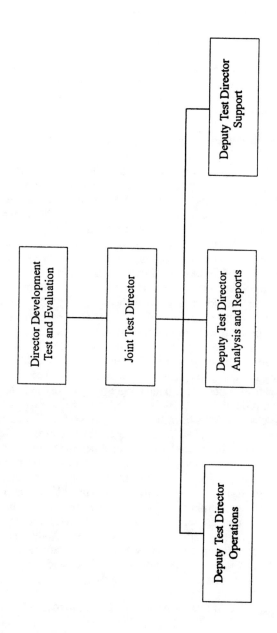

Figure 15-3. Example Joint Test Force Organization

by OSD for testing, a program test design must be developed. Historically, a support agent (i.e., generally a Federally Funded Research and Development Center) has been tasked to develop the program test design for the Joint Test Force. In some cases, the Joint Test Force has selected and tasked a support contractor directly to develop a program test design. Both of these methods require the Joint Test Force to be intimately involved in this process if an acceptable test design is to be produced. Although these support organizations have highly qualified technical personnel and can produce an excellent theoretical design, they lack the operational and Service experience to produce a meaningful JT&E design without major Service input and review.

15.6 Preparing the Test Plan

One of the most important JT&E documents is the program test plan. The program test plan helps each participating MAJCOM within the Services to know how it will benefit from the test and the specific cost of the test to each MAJCOM in terms of resources. This document depicts how the Joint Test Force will execute testing to accomplish the objectives of the JT&E. The plan includes resources; schedules; data collection, procedures, summarization, and analysis; test concept; test objectives; measures of effectiveness; and the reporting and briefing requirements. It should provide the overall framework to allow detail scheduling and identification of resource requirements and acquisition of those resources.

The program test plan should specify the methodology for the entire problem-solving effort. It should include the composition of the problem-solving team, the schedule of the major problem-solving actions, the scope of the research efforts to be conducted, the data sources to be searched, the general method for compiling the collected information, the composition of the information collection team, the method for acquiring the computer data base, the major resources required, the contribution of each effort to accomplish the program objectives, and the type and format of each output product.

The program test plan specifies how the test will be conducted (i.e., when and where testing will occur), what units will participate, the test scenarios, the general method of accomplishment, key data elements and collection techniques, the major resource requirements from each participating unit, the composition of each unit test team (during planning, execution, and reporting), and the expected contribution of each individual test to achieve the overall program objectives. Examples of key parts of the program test plan are the program design matrix, the program event schedule, the individual activity/test plans, the program level data management plan, and the budget and financial management plan. Additional planning activity includes such areas as security, safety, and administration. The program test plan is the "road map" for getting the test accomplished and it must be agreed to and coordinated by the participating Services, and approved by OSD.

15.7 Conducting the Joint Test and Evaluation

Conducting a successful large-scale JT&E is a challenge to OSD and the Services. The goal of getting the test planned, conducted, and reported in a reasonable time frame historically has been difficult to achieve. The Joint Test Director and his staff must maintain a disciplined approach to carrying out the approved program test plan and schedule. Additionally, they must be prepared to make judicious adjustments in test activities that are at variance with the approved plan but are necessary to successfully acquire the essential data and information to accomplish the program objectives. Test execution is normally a fast-paced and dynamic phase of the test. The Joint Test Director and his staff must understand and be committed to the concept of structured testing, have a complete grasp of the critical forces and systems needed to accomplish the test objectives, and have an understanding of the operations of the test facilities and ranges being used in the test. Continual communication and coordination among the many organizations and activities necessary to complete the test objectives are demanding and imperative. The Joint Test Force Director and his staff must continually get all of the involved activities to work together and to make the decisions and adjustments essential to carrying out a successful test.

One method that historically has helped in the successful conduct of JT&E is to make sure that the Joint Test Team and the supporting forces are trained and well prepared for their specific roles in the test. These activities include the military operators, instrumentation and equipment engineers, data collectors, computer data base operators, analysts, and report writers. Pretest preparation and mission rehearsal by the total Joint Test Force can often significantly reduce problems that occur during test conduct. These rehearsal activities can actually shorten the time and resources required for testing by increasing the success rate of collecting valid data and accomplishing the necessary number of trials for the test.

15.8 Joint Test and Evaluation Analysis and Reporting (See References 27–42)

Analysis and reporting are activities that continue throughout the life of the JT&E. Early analysis is conducted to verify that the planned test events are feasible to accomplish, and have a high likelihood of meeting, the objectives of the test. Analyses are conducted to refine and improve the test data collection, validation, and summarization processes; to assist in establishing realistic scenarios and test layouts; to investigate the impacts of changes from the expected tests and conditions; and to design quick look test procedures. Analyses cover a wide variety of fields and require many areas of expertise such as operations research analysis, engineering analysis, instrumentation analysis, financial analysis, and other forms of program analysis. Analytical modeling and simulation and special test facilities are always applied in some form to lessen the cost and burden of large-scale force-on-force testing.

Quick look analyses allow the Joint Test Force Director and staff to monitor and make adjustments to ensure that data collection goals are being successfully achieved. The data are reviewed during testing to confirm their appropriateness and quality, the expected quantities, and the performance of the collection systems. The goal of this effort is to provide timely feedback to the Joint Test Director so that equipment can be recalibrated or data collectors can be advised of changes to make in their data collection methods, and adjustments can be made to correct problems before the test has ended and too much data are lost.

Posttest analyses are conducted to ensure that the captured valid data are successfully processed into the analytical data elements essential to support the program level analysis. It is extremely important that test event data and information be valid, reconstructed, correlated, and put into the proper formats for later program level analysis. Validation analyses review the data to ensure that they were collected under the conditions specified in the test plan. Reconstruction analyses are often required to make corrections in the data by reconstructing what actually happened during the test trial. These corrections are sometimes required because of test range equipment failure, calibration corrections, etc. Both primary and backup data sources are often required to provide a complete data package for the program level analysis. The posttest analysis should also identify and explain any anomalies that occur in the data that could not be removed and that may affect later program level analyses.

Once the data are validated and put into the proper formats, they are entered into the program level analyses data bases. These data bases are used to support the analyses that address the specific objective of the JT&E. In complex analyses, it is sometimes necessary to combine the results of the statistical and operational analysis of the data with the results of other data generating activities such as computer simulations and laboratory tests. It is imperative that each analysis and data presentation clearly identify the basis and sources of data and information being used to support it. For example, the operational quality and credibility of results from a properly conducted field/flight test are different from those based solely on laboratory testing and analytical modeling and simulation. It is important that the decision maker and other users of this information know the origin of the results.

JT&E produces many levels and types of reports. These reports must contain data that are fundamental to the resolution of high level Joint Service problems. They should be carefully orchestrated and provide an effective presentation of the essential information to support action by decision makers. The audience for these reports ranges from the operators and supporters within the individual Services to the Secretary of Defense and Congress.

JT&E test reports are typically due 90 days after completion of a test and should describe the detailed results obtained for each objective as defined in the test plan, and the technical parameters of the systems and conditions at the time of the test. The reports should provide detailed explanations of any deviations from the planned activities, to include the reasons for the deviations and the impact on the data that were collected as a result of the deviations. The reports should identify all objectives initially stated in the test plan that were addressed during the test, and provide an explanation/description of the reasons for any objectives that were not addressed.

Test reports normally document all operational aspects of the test, to include the environmental conditions; terrain; weather; scenarios; participating units; equipment; and the state of training, data collection, and management methods. Sufficient data and information should also be provided for individual users to arrive at informed judgments on how the test results relate to their situations of interest. In some cases, the individual Services have performed independent analyses and reporting based on the data collected during JT&E. The test reports should strive to document precisely what occurred during the test, with detailed explanation of successes, failures, and conditions of the overall planned events. The presentations should be clear and easily understood, and all conclusions and recommendations should be objectively reached based on the valid data contained in the reports.

15.9 Disposition of Joint Test and Evaluation Resources

Once a JT&E is completed, the posttest disposition of the test instrumentation, threat simulators, facilities, and administrative support equipment acquired by OSD for the joint test is negotiated between OSD and the Services. Often, this special equipment contributes significantly to other efforts within the Services long after the joint tests have been disestablished.

15.10 Example Joint Test and Evaluation Programs

Joint Service Testing and Evaluation sponsored by the DDT&E began in about 1970 and since that time has resulted in over fifteen tests being completed. These JT&Es have contributed greatly to the interservice operations and the ability to conduct joint warfighting efforts. The following JT&Es depict the broad range of issues and problems investigated in Joint Service T&E activities.

15.10.1 Joint Test and Evaluation of Combat Air Support Target Acquisition (SEEKVAL)

The SEEKVAL series of T&Es were conducted during the mid-1970s to address direct and aided visual acquisition of combat targets by fixed- and rotary-wing aircraft. A series of experiments were conducted to validate techniques and instrumentation for recording and measuring characteristics of target and background scenes during airborne target acquisition flight tests. The overall objective was to evaluate alternative systems and techniques for acquiring targets in combat air support. Imagery was collected on targets with background and topographical conditions representative of a potential combat air support situation in Central Europe. A part of the T&E involved the use of the range measuring system (RMS-2) to vector aircraft over predetermined flight paths to gather information on target acquisition. The T&E also applied modeling and simulation to investigate factors that affect the target acquisition of electro-optical and other sensor systems.

15.10.2 Joint Test and Evaluation for Validation of Simulation Models (HITVAL)

The joint T&E for HITVAL was conducted during the early 1970s to address the validation of simulation models used in assessing the effectiveness of ground-to-air gun firings. These tests highlighted the difficulty that can be experienced with instrumentation and validation of computer models, as well as the importance of conducting an effective feasibility study prior to the commitment to physical testing.

15.10.3 Joint Test and Evaluation for Electronic Warfare Joint Test (EWJT)

The joint T&E for EWJT was conducted starting in approximately 1974 and addressed a large-scale field/flight test of joint electronic warfare activities. The primary method of investigation was field/flight testing making use of the range measurement system (RMS) for instrumentation. It examined various scenarios, jamming options, and threat penetration tactics.

15.10.4 Joint Test for A-7/A-10 Fly-Off

This joint T&E was conducted in approximately 1974 to assist the U.S. Air Force in selecting the most appropriate aircraft for the close air support mission. Realistic missions and scenarios were examined by pilots flying both aircraft under simulated combat and environmental conditions.

15.10.5 Joint Test and Evaluation of Tactical Aircraft Effectiveness and Survivability in Close Air Support Antiarmor Operations (TASVAL)

The joint T&E for TASVAL was initiated in approximately 1977 and addressed the effectiveness and survivability of both fixed- and rotary-wing aircraft operating in the close air support mission. A primary concern of the test was joint attack of armor targets. Field/flight testing and analytical modeling and simulation were the principal methods of the investigation.

15.10.6 Joint Test and Evaluation for Data Link Vulnerability Analysis (DVAL)

The joint T&E for DVAL was conducted from approximately 1979 through 1984 to address the vulnerability of radio frequency data links to jamming in a hostile electronic countermeasures environment. The T&E focused on the identification and quantification of problems and potential problems with data link communications.

15.10.7 Joint Test and Evaluation for Identification Friend, Foe, or Neutral (IFFN)

The joint IFFN T&E was conducted starting in approximately 1979 to assess the baseline capabilities of U.S. forces within the North Atlantic Treaty Organization (NATO) air defense command and control system to perform the IFFN function. The identification of targets and the employment of beyond visual range weapons was of interest. Both direct and indirect forms of identification were investigated. The joint T&E helped identify deficiencies in the performance of the IFFN function and proposed procedural and equipment modifications. The extensive test bed developed for the IFFN test has undergone numerous upgrades and continues to serve as an important tool in addressing target identification as well as other command and control issues.

15.10.8 Joint Test and Evaluation for Forward Area Air Defense (JFAAD)

The JFAAD T&E was initiated in the late 1970s to address the air defense of forward areas. These areas are characterized by the potential over flight of both friendly and enemy aircraft.

15.10.9 Joint Test and Evaluation for Command, Control, and Communications Countermeasures (C^3CM)

The joint T&E for C^3CM was a seven-year effort starting in approximately 1982 that addressed both the offensive and defensive aspects of C^3CM. It provided information that could be applied by combat commanders to improve the effectiveness of their forces by disrupting the enemy C^3 system. The test consisted of a combination of field tests, command post exercises, engineering studies, and analytical modeling and simulation.

15.10.10 Joint Test and Evaluation for Electromagnetic Interference (JEMI)

The JEMI T&E was conducted beginning in 1988 to identify and quantify the effects of electromagnetic interference (EMI) on friendly electronic systems employed in Joint Service operations. The T&E also developed and validated an improved methodology to identify, characterize, and prioritize potential joint EMI interactions and identify and develop solutions to reduce the effects of interference. The methodology identified potential EMI problems using results from field/flight operations, historical data bases, and analytical modeling and simulation. Expertise in operations, engineering, and testing was used to characterize and prioritize the potential EMI problems. Results were used by the Services to resolve potential EMI during the development acquisition and application of systems. Also, engineering fixes and operational workarounds were devised and employed. A combination of simulation, analysis, and field/flight testing was employed. The JEMI methodology developed during the T&E provided a tool that can be applied to maximize the potential of electronic systems and minimize their limitations in Joint Service operations.

15.10.11 Joint Test and Evaluation for Over-the-Horizon Targeting (JOTH-T)

The JOTH-T T&E was conducted in the late 1980s and addressed concepts and systems to share data and target information among various Service systems over large areas.

15.10.12 Joint Test and Evaluation of Camouflage, Concealment, and Deception (JCCD)

The JCCD T&E began in the 1990s to address the effectiveness of camouflage, concealment, and deception (CCD). It is designed to provide guidance to the Services for CCD application to enhance the survivability of critical fixed and relocatable assets against conventional manned aerial threats to include precision guided weapons. The JCCD is also analyzing the impact of CCD on target imagery used for intelligence analysis. A wide range of physical testing and analytical methods are being applied.

15.10.13 Joint Test and Evaluation for Air Defense Operations/Joint Engagement Zone (JADO/JEZ)

The JADO/JEZ T&E, begun in the 1990s, is investigating and evaluating a new concept for Joint Service air defense operations. The major thrust of the concept is to operate friendly surface-to-air missiles and fighter aircraft within the same airspace and engagement zones. Ideally, the air defense weapons of all the Services would be synergistically operated within the same airspace with standard plans and employment doctrine. With proper identification and command and control, this concept will allow friendly firepower to be efficiently applied in areas of concentrated enemy attack. Interoperability and command and control connectivity issues are being addressed. Computer models, simulations, and live field/flight tests are being used to examine the issues.

15.10.14 Joint Test and Evaluation for Logistics over the Shore (JLOTS)

The JLOTS T&E covers a long series of tests conducted beginning in approximately 1970 to address joint logistics over the shore operations. These tests were designed to aid in the development and demonstration of the container handling capability of assault follow-on echelon and logistics over the shore operations. The concept investigates the difficulties involved in moving military cargo across a shoreline without minimum port facilities. These tests employ a wide range of capabilities to include "roll-on/roll-off" and are collecting information to validate and refine operational techniques, planning factors, personnel, and equipment requirements. The T&E data also assist in determining deployment force structures and sea lift requirements.

15.10.15 Joint Test and Evaluation for Band IV (Infrared Countermeasures)

The joint T&E for Band IV (IR countermeasures) is an ongoing activity designed to increase the survivability of U.S. aircraft by providing countermeasures and counter-countermeasures against potential enemy threat systems. Testing includes both fixed- and rotary-wing aircraft and affects tactics, training, the development of countermeasures, and appropriate computer models and simulations.

15.10.16 Joint Test and Evaluation for Tactical Missile Signatures (JTAMS)

The JTAMS T&E is an ongoing activity directed at establishing improved methods and standards for acquiring and disseminating information on tactical missile signatures. The Joint T&E is addressing the processes and standards to institutionalize engineering practices for the collection, characterization, storage, and dissemination of missile signature data.

15.10.17 Joint Crisis Action Test and Evaluation (JCATE)

The JCATE is an ongoing activity to evaluate new Command, Control, Communications, Computers, and Intelligence (C4I) technologies and concepts. Information exchange and human decision processes are an important part of this investigation.

15.10.18 Joint Test and Evaluation for Smart Weapons Operability Enhancement (SWOE) Process

The SWOE joint T&E is an ongoing activity to provide a validated information and scene generation process for battlefield scenarios. Analytical modeling and simulation and field/flight testing are being applied in the investigation.

15.10.19 Joint Advanced Distributed Simulation (JADS)

The JADS T&E was chartered in 1994 and is designed to develop and test advanced distributed simulation (ADS) methodologies as they apply to T&E. The T&E will focus on how well ADS methodologies work in actual practice and how they can serve T&E in the future. Alternative cases will be examined to test and demonstrate the benefits of the technologies.

15.10.20 Joint Theater Missile Defense (JTMD)

The JTMD T&E was also chartered in 1994 and will address the problem of locating and destroying enemy ballistic and cruise missile targets. The focus will be on the testing of near-term systems and capabilities to locate, attack, and destroy mobile theater missile elements, missile stocks, and infrastructure.

References

[1] DoD Directive 5000.1, "Defense Acquisition," Office of the Secretary of Defense, Under Secretary of Defense for Acquisition, 23 February 1991.

[2] DoD Instruction 5000.2, "Defense Acquisition Management Policies and Procedures," Office of the Secretary of Defense, Under Secretary of Defense for Acquisition, 23 February 1991.

[3] DoD Manual 5000.2-M, "Defense Acquisition Management Documentation and Reports," Office of the Secretary of Defense, Under Secretary of Defense for Acquisition, 23 February 1991.

[4] DoDD 5000.3, Test and Evaluation. Establishes policy and guidance for the conduct of test and evaluation in the Department of Defense, provides guidance for the preparation and submission of a Test and Evaluation Master Plan and test reports, and outlines the respective responsibilities of the Director, Operational Test and Evaluation (DOT&E), and Deputy Director, Defense Research and Engineering (Test and Evaluation (DDDR&E (T&E)).

[5] *Joint Test and Evaluation Handbook.* Washington, D.C.: Pentagon, Office of the Director, Development Test and Evaluation, May 1992.

[6] DoD 5000.3-M-4 *Joint Test and Evaluation Procedures Manual.* Washington, D.C.: Pentagon, Office of Deputy Director Defense Research and Engineering (Test and Evaluation), August 1988.

[7] *DoD Budget Guidance Manual.* DoD 7110-1-M. Provides guidance on costs for Joint Tests.

[8] DoD 5010-12-L, *Acquisition Management Systems and Data Requirements*

Control List (AMSDL).

[9]DoD STD-963, *Data Item Descriptions (DID), Preparation of.*

[10]Air Force Regulation 80-20, Research and Development, Managing the Joint Test and Evaluation Programs, 22 August 1984.

[11]Meredith, J. R., and Mantel, S. J., Jr., *Project Management: A Manageria Approach* (2nd ed.). New York: John Wiley and Sons, Inc., 1989

[12]Kezsborn, D. S., Schilling, D. L., and Edward, K. A., *Dynamic Project - Management: A Practical Guide for Managers and Engineers.* New York: John W ley and Sons, Inc., 1989.

[13]AFR 310-1, *Management of Contractor Data.*

[14]AFSCR 310-1, *Management of Contractor Data.*

[15]AFLCR 310-1, *Management of Contractor Data.*

[16]Buffa, E. S., *Modern Production Management* (4th ed.). New York: John Wiley and Sons, Inc., 1973.

[17]Defense Systems Management College, *Systems Engineering Management Guide*, 1982.

[18]Chestnut, H., *Systems Engineering Methods.* New York: John Wiley and Sons, Inc., 1967.

[19]Cochran, W. G., and Cox, G. M., *Experimental Designs.* New York: John Wiley and Sons, Inc., 1957.

[20]Fisher, R. A., *The Design of Experiments.* New York: Hafner Publishing, 1960.

[21]Kempthorne, O., *Design and Analysis of Experiments.* New York: John Wiley and Sons, Inc., 1962.

[22]Chew, V. (ed.), *Experimental Designs in Industry.* New York: John Wiley and Sons, Inc., 1958.

[23]Parizan, E., *Modern Probability Theory and Its Applications.* New York: John Wiley and Sons, Inc., 1960.

[24]Wilson, E. B., Jr., *An Introduction to Scientific Research.* New York: McGraw-Hill, 1952.

[25]Biles, W. E., and Swain, J. J., *Optimization and Industrial Experimentation.* New York: John Wiley and Sons, Inc., 1980.

[26]Box, G. E. P., Hunter, W. G., and Hunter, J. S., *Statistics for Experimenters: An Introduction to Design Data Analysis and Model Building.* New York: John Wiley and Sons, Inc., 1978.

[27]Devore, J. L., *Probability and Statistics for Engineering and the Sciences.* Monterey, California: Brooks/Cole, 1982.

[28]ANSI Z39.18-198X, Scientific and Technical Reports: Organization, Preparation, and Production. (America National Standards Institute, 1430 Broadway, New York, NY 10018, Dec, 1986). The accepted reports standards for DoD as well as academia and the private sector.

[29]Hoshousky, A. G., *Author's Guide for Technical Reporting.* USAF Office of Aerospace Research, 1964.

[30]Rathbone, R., and Stone, J., *A Writer's Guide for Engineers and Scientists.* Englewood Cliffs, New Jersey: Prentice-Hall, 1961.

[31]Reisman, S. J. (ed.), *A Style Manual for Technical Writers and Editors.* New York: Macmillan, 1962.

[32]Freeman, L. H., and Bacon, T. R., *Shipley Associates Style Guide* (rev ed.). Bountiful, Utah: 1990.

[33] AFR 80-44, Defense Technical Information Center. Prescribes DTIC and outlines Air Force support responsibilities.

[34] DoDD 5230.25, Distribution Statements on Technical Documents. Establishes policies and procedures for marking technical documents to show that they are either releasable to the public or that their distribution must be controlled within the United States Government. This regulation applies to all newly created technical data.

[35] Jordan, S. (ed.), *Handbook of Technical Writing Practices*. New York: John Wiley and Sons, Inc., 1971.

[36] Duckworth, W. E., Gear, A. E., and Lockett, A. G., *A Guide to Operations Research*. New York: John Wiley and Sons, Inc., 1977.

[37] Beer, S., *Decisions and Control—The Meaning of Operational Research and Management Cybernetics*. New York: John Wiley and Sons, Inc., 1966.

[38] Battersby, A., *Network Analysis for Planning and Scheduling*. London, England: Macmillan, 1964.

[39] Brown, R. G., *Smoothing, Forecasting, and Prediction of Discrete Time Services*. Englewood Cliffs, New Jersey: Prentice-Hall, 1962.

[40] Lee, A. M., Applied Queuing Theory. London, England: Macmillan, 1966.

[41] Churchman, C. W., Ackoff, R. L., and Arnoff, E. L., *Introduction to Operations Research*. New York: John Wiley and Sons, Inc., 1957.

[42] Gillett, B. E., *Introduction to Operations Research, A Computer-Oriented Algorithmic Approach*. New York: McGraw-Hill Book Company, Inc., 1976.

Appendix 1
Example Test and Evaluation Quantification Problems and Solutions*

1. Estimate the arithmetic mean and standard deviation of a population based on a sample.

a) Suppose a random sample of $n = 20$ has been collected on the time required to complete a task when applying a new process. The sample times required to complete the task are recorded and depicted in Table 1.

Table 1. Measured Times in Minutes Required to Complete Task Using New Process

5.0	4.8
6.3	5.7
7.5	6.9
4.3	4.2
6.8	8.1
7.2	7.6
8.1	7.7
7.8	8.2
6.1	5.3
6.2	5.7

b) Calculate estimates of the population parameters $\bar{x}$ = average time to complete the task, S = standard deviation of time to complete the task, and $S_{\bar{x}}$ = standard deviation of the average time to complete the task.

$$\bar{x} = \frac{\sum x}{n} = 6.48 \text{ minutes}$$

$$S = \left(\frac{\sum(\bar{x}-x)^2}{n-1}\right)^{\frac{1}{2}} = 1.31 \text{ minutes}$$

$$S_{\bar{x}} = \frac{S}{\sqrt{n}} = .29 \text{ minutes}$$

2. Suppose, for the sample data in Problem 1, we wish to establish a 95% confident interval within which the average time to complete the task μ_o will fall.

---

*Please see References 1–30.

Procedure	Example
a. Choose the confidence interval ϕ, this sets α at $1 - \phi$.	a. $\phi = 95\%$ $\alpha = .05$
b. Since we are interested in both an upper and lower confidence limit, look up $t_{a/2}$ in a table for the t distribution for the appropriate n.	b. $t_{a/2} = 2.093$
c. Estimate confidence interval by $\mu_o = \bar{x} \pm t_{a/2} \dfrac{S}{\sqrt{n}}$	c. $6.48 - 2.093(.29) \leq \mu_o \leq 6.48 + 2.093(.29)$, thus we conclude that we are 95% confident that: $5.87 \leq \mu_o \leq 7.09$ minutes.

3. For the data collected in Problem 1, suppose we wish to be 95% confident that the task can be completed on the average (μ_o) in a standard time of 10 minutes or less. Does the sample data indicate that the standard has been met?

Procedure	Example
a. Set up the null hypothesis (H_0) and the alternative hypothesis (H_1).	a. $H_0: \mu = 10$ minutes $H_1: \mu < 10$ minutes
b. Choose α, the significance level of the test.	b. $\alpha = .05$
c. Look up $t_{1-\alpha}$ in table of the t distribution for the appropriate n.	c. $t_{.95} = -1.729$ $n = 20$
d. Compute value of t_c statistic where $t_c = \dfrac{\bar{x} - \mu_0}{S/\sqrt{n}}$	d. $t_c = \dfrac{6.48 - 10}{.29} = -12.14$
e. t_c must be $\leq t_{.95}$ to reject H_0 and, in turn, accept H_1.	e. Since $-12.14 \leq -1.729$, we accept H_1, and conclude that we are 95% confident that the standard of completing the task in 10 minutes or less (on the average) has been met.

4. Suppose for the sample data in Problem 1, we wish to estimate a one-sided upper 95% confidence limit $\left(\overline{X_L}\right)$ on the average time to complete the task. That is, we wish to establish an average time which 95 out of 100 sample means likely will not exceed.

APPENDIX 1

Procedure	Example
a. Choose the confidence interval ϕ, this sets α at $1 - \phi$.	a. $\phi = 95\%$ $\alpha = .05$
b. Since we are interested in only the upper confidence limit, look up t_α for the appropriate n.	b. $t_\alpha = 1.72$
c. Estimate upper confidence limit by $$\overline{x_L} = \overline{x} + t_\alpha \frac{S}{\sqrt{n}}$$	c. $\overline{x_L} = \overline{x} + t_\alpha \dfrac{S}{\sqrt{n}}$ $\overline{x_L} = 6.48 + (1.729)(.29)$ $\overline{x_L} = 6.98$ minutes We conclude that we are 95% confident that the average time required to complete the task will not exceed 6.98 minutes. $\overline{x_L} = 6.98$ minutes

5. Suppose we are interested in how low two sets of aircrews with different types of training can fly their aircraft. Each group was instructed to fly as low as possible and the average height of flight was recorded in feet for each within the group. Results of the flights are presented in Table 2.

Table 2. Average Height of Flight By Aircrews in Feet

Group A	Group B
105	125
140	160
130	140
160	130
120	170
115	120
150	200
180	160
130	135
125	130

a) Calculate $\overline{x}$, S, and $S_{\overline{x}}$, for each group.

b) Since $\overline{x_A} = 135.5$ feet, $\overline{x_B} = 147$ feet, $S_\alpha = 22.54$ feet, $S_B = 25.08$ feet, $S_{\overline{x_A}} = 7.13$ feet, and $S_{\overline{x_B}} = 7.93$ feet, it can be seen that group A can fly lower than group B. However, show that we are 95% confident that this finding did not occur due to chance.

Procedure	Example
a. Set up the null hypothesis (H_0) and the alternate hypothesis (H_1).	a. $H_0: \mu_A - \mu_B = D_0 = 0$ $H_1: \mu_A - \mu_B \neq 0$
b. Choose α, the significance level for the test.	b. Let $\alpha = .05$.
c. Look up $t_{\alpha/2}$ for two-tailed test with $\alpha = .05$ for appropriate m.	c. $t_{.025} = 2.101$ m = degrees of freedom = $n_1 + n_2 - 2$ $m = 18$
d. Compute value of $|t_c|$ statistic where $t_c = \dfrac{(\overline{x}_A - \overline{x}_B) - (\mu_A - \mu_B)}{\sqrt{n_A S_A^2 + n_B S_B^2}} \sqrt{\dfrac{n_A n_B (n_A + n_B - 2)}{n_A + n_B}}$	d. $t_c = \dfrac{(135.5 - 147)}{\sqrt{10(22.54^2) + 10(25.08^2)}} \sqrt{\dfrac{(10)(10)(10+10-2)}{10+10}}$ $t_c = -1.023$
e. Using a two-tailed test with $\alpha = .05$, we would reject H_0 if $|t_c| \geq 2.101$.	e. $|t_c| \geq 2.101$ Therefore we conclude that there is not good reason to believe that group A can fly lower on the average than group B.

6. For Problem 5, is there a reason to believe that the variance in how low group A can fly is significantly less than group B?

Procedure	Example
a. Set up a statistical hypothesis test of whether the two samples come from the same normal population.	a. $H_0: \sigma_A^2 = \sigma_B^2$ $H_1: \sigma_A^2 > \sigma_B^2$
b. Choose α, the significance level of the test.	b. Let $\alpha = .01$.
c. Look up F_α for $m = n_A - 1$, $n = n_B - 1$ degrees of freedom in a table of the F distribution.	c. $F_\alpha, (n_A - 1), (n_B - 1)$ $F_\alpha, 9, 9$ $F_\alpha = 5.35$
d. Calculate the F_c statistic: $F_c = \dfrac{\sigma_A^2}{\sigma_B^2}$	d. $F_c = \dfrac{\sigma_A^2}{\sigma_B^2}$ $F_c = \dfrac{22.54}{25.08}$ $= .899$
e. If $F_c < F_\alpha$, the test does not reject H_0.	e. If $F_c < F_\alpha$, accept H_0. $.899 < 5.31$ The test does not give good reason to doubt that the two samples come from the same normal population.

APPENDIX 1

7. Suppose that success criterion for a test has been established such that 50% of the bombs dropped must have radial miss distances of 100 feet or less from the target (i.e., CEP ≤ 100 feet). Further, suppose 32 bombs will be dropped during the test and it is desired to be 90% confident that the bombing accuracy meets this criterion. What is the minimum number of bombs that must score 100 feet or less to meet the criterion?

Procedure	Example
a. Let n = number of bombs to be dropped and r = number of successes (i.e., score 100 feet or less).	a. $n = 32$ $r = ?$
b. Let q = proportion of the bomb drop population that is not successful (i.e., score greater than 100 feet). Let p = proportion of the bomb drop population that is successful, and $q = 1 - p$.	b. Use a one-sided binomial table for a confidence level of 90% to look up r for a $q = 50\%$ with $n = 32$. From this table, we see, $r = 21$ for $q = 1 - .52790 = .4721$; therefore, we are 90% confident that the CEP will be 100 feet or less if at least 21 of the 32 bombs dropped score 100 feet or less.

8. Sampling from an infinite population. Suppose that we draw a random sample of $N = 10$ bombs from a population where 90% (p) are known to be good and 10% (q) are known to be defective. What is the probability that at most two bombs in the sample of 10 will be defective.

Procedure	Example
a. Assume that the population is infinite, the sample is random, and the binomial distribution is appropriate.	a. $N = 10$ $p = .9$ $q = 1 - p = .1$
b. Calculate the probability that at most two bombs in a random sample of 10 will be defective: $$P(x \leq 2) \approx \sum_{x=0}^{2} \frac{N!}{x!(N-x)!}(q)^2(p)^{N-x}$$	b. $$P(x \leq 2) \approx \sum_{x=0}^{2} \frac{10!}{x!(10-x)!}\left(\frac{1}{10}\right)^x \left(\frac{9}{10}\right)^{10-x}$$ $P(x \leq 2) \approx .93$ That is, the probability is equal to .93 that at most two bombs in the sample of 10 will be defective.

9. Sampling from a finite population. Suppose, for Problem 7, that the population of bombs is finite and equal to 100 bombs. What is the probability that at most two bombs in the sample of 10 will be defective when the defective proportion in the population $p = .1$?

Procedure	Example
a. Since the population (N) is finite, and the sample is random, the hypergeometric distribution is appropriate because we are sampling without replacement.	a. $N = 100$ $n = 10$ $p = .1$
b. Calculate the probability that at most two bombs in a random sample of 10 will be defective: $$p(x \leq 2) = \sum_{x=0}^{2} \frac{\binom{N_p}{x}\binom{N-N_p}{n-x}}{\binom{N}{n}}$$ where $\binom{y}{b} = \frac{y!}{b!(y-b)!}$	b. $p(x \leq 2) = \sum_{x=0}^{2} \frac{\binom{10}{x}\binom{90}{10-x}}{\binom{100}{10}}$ $p(x \leq 2) = .94$

c. Note that the answer for sampling from a finite population of 100 in this problem is not much different from that in Problem 8, where the population was assumed to be binomial and infinite. However, for situations in which the population is very small, a serious error could be introduced by using the binomial distribution when the population is finite.

---

10. Sampling from a finite population. Suppose, for Problem 7, the population of bombs is finite and equal to $N = 30$ bombs. What is the probability that at most two bombs in the sample of 10 will be defective when the defective proportion in the population is $p = .1$?

Procedure	Example
a. Since the population (N) is finite, and the sample is random, the hypergeometric distribution is appropriate because we are sampling without replacement.	a. $N = 30$ $n = 10$ $p = .1$
b. Calculate the probability that at most two bombs in a random sample of 10 will be defective. $$p(x \leq 2) = \sum_{x=0}^{2} \frac{\binom{N_p}{x}\binom{N-N_p}{n-x}}{\binom{N}{n}}$$	b. $p(x \leq 2) = \sum_{x=0}^{2} \frac{\binom{3}{x}\binom{27}{10-x}}{\binom{30}{10}}$ $p(x \leq 2) = .97$

---

11. Suppose a sample of 44 missiles are fired at a target in two modes of operation (i.e., mode A and mode B) and that the results are as presented in the table below. Further suppose that if a missile guides to within 30 feet of the target, a lethal kill is assumed to occur. From these data, it can be seen that the point estimates of P_K for modes A and B are .24 and .65, respectively. Test the hypotheses that we are 95% confident that mode B is indeed significantly better than mode A (i.e., α = significance level = .05).

	Delivery Mode A	Delivery Mode B	
Kill miss distance ≤ 30 feet	5	15	20
No kill miss distance > 30 feet	16	8	24
	21	23	44

Procedure	Example
a. Set up the null hypothesis H_0, that the proportion of successes are not significantly different, and the alternative hypothesis H_1.	a. $H_0 : P_A = P_B$ $H_1 : P_B > P_A$
b. Choose α, the significance level of the test.	b. $\alpha = .05$
c. Look up x_α^2 in a table for the chi squared distribution with one degree of freedom. Note, the x^2 distribution tends to be unreliable for this test when the sample size N is less than about 40 and the number in any cell is less than 5. Under such conditions, Fisher's Exact Probability Calculation can be used to estimate the exact probability of deviations from the null hypothesis as extreme as the observed given that the null hypothesis is true and the marginal totals are fixed (see Reference 20).	c. $x_{.05}^2 = 3.841$
d. Calculate value of x_c^2 where $$x_c^2 = \frac{N\left(|BC - AD| - \frac{N}{2}\right)^2}{(A+B)(C+D)(A+C)(B+D)}$$	d. $$x_c^2 = \frac{44\left(|15.16 - 5.8| - \frac{44}{2}\right)^2}{(20)(24)(21)(23)}$$ $x_c^2 = 6.0132$
e. If $x_c^2 \geq x_{.05}^2$ reject H_0 and accept H_1.	e. Since $x_c^2 \geq x_{.05}^2$ i.e., $6.0132 \geq 3.841$, we accept the hypothesis that the proportion of missile kills in mode B is indeed greater than in mode A.

12. Suppose an aircraft is flying at 100 feet altitude toward a ground radar site which is capable of detecting the aircraft at line of sight. What is the range at which you would expect to detect the aircraft because of the Earth's curvature?

Procedure	Example
a. The line of sight based on the Earth's curvature can be estimated by the following relationship: $D = \sqrt{2H}$ D = line of sight in nautical miles H = altitude of aircraft in feet	a. $D = \sqrt{2H}$ $D = \sqrt{2(100)}$ $D = 14.1$ nautical miles

13. Suppose the radar antenna is elevated to 50 feet for Problem 12. What is the expected line of sight?

Procedure	Example
a. The line of sight based on the Earth's curvature can be estimated by the following relationship: $D = \sqrt{2H} + \sqrt{2H_R}$ where D = line of sight in nautical miles H = altitude of aircraft in feet H_R = height of the radar antenna in feet	a. $D = \sqrt{2H} + \sqrt{2H_R}$ $D = \sqrt{2(100)} + \sqrt{2(50)}$ $= 24.1$ nautical miles

14. Suppose a target drone is being flown at 400 knots and will be turned at 4 g acceleration during missile launch. Approximately how long will it take to turn the drone around and what will be its turn radius?

Procedure	Example
a. Estimate turn radius using the following relationship: $R = \dfrac{V^2}{g\sqrt{G^2 - 1}}$ where R = drone turn radius V = drone velocity g = acceleration due to gravity = 32.2 ft/sec² G = load factor during the drone turn $t = \dfrac{\pi R}{V}$ where t = time to turn through 1/2 circle	a. $G = 4$ $g = 32.2$ ft/sec² $V = 400$ knots $\times 1.67 \dfrac{\text{ft/sec}}{\text{knot}}$ $= 667$ ft/sec $R = \dfrac{667^2}{32.2\sqrt{4^2 - 1}}$ $R = \dfrac{667^2}{32.2\sqrt{4^2 - 1}}$ $R = 3567.4$ feet $t = \dfrac{\pi R}{V}$ $t = \dfrac{3.14161(3567.4)}{667}$ $t = 16.8$ seconds

15. Estimate weapons delivery accuracy when data are normally distributed (see parts a and b).

a) Suppose a pilot delivers 32 bombs in a dive bomb mode and the bomb accuracy scores are shown in Table 3. Estimate the dive bombing accuracy.*

Table 3. Bomb Accuracy Scores

Bomb #	X, miss dist. in feet	Y, miss dist. in feet	R, radial miss dist. in feet	θ, angle to radial miss dist.	Bomb #	X, miss dist. in feet	Y, miss dist. in feet	R, radial miss dist. in feet	θ, angle to radial miss dist.
1	2.5	2.6	3.61	46.12	17	19.6	19.6	27.72	45.00
2	3.5	3.7	5.09	46.59	18	21.0	20.5	28.65	44.31
3	4.0	4.1	5.73	45.71	19	21.5	21.8	31.05	45.40
4	5.2	5.3	7.43	45.55	20	22.4	22.5	31.89	45.13
5	7.1	7.0	9.97	44.59	21	23.8	23.9	33.87	45.12
6	7.5	7.4	10.54	44.62	22	25.7	25.6	36.13	44.89
7	8.6	8.3	11.95	43.98	23	27.4	27.5	38.96	45.10
8	9.6	9.4	13.44	44.40	24	29.3	29.5	41.86	45.19
9	10.0	10.1	14.21	45.29	25	32.6	32.5	45.89	44.91
10	12.5	12.3	17.54	44.54	26	34.8	34.5	48.58	44.75
11	13.7	12.8	18.75	43.05	27	35.6	35.5	50.13	44.92
12	15.6	15.8	22.20	45.36	28	38.5	37.5	52.35	44.25
13	16.1	16.5	23.05	45.70	29	39.4	39.5	55.93	45.07
14	16.7	16.8	23.69	45.17	30	42.8	42.5	59.89	44.80
15	17.5	17.1	24.47	44.34	31	43.9	44.5	63.36	45.39
16	18.9	19.1	26.87	45.30	32	47.6	47.5	67.10	44.94

b) Median bomb CEP estimate: One nonparametric method of estimating the CEP that requires no assumptions regarding the basic distribution of the data is selecting the median bomb miss distance. When the sample size is even, this CEP can be estimated by averaging the two median bomb's miss distances (e.g., bombs 16 and 17 in this specific example).

$$\text{CEP} = \frac{26.87 + 27.72}{2} = 27.3 \text{ feet}$$

c) The circular error average (CEA) can be estimated by the following relationship:

$$\text{CEA} = \sum_{i=1}^{n} \frac{R_i}{n}$$

where
- R_i = radial miss distance of the ith bomb
- n = number of bombs in the sample

---

*It is quite common in problems of this nature to estimate weapons delivery accuracy in terms of circular error probable (CEP) and circular error average (CEA). CEP is defined as the radius of a circle within which 50% of the bomb impacts would be expected to occur. CEA is simply an estimate of the arithmetic average of the weapon radial miss distances.

CEA = 29.84 feet*

d) If one knows (or can show) that it is reasonable to assume that the miss distances are distributed bivariate normal, a parametric method of estimating CEP can be applied. For example, the Kolmogorov-Smirnov (KS) statistical test could be applied as described on page 62 of Reference 20 to ascertain whether it is reasonable to assume that the miss distances are indeed normally distributed. This statistical test is based on examining the absolute difference D between the sample and normal population cumulative percentages. When the largest observed difference is equal to or greater than a value extracted from a D table for a particular level of significance, the assumption that the population has the specific distribution form is rejected. Maximum D values were estimated for x, y, and R for the data in Table 3. An example calculation for a single D_x value is illustrated below:

Procedure	Example
a. Calculate cumulative percentage observed: $F(x) \times 100$	a. $F(x)$ observed = $\frac{1}{32} \times 100 = 3.13\%$
b. Estimate expected cumulative percentage. Use $z = \dfrac{x - \bar{x}}{S_x}$ to enter the table of normal distribution to find expected cumulative percentage.	b. $z = \dfrac{2.5 - 21.09}{12.97} = -1.43$ $F(x)$ expected from normal distribution for $(z = -1.43) = 7.64\%$.
c. Calculate D_x where $D_x = F(x)$ expected $- F(x)$ observed.	c. $D_x = 7.64 - 3.13$ $= 4.51$
d. Repeat the procedure above for all values of x, y, and R and select maximum values of D_x, D_y, and D_R.	d. Max $D_x = 8.5$ Max $D_y = 8.1$ Max $D_R = 8.7$

The D value extracted from the table for KS statistical test with a significance level of .01 is equal to 29. Since the maximum D_x, D_y, and D_R are not equal to or greater than 29, it is reasonable to assume that the sample values of x, y, and R are normally distributed. Consequently, the following CEP formula for normally distributed data are applied.

REP = range error probable = the radius of a circle within which 50% of the bombs are expected to impact along the range track of the aircraft (i.e., x axis)
S_x = the estimate of the standard deviation of the x axis (range) impacts
REP = .674 S_x = (.674)(12.97) = 8.74 feet
DEP = deflection error probable = the radius of a circle within which 50%

---

*Since the estimate of the median bomb CEP and the CEA do not differ greatly, this gives some indication that the data may be near normally distributed.

APPENDIX 1

of the bombs are expected to impact across the track of the aircraft (i.e., y axis)

S_y = the estimate of the standard deviation of the y axis impacts
DEP = .674 S_y = (.674)(12.98) = 8.75 feet
CEP = .873 (REP + DEP)
  = .873 (8.7 + 8.75)
  = 15.27 feet

16. Estimating weapons delivery accuracy when data are not normally distributed:

a) Suppose that a guided weapon system delivered 32 scores as shown in Table 4. Examine estimates of weapons delivery accuracy in terms of CEP (both nonparametric and parametric estimates), CEA, and percentage of bombs within lethal warhead distance (i.e., assume the lethal warhead distance to be 12 feet).

Table 4. Scores of Guided Weapon System

Bomb #	X, miss dist. in feet	Y, miss dist. in feet	R, radial miss dist. in feet	q, angle to radial miss dist.	Bomb #	X, miss dist. in feet	Y, miss dist. in feet	R, radial miss dist. in feet	q, angle to radial miss dist.
1	0	0	0	0	17	4.4	4.7	6.03	43.11
2	1.5	1.3	1.98	40.91	18	-4.5	-4.8	6.17	223.15
3	-1.5	-1.4	2.05	223.03	19	4.6	4.9	6.31	43.19
4	2.1	2.3	2.84	42.39	20	-4.7	-4.9	6.51	223.8
5	-2.2	-2.4	2.98	222.51	21	5.1	5.2	7.14	44.44
6	2.5	2.6	3.47	43.87	22	-5.2	-5.2	7.35	225.00
7	-2.6	-2.8	3.55	222.87	23	5.2	5.3	7.28	44.45
8	2.8	3.1	3.82	42.09	24	-5.3	-5.3	7.50	225.00
9	-3.4	-3.7	4.64	222.58	25	6.1	6.2	8.56	44.53
10	3.5	3.8	4.76	42.65	26	7.1	7.0	10.11	45.41
11	-3.6	-3.9	4.90	222.71	27	100.0	103.0	139.38	44.15
12	4.0	4.1	5.59	44.29	28	-350.0	-375.0	490.32	224.45
13	-4.1	-4.2	5.73	224.31	29	850.0	750.0	1284.72	48.58
14	4.2	4.3	5.87	44.33	30	-950.0	-1005.0	1307.26	223.39
15	-4.3	-4.3	6.08	225.00	31	980.0	1000.0	1372.14	44.42
16	4.3	4.6	5.89	43.07	32	-2000.0	-3000.0	3605.55	236.31

b) The median bomb CEP estimate is as follows:

$$CEP = \frac{5.89 + 6.03}{2} = 5.96 \text{ feet}$$

c) Confidence limits on the median bomb CEP can be established on the median bomb CEP estimate by use of the binomial distribution (see Reference 20). The procedure and example are illustrated below.

Procedure	Example
a. Choose $1 - \alpha$, the confidence level of the test.	a. For example, choose $1 - .05 = .95$.
b. Look up rank numbers that represent upper and lower limits on the median in Table D (Reference 20) for sample size $n = 32$ and confidence level .95.	b. For $n = 32$, $1 - \alpha = .95$ Lower limit = 10th value Upper limit = 23rd value
c. Establish confidence limits based on 10th and 23rd value.	c. Estimate of Median CEP = 5.96 feet Lower .95 CEP confidence limit = 4.76 feet Upper .95 CEP confidence limit = 7.28 feet

d) The circular error average (CEA) estimate is CEA = 260.52 feet.*

e) If one applies the Kolmogorov-Smirnov (KS) statistical test to ascertain whether the miss distance data are normally distributed, the results are as follows (see Problem 15 for discussion of KS test):

$$D_x = 42.89$$
$$D_y = 42.89$$
$$D_R = 57.93$$

The maximum D values estimated above for x, y, and R for the data in Table 4 are all larger than the D value extracted from the table for a KS statistical test with a significance level of .01 (i.e., $D = 29$). Therefore, it is not reasonable to assume that the data come from a population that is normally distributed. Although it is not good practice and we would not normally compute the parametric CEP estimates for these data, we will do so in this example problem for comparative and illustrative purposes.

$$\text{REP} = .674\, S_x = (.674)(463.54) = 312.43 \text{ feet}$$
$$\text{DEP} = .674\, S_y = (.674)(609.79) = 410.99 \text{ feet}$$
$$\text{CEP} = .873\, (\text{REP} + \text{DEP}) = 631.55 \text{ feet**}$$

f) Percentage weapons impacting within the lethal miss distance. From Table 3, it can be seen that 26 of the 32 weapons impacted equal to or less than 12 feet from the target. Therefore,

$$\text{Percentage Weapons Impacting Within Lethal Miss Distance} = \frac{26}{32} \times 100 = 81.25\%$$

*The vast difference between the value of the CEA estimate and the value estimated in part b of this problem for the CEP is a strong indicator that the data are not normally distributed.

**Note, the vast difference in this parametric estimated value of CEP and the nonparametric estimate obtained in paragraph b as well as the CEA estimate obtained in part c of this problem.

From these estimates, it can be seen that the parametric estimates of CEP and CEA (i.e., 631.55 feet and 260.52 feet, respectively) are not very good indicators of this particular weapon's accuracy because over 81% of the weapons scored less than 12 feet. Consequently, one would usually choose either a nonparametric method of estimating CEP or a percentage of weapons impacting within the lethal miss distance as the best indicator of weapons delivery accuracy for weapons systems with distributions of this nature.

17. Comparing the averages of pilot reaction times for three different data display methods: Suppose samples were taken of the times required for a pilot to recognize and react to a given situation. Three different display systems (system 1, system 2, and system 3) were applied and the data shown in Table 5 represent the time in seconds for the pilot to recognize the situation of interest and to react. Are the averages for reaction times by the pilots to the different displays significantly different at the .05 level of significance?

Table 5. Pilot Reaction Time in Seconds

System 1	System 2	System 3
1.7 (1)*	4.0 (7)	13.1 (13)
1.9 (2)	5.0 (8)	15.5 (15)
2.3 (3)	10.1 (12)	16.8 (17)
2.8 (4)	14.2 (14)	17.1 (18)
3.5 (5)	16.2 (16)	17.5 (19)
3.9 (6)		18.0 (20)
7.0 (9)		
8.3 (10)		
9.5 (11)		
$R_1 = 51$	$R_2 = 57$	$R_3 = 102$
$n_1 = 9$	$n_i = 5$	$n_i = 6$
$\dfrac{R_i^2}{n_i} = 289$	$\dfrac{R_i^2}{n_i} = 649.8$	$\dfrac{R_i^2}{n_i} = 1734$

*The numbers shown in parentheses are the ranks, from lowest to highest, for all measurements combined.

a) This problem can be addressed by use of the Kruskal-Wallis sum of ranks or H test. It is based on having independent samples of measurements which can be combined into an ordered series and ranked from 1, smallest, to N, largest [N being the total number of measurements (see page 109 of Reference 20)].

b) The procedure and example for the test are illustrated below.

Procedure	Example
a. Choose α, the significance level of the test.	a. Let $\alpha = .05$ as stated for the problem.
b. Look up $x^2_{1-\alpha}$ for $t - 1$ degrees of freedom in a x^2 table, where t is the number of display systems to be compared.	b. $x^2_{.95}$ for 2 degrees of freedom (i.e., $t - 1 = 3 - 1 = 2$) = 5.991.

(Continued on next page.)

(Continued)

Procedure	Example
c. We have $n_1, n_2, \ldots, n_t$ measurements on each of the display systems 1, 2, …t: $N = n_1 + n_2, \ldots, n_t$	c. From Table 4, $N = 9 + 5 + 6 = 20$.
d. Compute R_i, the sum of the ranks of the measurements on the ith reaction time, for each of the display systems.	d. $R_1 = 51$ $R_2 = 57$ $R_3 = 102$
e. Compute $$H = \frac{12}{N(N+1)} \sum_{i=1}^{t} \frac{R_i^2}{n_i} - 3(N+1)$$	e. $H = \dfrac{12}{20(20+1)}(2672.8) - 3(20+1)$ $= 13.37$
f. If $H > x_{1-\alpha}^2$, conclude that the averages of the t reaction times differ; otherwise, there is no reason to believe that the averages differ.	f. Since H (i.e., 13.37) is larger than $x_{.95}^2 = 5.991$, there is good reason to believe that the averages for the reaction times differ significantly at the .05 level of significance.

18. Suppose that an aircraft penetrates the airspace of an enemy missile defense system and that when no electronic countermeasures are applied, is fired at three times by a missile with a single shot probability of kill $(P_K) = .6$. Also, suppose that when electronic countermeasures are applied, only two missiles are fired with each having a $P_K = .3$. Estimate the increase in the probability of aircraft survival when electronic countermeasures are applied.

a) Case 1, no electronic countermeasures: If one assumes that the missile firings are independent, a binomial estimate of the probability of survival can be made as follows:

Procedure	Example
a. $P_s = (1 - P_k)^n$ where P_s = probability of survival of n shots with no countermeasures P_k = single shot probability of kill with no countermeasures n = number of shots or missiles fired	a. $P_s = (1 - .6)^3$ $P_s = .064$

b) Case 2, with electronic countermeasures: same assumption as above.

Procedure	Example
a. $P_s^* = (1 - P_k^*)^n$	a. $P_s^* = (1 - .3)^2$

(Continued on next page)

APPENDIX 1

(Continued)

Procedure	Example
where P_s^* = probability of survival of n shots with countermeasures P_k^* = single shot probability of kill with countermeasures	$P_s^* = .49$

c) Delta increase in probability of survival resulting from countermeasures is as follows:

$$\Delta P_s = P_s^* - P_s = .49 - .064 = .426$$

19. Suppose an aircraft penetrates the airspace of five different enemy missile defense systems with single shot probabilities of kill of .001, .005, .01, .015, and .02. Further suppose that the missile systems are able to fire 3, 2, 1, 1, and 1 missiles respectively. Estimate the aircraft's probability of survival against the defense system.

a) If one assumes that the missile firings are independent, a binomial estimate of the probability of survival can be made as follows:

Procedure	Example
a. $P_s = \prod_{i=1}^{N}(1-P_{ki})^{n_i}$ where P_s = estimate of the probability of survival against the total defense system N = total number of individual missile systems (i.e., $i = 1, 2, ..., N$) N = 5 P_{ki} = single shot probability of kill of the ith defense missile system n_i = number of shots taken by the ith defense missile system	a. $P_s = (1 - .001)^3 (1 - .005)^2 (1 - .01)^1$ $(1 - .015)^1 (1 - .02)^1$ $P_s = (.997)(.99)(.985)(.98)$ $P_s = (.943)$

20. Suppose that a single communications system has a reliability of .8. How many independent systems would need to be installed to have an overall reliability of .95 or greater?

Procedure	Example
a. $R = 1 - (1 - R_1)^n$ R = probability that at least one system works (i.e., desire $R = .95$ as a minimum)	a. $R = 1 - (1 - .8)^n$ $.95 = 1 - (1 - .8)^n$ $1 - .95 = .2^n$ $\ln .05 = n \ln .2$

(Continued on next page.)

(Continued)

Procedure		Example
R_1	= probability that a single system works	$n = \dfrac{\ln .05}{\ln .2}$
n	= number of single systems	$n = 1.86 \approx 2$
$(1 - R_1)^n$	= probability that none of the n systems work	

21. The reliability, maintainability, and availability (RMA) of a weapon system were subjected to 3000 hours of operation, which resulted in the data shown in Table 6. Estimate values for the mean time between failure (MTBF), mean time between critical failure (MTBCF), mission reliability (R_m) (i.e., the probability of successfully being able to complete a 2 hour mission), mean time to repair (MTTR), mean downtime (MDT), inherent availability (A_1), and operation availability (A_o).

Table 6. Data for Problem 21

Total system operating hours (T) = 3000 hours
Total number of maintenance actions (M) = 20
Total number of failures (F) = 15 failures
Total number of critical failures (F_c) = 3 failures
Total number of equipment repair hours (H_R) = 6 hours
Total down hours of system to correct 20 maintenance actions (H_D) = 12 hours

a) Estimation of MTBF, MTBCF, and R_m:

Procedure	Example
a. $\text{MTBF} = \dfrac{\text{Total operating hours}}{\text{Total number of failures}}$	a. $\text{MTBF} = \dfrac{3000}{15} = 200$ hours
b. $\text{MTBCF} = \dfrac{\text{Total operating hours}}{\text{Total number of critical failures*}}$	b. $\text{MTBCF} = \dfrac{3000}{3} = 1000$ hours
c. $R_m = e^{-\dfrac{t}{\text{MTBCF}}}$	c. $R_m = e^{-\dfrac{2}{1000}}$ $= .998$

where
R_m = mission reliability and is assumed to be a function of the exponential failure distribution
t = mission duration in hours

*Critical failures prevent mission accomplishment.

b) Estimation of MTTR and MDT:

Procedure	Example
a. $\text{MTTR} = \dfrac{\text{Total repair hours}}{\text{Total number of repairs}}$	a. $\text{MTTR} = \dfrac{6}{15} = .4$ hours
b. $\text{MTBM} = \dfrac{\text{Total operating hours}}{\text{Total number of maintenance actions}}$	b. $\text{MTBM} = \dfrac{3000}{20} = 150$ hours
c. $\text{MDT} = \dfrac{\text{Total down time}}{\text{Total number of maintenance actions}}$	c. $\text{MDT} = \dfrac{12}{20} = .6$ hours

c) Estimation of A_i and A_o:

Procedure	Example
a. $A_i = \dfrac{\text{MTBF}}{\text{MTBF} + \text{MTTR}}$ A_i = inherent availability, a measure of system availability with respect only to operating time and corrective maintenance	a. $A_i = \dfrac{200}{200 + .4} = .998$
b. $A_o = \dfrac{\text{MTBM}}{\text{MTBM} + \text{MDT}}$ A_o = operational availability, a measure of system availability with respect to all segments of time during which the equipment is intended to be operational.	b. $A_o = \dfrac{150}{150 + .6} = .996$

References

[1] Hoel, P.G., *Introduction to Mathematical Statistics*. New York: John Wiley and Sons, Inc., 1962.

[2] Dixon, W.J., and Massey, F.J., Jr., *An Introduction to Statistical Analysis* (2nd ed.). New York: McGraw-Hill Book Company, 1957.

[3] Ostle, B., *Statistics in Research*. Ames, Iowa: Iowa State University Press, 1963.

[4] Mendenhall, W., *Introduction to Probability and Statistics* (2nd ed.). Belmont, California: Wadsworth Publishing Company, Inc., 1967.

[5] Freund, J. E., *Mathemtical Statistics*. Englewood Cliffs, New Jersey: Prentice Hall, Inc., 1960.

[6] Li, J.C.R., *Introduction to Statistical Inference*. Ann Arbor, Michigan: J.W. Edwards Publisher, Inc., 1961.

[7] Mendenhall, W., *Introduction to Linear Models and Design and Analysis of Experiments*. Belmont, California: Wadsworth Publishing Company, Inc., 1968.

[8] Johnson, P.O., *Statistical Methods in Research*. New York: Prentice-Hall, Inc., 1950.

[9] Brownlee, K.A., *Statistical Theory and Methodology in Science and Engineering*. New York: John Wiley and Sons, Inc., 1960.

[10] Burington, R., and May, D., *Handbook of Probability and Statistics With Tables* (2nd ed.). New York: McGraw-Hill, 1970.

[11] Dixon, W.J., and Massey, F.J., Jr., *Introduction to Statistical Analysis*. New York: McGraw-Hill Book Company, 1957.

[12] Fisher, R.A., *Statistical Methods for Research Workers*. New York: Hafner Publishing, 1958.

[13] U.S. Army Materiel Command, *Experimental Statistics*, Sections 1-5 (pamphlets 110-114). Washington, D.C.: USAMC.

[14] Draper, N., and Smith, H., *Applied Regression Analysis* (2nd ed.). New York: John Wiley and Sons, Inc., 1981.

[15] Sheffe, H., *The Analysis of Variance*. New York: John Wiley and Sons, Inc., 1959.

[16] Mace, A.E., *Sample-Size Determination*. New York: Reinhold Publishing Corporation, 1964.

[17] Brownlee, K.A., *Industrial Experimentation*. New York: Chemical Publishing Co., Inc., 1948.

[18] Cooke, J.R., Lee, M.T., and Vanderbeck, *Binomial Reliability Table*, NAVWEPS REPORT 8090. Springfield, Virginia: National Technical Information Service.

[19] Cochran, W.G., and Cox, G.M., *Experimental Designs*. New York: John Wiley and Sons, Inc., 1957.

[20] Tate, M.W., and Clelland, R.C., *Nonparametric and Shortcut Statistics*. Danville, Illinois: Interstate Printers and Publishers, Inc., 1957.

[21] Siegel, S., *Nonparametric Statistics for the Behavioral Sciences*. New York: McGraw-Hill Book Company, 1956.

[22] Conover, W.J., *Practical Nonparametric Statistics*. New York: John Wiley and Sons, Inc., 1971.

[23] Hollander M., and Wolfe, D.A., *Nonparametric Statistical Methods*. New York: John Wiley and Sons, Inc., 1973.

[24] Hald, A., *Statistical Tables and Formulas*. New York: John Wiley and Sons, Inc., 1952.

[25] Fisher and Yates, *Statistical Tables for Biological, Agricultural, and Medical Research*. New York: Hafner Publishing, 1963.

[26] Wald, A., *Sequential Analysis*. New York: John Wiley and Sons, Inc., 1948.

[27] Graybill, F.A., *An Introduction to Linear Statistical Models*, Vol I. New York: McGraw-Hill Book Company, 1961.

[28] Fisher, R.A., *The Design of Experiments*. New York: Hafner Publishing, 1960.

[29] Chew, V. (ed.), *Experimental Designs in Industry*. New York: John Wiley and Sons, Inc., 1958.

[30] Kempthorne, O., *Design and Analysis of Experiments*. New York: John Wiley and Sons, Inc., 1962.

Appendix 2
Example Department of Defense Test and Evaluation Facilities

Numerous major facilities in the U.S. defense establishment are capable of providing one or more of the five types of test and evaluation (T&E) environments identified earlier in Chapter 4. Historically, these environments have required continuous upgrades to keep pace with advanced technology developments and modern systems. Numerous studies and reviews have recently taken place and ongoing efforts are continuing to address how best to downsize, consolidate, and provide the essential capabilities in a more efficient manner within the U.S. Department of Defense. Although the brief description of T&E facilities provided in this appendix may change somewhat from an organizational perspective, it should remain conceptually and functionally representative of U.S. Defense T&E capabilities.

1) Air Force Development Test Center (AFDTC), Eglin AFB, Florida. This is the Air Force Materiel Command (AFMC) center for development and testing of conventional munitions, flight certification, and testing of some electronic systems. Land range and restricted airspace total approximately 724 square miles, and restricted airspace over water totals approximately 86,500 square miles. The Electromagnetic Test Environment range consists of multiple threat simulators, emitters, and time-space-position-information (TSPI) used for testing, training, and exercises. Real time mission monitoring, aircraft vectoring, and data readout are available at the central control facility. Mission monitoring is also available at Site A-20. The Preflight Integration of Munitions and Electronic Systems (PRIMES) facility anechoic chamber enables ground intra-aircraft electromagnetic interference (EMI) checkout in a simulated flight environment. The Airborne Seeker Evaluation Test System (ASETS) and Guided Weapons Evaluation Facility (GWEF) enable checkout of individual systems in flight and simulated flight conditions. The McKinley Climatic Laboratory produces global environmental testing conditions that range from the extremes of windswept arid deserts to the cold northern regions. Conditions include sand and dust, moisture, precipitation, snow, salt spray, and high humidity. Four runways are available at Eglin, Hurlburt, and Duke Fields; the longest is 12,500 feet. Extensive instrumentation, data reduction, aviation servicing, maintenance, ordnance handling, storage, and machine shop facilities are available.

2) USAF Weapons and Tactics Center (USAFWTC), Nellis AFB, Nevada. This Air Force center provides an operationally oriented range where multiple air and ground participants can perform integrated air-to-air and air-to-ground training and testing. Mobile and fixed, manned and unmanned threat simulators equipped with the modular instrumentation package system (MIPS) transmit data through the range instrumentation system (RIS). RIS, meshed with the Red Flag measure-

ment and debriefing system (RFMDS), displays command, control, and communications (C^3), training, and exercise data in real time at the range control and combined operations centers and the RFMDS consoles. Two MSRT4s and a mobile signal analysis system perform frequency analysis functions. Video, TSPI, and precision tracking radar are located on the munitions training ranges south of the electronic combat (EC) ranges. Two 10,000-foot runways are located on base with an alternate at Indian Springs AFB on the southern edge of the ranges. Aviation servicing, maintenance, ordnance handling, storage, and machine shops are available. Heavy utilization of ranges for Red Flag, Green Flag, and the Fighter Weapons School limits the availability for testing. The Nellis ranges, combined with ranges in Utah, Ft. Irwin, China Lake, Edwards AFB, and Pt. Mugu, make up the U.S. western test range complex.

3) Utah Test and Training Range (UTTR), Hill AFB, Utah. This major range and test facility base (MRTFB), operated for the DoD by the 545th Test Group, is detached from the Air Force Flight Test Center at Edwards AFB. It provides range facilities for all phases of T&E of manned and unmanned aircraft systems. Land ranges (2,136 square miles) and airspace (9,260 square miles) are spread over sparsely populated desert and are instrumented for training, exercises, and testing. Limited pulsed, portable emitter systems are available. Tracking radars, cinetheodolites, the high accuracy multiple object tracking system (HAMOTS), the television ordnance scoring system (TOSS), and flight test television are displayed for data and C3 information at the Range Mission Control Center (RMCC) at Hill AFB. Data processing and reduction are available. Remoteness of the ranges from the RMCC, widely varied terrain, and seasonally adverse weather can impact T&E at UTTR. Runways are available at Hill AFB (13,500 feet), Wendover Field on the Utah/Nevada border (9,100 feet), and Michael Army Airfield at Dugway Proving Ground (13,500 feet). UTTR has the potential to be upgraded to a national electronic combat test range.

4) Air Force Electronic Warfare Environment Simulator (AFEWES), Carswell AFB, Texas. The AFEWES is an AFMC secure laboratory for evaluation of electronic warfare (EW) hardware in a high density simulated threat environment. Radio frequency (RF) signal propagation between transmitter and receiver is contained within waveguides and coaxial cables with signals adjusted to produce the effects of various range, aircraft movement, antenna scanning, and other real world environmental factors. Over 200 simulated threat signals can be generated from 175 separate sources with 0.5- to 18.0-GHz coverage by the multiple emitter generator (MEG). Electronic Countermeasures (ECM) and C^3 equipment and techniques can be tested against anti-aircraft, surface-to-air, acquisition, and ground-controlled intercept/air intercept radars. Jammer techniques, infrared countermeasures, and chaff and clutter effects can also be simulated. The real time data reduction system can interface the 1553B avionics bus with the VAX 11/785 Evaluation and Analysis System for Simulation Testing (EASST) to provide test design, dry run operation, real time and quick look analyses, presentations, and playback.

5) Rome Laboratory, Griffiss AFB, New York. Rome Laboratory, an AFMC Electronic Systems Center facility scattered at nine separate locations in New York and Massachusetts, specializes in dynamic and static radar cross section and antenna pattern measurements at its precision antenna measurement system (PAMS). The Electromagnetic Compatibility Analysis Facility (EMCAF) contains three shielded anechoic chambers; the largest is 48 by 40 by 32 feet. Its tactical and

strategic aircraft static antenna measurement test-beds contain F-15, F-4, F-111, A-10, B-1, C-130, B-52, and KC-135 airframes for hardware mounting. The available runways at Griffiss AFB are 11,820 feet by 300 feet; and those at Hanscom AFB, Massachusetts, are 7,001 feet by 150 feet and 5,106 feet by 150 feet.

6) Naval Air Station (NAS), Fallon, Nevada. This is the location of the Naval Strike University, whose mission is to provide an operationally oriented range where multiple air and ground participants can perform integrated training. Its EW range is interfaced with the tactical air combat training system (TACTS), giving it a capability similar to, but more limited than, the RFMDS and Green Flag exercises at Nellis AFB. Fallon has two runways; the longest is almost 14,000 feet. It has approximately 235 square miles of land range.

7) Naval Surface Warfare Center Dahlgren Division (NSWC), Dahlgren, Virginia. The NSWC is the principal Navy Research Development Test and Evaluation (RDT&E) center for surface ship weapons, ordnance, mines, and strategic systems, as well as the center for hazards of electromagnetic radiation to ordnance (HERO) and missile electromagnetic vulnerability (EMV) testing. On site are two outdoor ground test facilities (each 100 feet by 240 feet) for EMV/HERO testing; a third one is operated by NSWC at the Naval Air Warfare Center, Patuxent River, Maryland. The Electromagnetic Vulnerability Assessment Program (EMVAP) operates an anechoic chamber and a mode-stirred reverberation chamber for systems testing. Each has its own control station. A runway capable of accommodating propeller aircraft is located on site. NSWC is the monitor for the electromagnetic compatibility assessment and fleet frequency management program for automatic ship radar and satellite communications frequency deconfliction.

8) Naval Air Warfare Center-Aircraft Division (NAWC-AD), Patuxent River NAS, Maryland. The NAWC-AD is the principal Navy center for aircraft weapons systems T&E for all phases of acquisition and the life cycle process. NAWC-AD provides range and engineering support for the fleet and other DoD customers. Inland and offshore restricted airspace (50,000 square miles) with the Chesapeake Test Range (CTR) and Mid Atlantic Tracking System (MATS) is available and instrumented for testing and training. Real time readout of C^3 and data is displayed at the range control center. An anechoic chamber capable of housing fighter-sized aircraft can be linked to inflight or offshore platforms as well as other ground test facilities. These other facilities are all linked under the Integrated Laboratory System which consists of the Tactical Avionics/Software T&E Facility (TASTEF), Manned Flight Simulator (MFS), Electronic Warfare Integrated Test Systems Laboratory (EWISTL), Aircrew Systems Evaluation Facility (ASEF), Electromagnetic Environment Generating System (EMEGS), and the Electromagnetic Environmental Effects Test Laboratory (E3TL). Extensive test instrumentation and data reduction facilities are available. The NAS has two runways; the longest is 9,800 feet. Limited land test area is available; however, Ft. A.P. Hill, Virginia, and Aberdeen Proving Ground, Maryland, are nearby. Dover AFB, Delaware, and Langley AFB, Virginia, are also nearby.

9) Naval Air Warfare Center-Weapons Division (NAWC-WD), China Lake, California. This is the Navy center for development and testing of air warfare, missiles, and electronic warfare systems. Land ranges (approximately 1,800 square miles) and restricted airspace are instrumented for testing or training. Its weapons ranges have EW test capability for live firing against emitting targets with real time C^3 and data readout at the range control center. A separate, dedicated EW

threat environment simulation (EWTES) range is available for nonlethal testing and training. Michelson and Thompson Laboratories maintain simulation facilities which include infrared (IR) and RF hardware-in-the-loop and anechoic chambers. Armitage Field is on site and has three runways, the longest being 10,000 feet. Aviation servicing, maintenance, ordnance handling, storage, machine shop, and tooling support are available. Nellis AFB, Nevada, and Ft. Irwin and Edwards AFB, California, are closeby. The National Aeronautics and Space Agency (NASA) Goldstone Tracking Station adjacent to the EWTES limits frequency allocation and jamming.

10) Naval Air Warfare Center-Weapons Division (NAWC-WD/P), Pt. Mugu NAS, California. NAWC-WD/P is a Navy field activity for software support of all naval airborne EW systems. Several laboratories are located on site which use IR, laser, radar, and RF sensor simulations. Controlled airspace (approximately 35,000 square miles) over deep ocean can be monitored by the extended area tracking system (EATS) with real time C^3 and data readout at the range operations building. NAWC-WD/P is instrumented to support fleet training as well as testing. The Electronic Combat Simulation and Evaluation Laboratory (ECSEL) contains RF simulation of threat radars used for development, integration, software modification, and evaluation of radar warning receivers (RWRs), electronic support measures (ESM), and jammers in various configurations and scenarios. Two microwave anechoic chambers are available. Pt. Mugu NAS has an 11,000-foot runway. No land test ranges are available. Edwards AFB, NAWC-WD China Lake, and Ft. Irwin are also within the lower California area.

11) Marine Corps Air Station (MCAS), Yuma, Arizona. An EW training range for West Coast Marine air units is located here. Presently a TACTS system is in operation that is interfaced with the ground emitters. Yuma has four runways, with the longest being 13,299 feet by 200 feet.

12) Marine Corps Air Station (MCAS), Cherry Point, North Carolina. An EW training range for East Coast Marine air units is located here. It also has a TACTS system installed with limited threats similar to a small EW exercise. Four runways, with the longest being 9,000 feet by 200 feet, are located at Cherry Point.

13) Air Defense Artillery Center (ADAC), Ft. Bliss, Texas. ADAC is a U.S. Army Training and Doctrine Command (TRADOC) installation with responsibility for the ADAC and Air Defense Artillery School. The school includes the Army Development and Acquisition of Threat Simulators (ADATS) activity which provides mobile replicas, operators, and instrumentation for training and testing to customers worldwide. Land ranges (approximately 1,700 square miles) and restricted airspace include projectile, missile firing, and maneuver areas primarily designed to support training. A 13,572-foot runway is located on the main post. Instrumentation and experience for testing on the Ft. Bliss ranges are provided by the adjacent White Sands Missile Range.

14) U.S. Army Electronic Proving Ground (USAEPG), Ft. Huachuca, Arizona. USAEPG is a center for U.S. Army T&E of communications electronics (C-E) systems. It has three ranges instrumented for testing, with an operations control center and data processing facilities. Total land test area available is about 115 square miles. Restricted airspace (approximately 1,170 square miles) is available. Its electromagnetic environmental test facility (EMETF) uses simulation models and RF hardware to evaluate C^3 performance and EMI between systems. An EMI test facility with an anechoic chamber, antenna pattern measurement facility, com-

munications test facility, and radar and avionics test facility is also available. HQ USA Information Systems Command, USA Intelligence Center and School, and the Joint Test Element (JTE) of the Joint Tactical Command, Control, and Communications Agency (JTC3A) are also located at Ft. Huachuca. A 15,000-foot runway is on site at Libby Army Airfield (AAF) to support training and testing. Davis-Monthan AFB is nearby.

15) National Training Center (NTC), Ft. Irwin, California. This is the U.S. Army center for combined and joint training at the armored and mechanized infantry battalion task force level. The 1,000 square miles of land range and airspace are instrumented almost exclusively for training. The multiple integrated laser engagement system (MILES) and NTC Instrumentation System (NTC-IS) provide real time readout of ground vehicle position and status at the core instrumentation subsystem (CIS) building. JCS Initiative 38 should integrate the Red Flag Measurement and Debriefing System (RFMDS) from the U.S. Air Force Weapons and Tactics Center at Nellis AFB, Nevada, with NTC-IS to enable real time tracking of joint exercises with air-to-ground and surface-to-air engagements. NTC frequently uses ADATS. Proximity of Nellis AFB and China Lake Naval Air Warfare Center, California, makes joint operations feasible. NTC requires that any testing on its ranges be "transparent" to its primary mission of training. The NASA Goldstone Tracking Station collocated on site restricts frequency control allocation and jamming.

16) White Sands Missile Range (WSMR), New Mexico. WSMR is a U.S. Army center for many categories of weapons system testing. It operates and maintains numerous laboratories and facilities to examine high explosive, shock, electromagnetic, optical, and laser effects. Land range (approximately 5,600 square miles) and airspace are available primarily for testing with real time C^3 and data readout at the range control center. The Vulnerability Assessment Laboratory (VAL) contains three anechoic chambers and facilities for Electromagnetic Vulnerability (EMV), antenna pattern, and radar cross section measurement. The Army Test and Evaluation (ARMTE) Directorate operates the Electromagnetic Radiation Effects (EMRE) facility for Electromagnetic Pulse (EMP) effects, C^3 data link, ECM, and software testing using hardware and simulations. A C^3 branch specializes in Army tactical battlefield air defense fire support C^3 systems interoperability. Extensive testing, instrumentation, and data reduction facilities are available. A runway capable of accommodating propeller aircraft is located on post. Holloman AFB, New Mexico, with a 12,100-foot runway, and Ft. Bliss, Texas, are adjacent.

17) USA Armament Research Development and Engineering Center (ARDEC), Picatinny Arsenal, New Jersey. This Army Armament, Munitions, and Chemical Command (AMCCOM) facility provides EMI/EMC specifications for purchase orders, contracts, and other documents. Its electromagnetic test facility evaluates the susceptibility of munitions to RF, lightning, EMP, and static using a 25- by 11- by 7-foot EMI chamber. Portable instrumentation and telemetry vans are available. Small ammunition and gun test ranges are also located on site.

18) USA Combat Systems Test Activity (CSTA), Aberdeen Proving Ground (APG), Maryland. This Army Test and Evaluation Command (TECOM) facility performs electromagnetic compatibility and interference tests on weapons, vehicles, and support equipment using a 94- by 60- by 28-foot shielded enclosure. A mobile system is available for field testing. Phillips AAF is located on site; its longest runway is approximately 8,000 feet.

19) USA Communications Electronics Command (CECOM), Ft. Monmouth, New Jersey. This is an Army Materiel Command (AMC) facility responsible for RDT&E and acquisition of communications electronics and electronic systems. It uses an indoor antenna test range, with a 52- by 76-foot chamber, and the Earle Wayside Test Site, 60 acres of cleared land and a 10-acre communications area, to set up antennas and direction-finding equipment for test. The EW/Reconnaissance, Surveillance and Target Acquisition (EW/RSTA) center conducts R&D of EW jamming and detection systems.

20) USA Belvoir Research Development and Engineering Center, Ft. Belvoir, Virginia. This Army facility tests countermeasures and tactical sensors for target classification, electronic power generation and distribution, and physical security. A 30- by 17- by 14-foot anechoic chamber and a 20- by 24- by 13-foot enclosed EMI facility are located here. Davis AAF is on site with a 4,010- by 75-foot runway.

21) USA Harry Diamond Laboratories (HDL), Adelphi, Maryland. This Army Laboratory Command (LABCOM) facility investigates the survivability of weapons systems and components in high power RF fields. A 20- by 20- by 60-foot tapered anechoic chamber interfaced with automated test and data collection equipment is on site.

22) USA Missile Command (MICOM), Redstone Arsenal, Alabama. This Army AMC facility evaluates electromagnetic capability, interference, and vulnerability problems of missile systems using a 75- by 52- by 30-foot anechoic chamber in its millimeter wave simulation system and a 30- by 150- by 150-foot chamber in its electromagnetic test facility. Redstone AAF has a 7,300- by 150-foot runway.

23) USA Tank-Automotive Command (TACOM), Warren, Michigan. This Army AMC center for RD&E of tank-automotive equipment uses a 6- by 19.5-foot shielded chamber.

24) Aerospace Engineering Test Establishment (AETE), Cold Lake Canadian Forces Base (CFB), Alberta, Canada. The Cold Lake weapons ranges are used by AETE for flight tests and training with aircraft, weapons, and ground systems. The U.S. Maple Flag Exercise was periodically conducted here. No threat simulators are resident. Limited TSPI is available on the 135-square mile Primrose Lake Evaluation Range (PLER). The isolated location provides an uncluttered, less restricted RF environment. Cold Lake has three runways, the longest being 12,600 feet by 200 feet.

25) Department of National Defense, Goose Bay CFB, Labrador. Goose Bay possesses over 38,000 square miles of air ranges with limited land and target capabilities. It has no TSPI except normal air traffic control and communications equipment. The North Atlantic Treaty Organization (NATO) has used it for training exercises. Its large land area and isolated location provide an uncluttered, less restricted RF environment. USAF Air Mobility Command has maintained a detachment at this location. Two runways, with the longest being 11,050 feet by 200 feet, are available. This location has the potential to become a NATO forces training center.

Appendix 3
Glossary of Abbreviations and Acronyms

A_i	Inherent (Intrinsic) Availability
A_o	Operational Availability
AAF	Army Air Field
ABCCC	Airborne Command, Control, and Communications
ACAT	Acquisition Category
ACMI	Air Combat Maneuvering Instrumentation
ADAC	Air Defense Artillery Center
ADATS	Army Development and Acquisition of Threat Simulators
ADM	Acquisition Decision Memorandum
ADPE	Automatic Data Processing Equipment
ADS	Advanced Distributed Simulation
ADS	Automatic Diagnostic System
AETE	Aerospace Engineering Test Establishment
AF	Air Force
AFB	Air Force Base
AFDTC	Air Force Development Test Center
AFEWES	Air Force Electronic Warfare Environment Simulator
AFMC	Air Force Material Command
AFOTEC	Air Force Operational Test and Evaluation Center
AIAA	American Institute of Aeronautics and Astronautics
AMC	Army Material Command
AMCCOM	Army Armament, Munitions, and Chemical Command
ANSI	American National Standards Institute
APB	Acquisition Program Baseline
APG	Aberdeen Proving Ground
ARDEC	United States Army Research Development and Engineering Center
ARMTE	Army Test and Evaluation Directorate
ASEF	Aircrew Systems Evaluation Facility
ASETS	Airborne Seeker Evaluation Test System
ASP	Acquisition Strategy Panel

ATD	Advanced Technology Demonstration
ATE	Automatic Test Equipment
AWACS	Airborne Warning and Control System
AWS	Air Weather Service
BDA	Bomb Damage Assessment
BIT	Built-In-Test
Bug	Error in software which is consistent and repeatable
C	Cost
CAS	Close Air Support
C3	Command, Control, and Communications
C3I	Command, Control, Communications, and Intelligence
CCB	Configuration Control Board
C3CM	Command, Control, and Communications Countermeasures
C4I	Command, Control, Communications, Computers, and Intelligence
CCD	Camouflage, Concealment, and Deception
CCM	Counter-Countermeasures
CCP	Contract Change Proposal
CDR	Critical Design Review
CECOM	United States Army Communications Electronics Command
CED	Contract Exploration and Definition
CEM	Communications, Electronics, and Meteorological
CEP	Circular Error Probable
CERs	Cost Estimating Relationships
CFB	Canadian Forces Base
CI	Critical Issue
CINC	Commander-In-Chief of the Unified or Specified Command
CIS	Core Instrumentation Subsystem
CM	Countermeasures
CND	Can Not Duplicate
COEA	Cost and Operational Effectiveness Analysis
COMSEC	Communications Security
C_n^*	Present value cost of program at year n
C_t	Cost value at time t
CPU	Central Processing Unit
CSC	Computer Software Component
CSTA	United States Army Combat Systems Test Activity
CTR	Chesapeake Test Range
DAB	Defense Acquisition Board

DAE	Defense Acquisition Executive
DCP	Decision Coordinating Paper
DDR&E	Director of Defense Research and Engineering
DDT&E	Director of Development Test and Evaluation
DEF CAP	Defensive Captive Air Patrol
DEM/VAL	Demonstration and Validation
DEP	Deflection Error Probable
DEPSECDEF	Deputy Secretary of Defense
DIS	Distributed Interactive Simulation
DoD	Department of Defense
DoDD	Department of Defense Directive
DoDI	Department of Defense Instruction
DoDM	Department of Defense Manual
DPG	Defense Planning Guide
DPRB	Defense Planning and Resources Board
DR	Deficiency Report
DSSP	Defense Standardization and Specifications Program
DT&E	Development Test and Evaluation
DVAL	Joint Test and Evaluation for Data Link Vulnerability Analysis
e	Base of the natural logarithms
EASST	Evaluation and Analysis System for Simulation Testing
EATS	Extended Area Tracking System
EC	Electronic Combat
ECM	Electronic Countermeasures
ECP	Engineering Change Proposal
ECR	Embedded Computer Resources
ECSEL	Electronic Combat Simulation and Evaluation Laboratory
EEI	Essential Elements of Information
E2TL	Electromagnetic Effects Test Laboratory
EMCAF	Electromagnetic Compatibility Analysis Facility
EMD	Engineering and Manufacturing Development
EMEGS	Electromagnetic Environment Generating System
EMETF	Electromagnetic Environmental Test Facility
EMI	Electromagnetic Interference
EMP	Electromagnetic Pulse
EMRE	Electromagnetic Radiation Effects
EMV	Electromagnetic Vulnerability
EMVAP	Electromagnetic Vulnerability Assessment Program

ERM	Environmental Resources Management
ESM	Electronic Support Measures
EW	Electronic Warfare
EWISTL	Electronic Warefare Integrated Test Systems Laboratory
EWJT	Electronic Warfare Joint Test
EW/RSTA	Electronic Warfare/Reconnaissance, Surveillance, and Target Acquisition
EWTES	Electronic Warfare Threat Environment Simulation
FA	False Alarm
F	F statistic
F_c	F statistic value as calculated
F_α	F statistic value as obtained from table of distribution
FCA	Frequency Control Analysis
FD	Fault Detection
FEBA	Forward Edge of the Battle Area
FI	Fault Isolation
FOT&E	Follow-On Operational Test and Evaluation
FSD	Full Scale Development
FY	Fiscal Year
g	Acceleration due to gravity = 32.2 ft/sec^2
GBU	Guided Bomb Unit
GCI	Ground Control Intercept
Glitch	An unforeseen feature that renders software inefficient, inelegant, or clumsy
GPS	Global Positioning System
GWEF	Guided Weapons Evaluation Facility
HARM	High Speed Antiradiation Missile
HAMOTS	High Accuracy Multiple Object Tracking System
HDL	Harry Diamond Laboratories
HERO	Hazards of Electromagnetic Radiation to Ordnance
HITVAL	Joint Test and Evaluation for Validation of Simulation Models
H_0	Null hypothesis
H_1	Alternative hypothesis
IADS	Integrated Air Defense System
IC	Integrated Circuit
ICASE	Integrated Computer-Aided Software Engineering
IFF	Identification Friend or Foe
IFFN	Joint Test and Evaluation for Identification Friend, Foe, or Neutral

APPENDIX 3

IFR	Instrument Flight Rules
ILS	Integrated Logistics Support
ILSP	Integrated Logistics Support Plan
IOC	Initial Operational Capability
IOT&E	Initial Operation of Test and Evaluation
IP	Initial Point
IPA	Integrated Program Assessment
IPP	Implementation Program Plan
IR	Infrared
IRS	Interface Requirement Specification
IWSM	Integrated Weapon System Management
JADS	Joint Advanced Distributed Simulation
JADO/JEZ	Joint Air Defense Operations/Joint Engagement Zone
JCATE	Joint Crisis Action Test and Evaluation
JCCD	Joint Camouflage, Concealment, and Deception
JCS	Joint Chiefs of Staff
JEMI	Joint Electromagnetic Interference
JFAAD	Joint Forward Area Air Defense
JFS	Joint Feasibility Study
JLOTS	Joint Logistics Over the Shore
JOTH-T	Joint Over the Horizon-Targeting
JTAMS	Joint Test and Evaluation for Tactical Missiles Signatures
JTC3A	Joint Tactical Command, Control, and Communications Agency
JTE	Joint Test Element
JTMD	Joint Theater Missile Defense
JROC	Joint Requirements Oversight Council
JT&E	Joint Test and Evaluation
JTF	Joint Test Force
k	End year of program
L	Lagrangian function
LABCOM	Laboratory Command
LCC	Life-Cycle Cost
LFT&E	Live Fire Test and Evaluation
$\ln$	Natural logarithm
LRIP	Low-Rate Initial Production
LSA	Logistics Support Analysis
MAA	Mission Area Analysis (or Assessment)
MAJCOM	Major Command
MATS	Mid Atlantic Tracking System

MCAS	Marine Corps Air Station
MDT	Mean Downtime
MEG	Multiple Emitter Generator
MFS	Manned Flight Simulator
MICAP	Mission Capability
MICOM	United States Army Missile Command
MILES	Multiple Integrated Laser Engagement System
MIL-STD	Military Standard
MIPS	Modular Instrumentation Package System
MNS	Mission Need Statement
MOE	Measure of Effectiveness
MOP	Measure of Performance
MRTFB	Major Range and Test Facility Base
M&S	Modeling and Simulation
MTBCF	Mean Time Between Critical Failure
MTBF	Mean Time Between Failure
MTBM	Mean Time Between Maintenance
MTTR	Mean Time To Repair
N	Finite population size
n	Sample size
NAS	Naval Air Station
NASA	National Aeronautics and Space Agency
NATO	North Atlantic Treaty Organization
NAWC-AD	Naval Air Warfare Center—Aircraft Division
NAWC-WD	Naval Air Warfare Center—Weapons Division
NCA	National Command Authority
NSC	National Security Council
$\binom{N}{n}$	Number of possible combinations of N things taken n at a time
NSWC	Naval Surface Warefare Center
NTC	National Training Center
NTC-IS	National Training Center-Instrumentation System
O&M	Operations and Maintenance
O	Output
OA	Operational Assessment
OA	Operations Analysis
OCD	Operational Concept Demonstration
OD	Operational Demonstration
OFP	Operational Flight Program

APPENDIX 3

OPR	Office of Primary Responsibility
ORD	Operational Requirements Document
OSD	Office of the Secretary of Defense
OT&E	Operational Test and Evaluation
P	Probability
P_k	Probability of kill
P_s	Probability of survival
p	Proportion of successes (or estimate of probability of success)
PAMS	Precision Antenna Measurement System
PAT&E	Production Acceptance Test and Evaluation
PC	Planning Committee
PDM	Program Decision Memorandum
PDR	Preliminary Design Review
PE	Program Element
PLER	Primrose Lake Evaluation Range
PMB	Performance Measurement Baseline
PMD	Program Management Directive
POA&M	Plan of Action and Milestones
POL	Petroleum, Oil, and Lubricants
POM	Program Objective Memorandum
PRIMES	Preflight Integration of Munitions and Electronic Systems
q	Proportion of failures (or estimate of probability of failure)
QT&E	Qualification Test and Evaluation
QOT&E	Qualification Operational Test and Evaluation
R	Reliability
r	Interest rate
R&D	Research and Development
R&M	Reliability and Maintainability
RAM	Random Access Memory
RCM	Requirements Correlation Matrix
RDT&E	Research, Development, Test, and Evaluation
RECCE	Reconnaissance
REP	Range Error Probable
RF	Radio Frequency
RFMDS	Red Flag Measurement and Debriefing System
RFP	Request for Proposal
RIS	Range Instrumentation System
RM&A	Reliability, Maintainability, and Availability
RMCC	Range Mission Control Center

RML&A	Reliability, Maintainability, Logistics Supportability, and Availability
RMS-2	Range Measurement System-2
RTO	Responsible Test Organization
RTOK	Retest OK
RWR	Radar Warning Receiver
S&T	Science and Technology
SA	Supportability Assessment
SAC	Senior Advisory Council
SE	Support Equipment
SEAD	Suppressions of Enemy Air Defense
SECDEF	Secretary of Defense
SEEKVAL	Joint Test and Evaluation of Combat Air Support Target Acquisition
SOW	Statement of Work
SPO	System Program Office
SRD	System Requirements Document
SRS	Software Requirements Specification
SSA	Source Selection Authority
SSR	Software Specification Review
STD	Standard
ST&E	Security Test and Evaluation
STINFO	Scientific and Technical Information
SWOE	Smart Weapons Operability Enhancement
t	Time period
t	t statistic
TAB	Technical Advisory Board
TACOM	United States Army Tank-Automotive Command
TACTS	Tactical Air Combat Training System
TAG	Technical Advisory Group
TASTEF	Tactical Avionics/Software Test and Evaluation Facility
TASVAL	Joint Test and Evaluation for Tactical Aircraft Effectiveness in Close Air Support Anti-Armor Operations
TCTO	Time Compliance Technical Order
TD&E	Tactics Development and Evaluation
T&E	Test and Evaluation
TECOM	United States Army Test and Evaluation Command
TEMP	Test and Evaluation Master Plan

APPENDIX 3

TEMPEST	Control of Compromising Emanations
TEREC	Tactical Electronic Reconnaissance
TM	Technical Manuals
TO	Technical Orders
TOT	Time Over Target
TQM	Total Quality Management
TRADOC	United States Army Training and Doctrine Command
TSPI	Time-Space-Position Information
TTP	Technology Transition Plan
USAEPG	United States Army Electronic Proving Ground
USAFWTC	United States Air Force Weapons and Tactics Center
USD(A)	Under Secretary of Defense for Acquisition
UTTR	Utah Test and Training Range
VFR	Visual Flight Rules
VLSI	Very Large Scale Integration
V&V	Verification and Validation
VV&A	Verification, Validation, and Accreditation
WSMR	White Sands Missile Range
WSR	Weapon System Reliability
WWV	National Bureau of Standard Time Broadcasts
VAL	Vulnerability Assessment Laboratory
α	Level of significance (i.e., probability of rejecting a true hypothesis, Type I error)
β	Probability of accepting a false hypothesis, Type II error)
ϕ	Confidence interval $(1 - \alpha)$
χ^2	Chi square statistic
Σ	Sign for mathematical summation
Π	Sign for mathematical product
λ	Lagrangian multiplier
μ	Arithematic average or mean of distribution
$\hat{\mu} = \bar{x}$	Estimate of arithematic average or mean of distribution
σ	Standard deviation of distribution
$\hat{\sigma} = s$	Estimate of standard deviation of distribution
σ^2	Variance of distribution
$\hat{\sigma}^2 = s^2$	Estimate of variance of distribution

Selected Bibliography

Adamson, D. S., Andreani, E. C., Archer, G. W., et al., *Mathematical Model User's Manual Combined Arms Tactical Training Simulator (CATTS) Device 16A3,* Volumes I-IV. Redondo Beach, California: TRW, Inc. (AD A038 796; AD A038 797; AD A038 798; AD A038 799), 1977.

Anderson, L. B., et al., *Revised OPTSA Model Volume I: Methodology.* Arlington, Virginia: Institute for Defense Analysis (LD 35385MA), 1975.

Anderson, L. B., et al., *Maxmin and Minmax Strategies in Multistage Games and ATACM.* Arlington, Virginia: Institute for Defense Analysis (AD A030 246), 1976.

Anderson, T. W., *An Introduction to Multivariate Statistical Analysis.* New York: John Wiley and Sons, Inc., 1958.

Arkin and Colton, *Tables for Statisticians.* New York: Barnes and Noble, 1961.

Bailey, R. W., *Human Performance Engineering.* Englewood Cliffs, New Jersey: Prentice-Hall, 1982.

Bailey, T. J., and Martin, G. A., *CACDA Jiffy War Game. Technical Manual. Part I: Methodology.* Fort Leavenworth, Kansas: Army Combat Development Activity (LD 39809MA), 1977.

Barbieri, W. A., *Countermilitary Potential: A Measure of Strategic Offensive Force Capability.* Santa Monica, California: Rand Corporation (R-4131-PR), 1973.

Barlow's Tables. New York: Chemical Publishing, 1962.

Battilega, J. A., Blackwell, L. M., and Phelps, M. W., *An Introduction to Quantitative Strategic Force Analysis.* Denver, Colorado: Science Applications, Inc. (SAI-77-011-DEN), 1977.

Battilega, J. A., *Selected Mathematical Programming Techniques for Force Allocation Problems.* Denver, Colorado: Science Applications, Inc. (SAI-74-017-DEN), 1974.

Battilega, J. A., *Multi-Goal Optimization With the Arsenal Exchange Model.* Denver, Colorado: Science Applications, Inc. (SAI-75-30-DEN), 1975.

Battilega, J. A., and Cotsworth, W. L., *Multiple Criteria, Strategic Force Analysis, and the Arsenal Exchange Model.* Denver, Colorado: Science Applications, Inc. (SAI-77-058-DEN), 1977.

Beach, E. F., *Economic Models: An Exposition.* New York: John Wiley and Sons, Inc., 1957.

Benson, K. T., and Miller, C. E., *User's Manual for QUANTO—A Weapon Allocation Code.* Kirtland AFB, New Mexico: Air Force Weapons Laboratory (AD 779 754), 1974.

Benson, K. T., et al., *QUANTO—A Code to Optimize Weapon Allocations.* Kirtland AFB, New Mexico: Air Force Weapons Laboratory (AD 773 801), 1974.

Bergstrom, A. R., "Nonrecursive Models as Discrete Approximations to Systems of Stochastic Difference Equations." *Econometrica*, 34(1) (January), 1966, pp. 173-182.

Bishop, A. B., and Stollmack, S., *The Tank Weapon System*, Final Report. Columbus, Ohio: Ohio State University (AD 850 367), 1968.

Blackett, P. M. S., *Studies of War*. New York: Hill and Wang, 1962.

Blackwell, L. M., et al., *UNICORN (Version III) Methodology*. Denver, Colorado: Science Applications, Inc. (SAI- 76-048-DEN), 1976.

Bode, J. R., *Indices of Effectiveness in General Purpose Force Analysis*. Vienna, Virginia: Braddock, Dunn, and McDonald (BDM/W-74-070-TR), 1974.

Bonder, S., *A Summary Descripton of the VECTOR-II Theater-Level Campaign Model*. Ann Arbor, Michigan: Vector Research, 1976.

Bonder, S., and Farrell, R. *Development of Analytical Models of Battalion Task Force Activities*. Ann Arbor, Michigan: Systems Research Laboratory, University of Michigan (AD 714 677), 1970.

Bozovich, J. F., et al., *Mathematical Formulation of the Arsenal Exchange Model* (Revision 7). Denver, Colorado: Martin Marietta Corporation, 1973.

Bracken, J., *Two Optimal Sortie Allocation Models*. Arlington, Virginia: Institute for Defense Analysis (IDA Paper P-993), 1973.

Bracken, J., et al., *Quantity-Quality Tradeoffs*. Arlington, Virginia: Institute for Defense Analysis (HQ 74-16712, Study 5-443), 1975.

Bracken, J., et al., *A Theater-Level Analysis of NATO Aircraft Shelters and Air-Defense Attrition Rates (U)*. Arlington, Virginia: Institute for Defense Analysis (IDA Study 5-447), 1975.

Brode, H. L., *Height of Burst Effects at High Overpressure*. Los Alamos, New Mexico: Defense Atomic Support Agency (AEC)(DASA 2506), 1970.

Brown, R. A., et al., *SMOBSMOD (Strategic Mobility Simulation Model): User's Manual*. Bethesda, Maryland: U.S. Army Concepts Analysis Agency (LD-31060), 1974.

Christ, C. F., *Econometric Models and Methods*. New York: John Wiley and Sons, Inc., 1966.

Close, A. F., and Gillen, C. A., *Description of the Computer Program for METRIC-A Multi-Echelon Technique for Recoverable Item Control*. Santa Monica, California: Rand Corporation (AD 686056), 1969.

Cochran, W. G., *Sampling Techniques*. (2nd ed.), New York: John Wiley and Sons, Inc., 1963.

Cole, J. C., and Dorr, A. J., *The OASIS-II Program, Volume V: OASIS-II Listings*. La Jolla, California: Science Applications, Inc. (SAI-74-251-LJ), 1974.

Cole, J. C., and Dorr, A. J., *CARMONETTE: General Description*. McLean, Virginia: General Research Corporation (AD A007 843), 1974.

Coleman, H. W., and Steele, W. G., Jr., *Experimentation and Uncertainty Analysis for Engineers*. New York: John Wiley and Sons, Inc., 1989.

Combat II Model: A Step Toward an Integrated Combat Assessment. Vienna, Virginia: Braddock, Dunn, and McDonald (LD- 36936MA), 1974.

Cramer, H., *Mathematical Methods of Statistics*. Princeton, New Jersey: Princeton University Press, 1946.

Cramer, H., *The Elements of Probability Theory*. New York: John Wiley and Sons, Inc., 1955.

Crow, E. L., Davis, F. A., and Maxfield, M. W., *Statistics Manual*. New York:

Dover Publications, 1960.

Curtis, W., and Smith, K. L., *Preliminary Test Design Procedures*. Albuquerque, New Mexico: Sandia Corporation, 1968.

Department of Defense, *Life Cycle Costing Guide for System Acquisition (Interim)*. Washington, D.C.: U.S. Government Printing Office (008-000-00184-7), 1973.

DIVWAG Model Documentation. Fort Leavenworth, Kansas: Army Combat Development Activity (LD 40635MA), 1976.

Dixon, W. J., and Massey, F. J., Jr., *An Introduction to Statistical Analysis* (2nd ed.). New York: McGraw-Hill Book Company, 1957.

Dockery, J., Leiser, M., and Aitben, M., "A Theater-Level Simulation With COMO-III," in *Computers and Operations Research*, Volume 3, pp. 57-71. New York: Pergamon Press, 1976.

Dolins, L. P., *The IDA Level Defense Economic Model: An Interim Summary*. Arlington, Virginia: Institute for Defense Analysis (AD 711 553), 1970.

Donelson, J., et al., *IDA Strategic Nuclear Exchange Model (IDASNEM)*. Washington, D.C.: Institute for Defense Analysis (AD B020 653L), 1976.

Drake, M. K., and Frike, M. P. (eds.), *Nuclear Weapon Effects Handbook for Long Range Research and Development Program*. La Jolla, California: Science Applications, Inc., 1974.

Durbin, J., and Watson, G. S., "Testing for Serial Correlation in Least Squares Regression. Part I." *Biometrika*. Volume 37, December 1950, pp. 409-428.

Durbin, J., and Watson, G. S., "Testing for Serial Correlation in Least Squares Regression. Part II." *Biometrika*. Volume 38, June 1951, pp. 159-178.

Durbin, J., "Estimation of Parameters in Time-series Regression Models." *Journal of the Royal Statistical Society, Series B* (Methodological), Volume 22, No. 1, January 1960, pp. 139-153.

Everett, H. M., "Generalized Lagrange Multiplier Method for Solving Problems of Optimum Allocation of Resources." *Operations Research*. Volume 11, 1973, pp. 339-425.

Fain, J., "The Lanchester Equations and Historical Warfare: An Analysis of Sixty World War II Land Engagements," in *Proceedings of the 34th Military Operations Research Symposium*. Alexandria, Virginia: Military Operations Research Society (AD C002 913), 1975.

Farrell, R. L., Cherry, W. P., and Ahrens, R. A., *A Preliminary Design and Resource Estimate for a Combined-Arms Simulation Model (CASM)*. Volume I: Summary Report. Ann Arbor, Michigan: Vector Research (AD B010 532L), 1976.

Forsythe, G. E., "Generation and Use of Orthogonal Polynomials for Data Fitting With a Digital Computer," *Journal of the Society for Industrial and Applied Mathematics*. No. 5, 1957, pp. 74-88.

FOURCE, Division Level Combat Model. Working papers from TOS Cost and Operational Effectiveness Analysis. White Sands Missile Range, TRADCO Systems Analysis Activity (ATAA-TCA).

Fry, F. C., *Probability and Its Engineering Uses*. Princeton, New Jersey: Van Nostrand, 1948.

Function FEAR: A Set of Subroutines for the UNICORN Model for Rapid Calculation of Lethal Offsets From Nuclear Weapons (U). La Jolla, California: Science Applications, Inc. (SAI-75-125-LJ), 1975.

Green, D. W., et al., *The SRI-WEFA Soviet Econometric Model: Phase Three Documentation*. Volume I. Arlington, Virginia: SRI International (AD A043 287), 1977.

Groover, P. L., et al., *Assumptions and Rationale Used to Calculate an Arsenal Relative Force Size*. Springfield, Virginia: Science Applications, Inc. (SAI-78-03-POST), 1978.

Handbook of Chemistry and Physics. Cleveland, Ohio: Chemical Rubber Publishing, latest edition.

Hannan, E. J., "Testing for Serial Correlation in Least Squares Regression." *Biometrika*, Vol. 44, Parts 1 and 2, June 1957, pp. 57-66.

Hannan, E. J., "The Estimation of Relationships Involving Distributed Lags." *Econometrica*, Volume 33, No. 1, January 1965, pp. 206-224.

Hartley, D. S., and Cockrell, A. A., *An Examination of a Distribution of TAC Contender Solutions*. Washington, D.C.: National Military Command System Support Center (AD A014 426), 1975.

Hawkins, J. H., *The AMSAR War Game (AMSWAG) Computer Combat Simulation*. Aberdeen Proving Ground, Maryland: Army Material Systems Analysis Activity (AD A028 052), 1976.

Hengle, J. E., "Quantitative Methodology for Evaluating Strategic Command Control and Communications Systems," in *Proceedings of the 37th Military Operations Research Symposium (MORS)*. Alexandria, Virginia: Military Operations Research Society (AD C009 755L), 1976.

Hess, R. W., *BOMTAN: A Model for Estimating the Annual Cost of Bomber and Tanker Squadrons*. Santa Monica, California: Rand Corporation (LD 31844MA), 1974.

Hillerman, H. H., *The Theoretical Basis of the CODE 50 Nuclear Exchange Model*. Arlington, Virginia: Center for Naval Analysis (AD A043 377), 1972.

Hitch, C. J., and McKean, R. N., *The Economics of Defense in the Nuclear Age*. New York: Atheneum, 1978.

Hoaglin, D. C., and Vellamin, P. F., *Applications, Basics, and Computing of Exploratory Data Analysis*. Boston, Massachusetts: Duxbury Press, 1981.

Honig, J., et al., *Review of Selected Army Models*. Washington, D.C.: Assistant Vice Chief of Staff (AD 887 175L), 1971.

Hood, J. W., and Terry, W. E., *Strategic Arms Interactions: 1945-1961*. Monterey, California: U.S. Naval Postgraduate School (AD A001 264), 1974.

Hooper, J. W., and Zellner, A., "The Error of Forecast for Multivariate Regression Models." *Econometrica*, Volume 29, No. 4, October 1961, pp. 544-555.

Hopgood, R. E., *Military Planning Game*. Monterey, California: U.S. Naval Postgraduate School (AD 475 372), 1965.

Huff, D., *How to Lie With Statistics*. New York: Norton, 1954.

Hugus, D. K., *A Review of the Compilation of the DYNTACS(x) Data Base for the HELLFIRE COEA*. Fort Leavenworth, Kansas: Army Combat Development Activity (AD A034 919), 1976.

Ignizio, J., *Goal Programming and Extensions*. New York: Lexington, 1977.

"Integrated Nuclear and Conventional Theater Warfare Simulation (INWARS)," Level III Specifications, Volume I: Introduction. McLean, Virginia: BDM Corporation (BDM/W-78-402-TR).

Kahan, J. H., *Security in the Nuclear Age—Developing U.S. Strategic Arms Policy*. Washington, D.C.: The Brookings Institute, 1975.

Karr, A. F., *Adaptive and Behavioral Strategies in Two-Person, Two-Move, Two-Action, Zero-Sum Games*. Arlington, Virginia: Institute for Defense Analysis (IDA Paper P-993), 1973.

Karr, A. F., *On the CONAF Evaluation Model*. Arlington, Virginia: Institute for Defense Analysis (LD 37409MA), 1976.

Karr, A. F., *Review and Critique of the VECTOR-II Combat Model*. Arlington, Virginia: Institute for Defense Analysis (IDA Paper P-1315), 1977.

Kelleher, G. J., *A Damage Limiting Shelter-Allocation Strategy*. Arlington, Virginia: Institute for Defense Analysis (AD 615779), 1965.

Kendall, M. G., and Smith, B. B., *Tables of Random Sampling Numbers*. Cambridge, England: Cambridge University Press, 1939.

Kennedy, W. R., et al., *DIVLEV War Game Model Computer Program*. Aberdeen Proving Ground, Maryland: U.S. Army Material Systems Analysis Activity, 1977.

Keynes, J. M., *The General Theory of Employment, Interest, and Money*. London, England: Macmillan, 1936.

Keynes, J. M., *The Scope and Method of Political Economy* (4th ed.). London, England: Macmillan, 1930.

Kish, J. L., Jr., *Business Forms: Design and Control*. New York: The Ronald Press Company, 1971.

Kline, M., *Mathematics in Western Culture*. New York: Oxford, 1953.

Knox, F. M., *Design and Control of Business Forms*. New York: McGraw-Hill Book Company, Inc., 1952.

Koopmans, T. C. (ed.), *Statistical Inference in Dynamic Economic Models*. Cowles Commission Monograph 10. New York: John Wiley and Sons, Inc., 1950.

Lanchester, F. W., "Aircraft in Warfare: The Dawn of the Fourth Arm, No. 5: The Principle of Concentration." Reprinted 1956 in *The World of Mathematics*, Volume 4, Newman, J. (ed.). New York: Simon and Schuster, 1914, pp. 2138-2148.

Larson, H. J., and Bancroft. T. A., "Biases in Prediction by Regression for Certain Incompletely Specified Models." *Biometrika*, Volume 50, Parts 3 and 4, December 1963, pp. 391-402.

Leake, L. A., Tiede, R. V., and Whipple, S., Jr., *Information Flow and Combat Effectiveness. Part II: Methodology for Evaluating the Effectiveness and Cost of the Tactical Operations System (TOS)*. McLean, Virginia: Research Analysis Corporation (RAC-R-100), 1970.

Linstone, H. A., and Turoff, M. (eds.), *The Delphi Method*. Reading, Massachusetts: Addison-Wesley Publishing Company, 1975.

Livingston, I., and Groover, P., *The HALL Model for the Analysis of Aircraft Survivability Prior to Penetration*. Denver, Colorado: Science Applications, Inc. (SAI 74-001-DEN), 1974.

Low, L. J., and Means, E. H., *BALFRAM (Balanced Force Requirements Analysis Methodology*, Volume I: Seminar Guide. Menlo Park, California: Stanford Research Institute (AD B013 416L), 1976.

Mailart, J. D., et al., *Navy Modified Logistics Composite Model (LCOM) User's Manual*. Los Angeles, California: CACI, Incorporated-Federal (AD A006 975), 1974.

March, J. G., and Simon, H. A., *Organizations*. New York: John Wiley and

Sons, Inc., 1958.

Massey, H. G., *Introduction to the USAF Total Force Cost Model*. Santa Monica, California: Rand Corporation (AD A042 460), 1977.

Mathematical Formulations of the Arsenal Exchange Model: Revision 7. Denver, Colorado: Martin Marietta Corporation, 1975.

Mathematical Tables from the Handbook of Chemistry and Physics. Cleveland, Ohio: Chemical Rubber Publishing, latest edition.

MEECN System Simulator Capabilities Description. Washington, D.C.: Defense Communications Agency (TP 960-TP-74- 40), 1974.

Mobile Missile Survival Against a Ballistic Salvo—The Effects of Position Uncertainty and Other Parameters. Santa Monica, California: Rand Corporation (RM-2743), 1958.

Molina, E. C., *Poisson's Exponential Binomial Limit*. New York: Van Nostrand, 1942.

Mood, A. M., Graybill, F. A., and Bues, D. C., *Introduction to the Theory of Statistics* (3rd ed.). New York: McGraw-Hill Book Company, 1974.

Moroney, M. J., *Facts from Figures*. Baltimore, Maryland: Penguin Books, 1956.

National Bureau of Standards, *Tables of the Binomial Distribution*, Applied Mathematics Series 6. Washington, D.C.: U.S. Government Printing Office, 1949.

National Bureau of Standards, *Tables of Normal Probability*, Applied Mathematics Series 23. Washington, D.C.: U.S. Government Printing Office, 1953.

Neter, Wasserman, and Whitmore, *Applied Statistics*. Boston, Massachusetts: Allyn and Bacon Publishing Co., 1978.

Novick, D., *Program Budgeting in the Department of Defense*. Santa Monica, California: Rand Corporation (605 388), 1974.

Patinkin, D., *Money, Interest, and Prices* (2nd ed.). New York: Harper and Row, 1965.

Paulson, R. M., et al., *Using Logistics Models in System Design and Early Support Planning*. Santa Monica, California: Rand Corporation (AD 724 690), 1971.

Phelps, M. W., et al., *Generalized Decomposition Procedures for Strategic Balance Calculations*. Denver, Colorado: Science Applications, Inc. (SAI-76-084-DEN), 1976.

Pietrewicz, P. A., et al., *Development of a Methodology for Solving the Combat Potential Delivery Problem: A Final Research Report*. Denver, Colorado: Science Applications, Inc. (SAI-78-115-DEN), 1978.

Poulton, E., *The Environment at Work*. Springfield, Illinois: Charles C. Thomas, 1979.

Preliminary Design and Resource Estimate for Combined Arms Simulation Model (CASM), Volume I: Executive Summary. Vienna, Virginia: BDM Corporation (AD B010 526L), 1976.

Quade, E. S., *Analysis for Military Decision*. New York: American Elsevier Publishing Co., 1970.

Raiffa, H., and Schlaifer, R., *Applied Statistical Decision Theory*. Boston, Massachusetts: Division of Research, Graduate School of Business Administration, Harvard University, 1961.

The Rand Corporation, *A Million Random Digits with 100,000 Normal Deviates*. Glencoe, Illinois: Free Press of Glencoe, 1955.

Roebuck, J., Kroemer, K., and Thomson, W., *Engineering Anthropometry Techniques*. New York: John Wiley and Sons, Inc., 1975.

Roscoe, S. (ed.), *Aviation Psychology*. Ames, Iowa: Iowa State University Press, 1980.

Savage, L. J., *The Foundations of Statistical Inference*. New York: John Wiley and Sons, Inc., 1962.

Seminara, J., Gonzalez, W., and Parsons, S., *Human Factors Review of Nuclear Power Plant Control Room Design*. Palo Alto, California: Electric Power Research Institute, 1977.

Shackel, B. (ed.), *Applied Ergonomics Handbook*. Surrey, England: IPC Science and Technology Press, 1974.

Sheridan, T., and Ferrell, W., *Man-Machine Systems: Information, Control, and Decision Models of Human Performance*. Cambridge, Massachusetts: MIT Press, 1974.

Shubik, M., *Games for Society, Business, and War: Towards a Theory of Gaming*. New York: Elsevier Scientific Publishing Company, 1975.

Shubik, M., *The Uses and Methods of Gaming*. New York: Elsevier Scientific Publishing Company, 1975.

Shurleff, E., *How to Make Displays Legible*. La Mirada, California: Human Interface Design, 1980.

Simon and Grubles, *Tables on the Cumulative Binomial Probabilities*. Washington, D.C.: U.S. Government Printing Office, 1952.

Simon, H. A., "Prediction and Hindsight as Confirmatory Evidence." *Philosophy of Science*, Volume 22, No. 3, July 1955, pp. 227-230.

Singleton, W., *Introduction to Ergonomics*. Geneva, Switzerland: World Health Organization, 1972.

Singleton, W., Easterby, R., and Whitfield, D. (eds.), *The Human Operator in Complex Systems*. London, England: Taylor and Francis, 1971.

Slonim, M. J., *Sampling in a Nutshell*. New York: Simon and Schuster, 1960.

Spaulding, S. L., *User's Manual for Bonder/IUA Combat Model*. Washington, D.C.: Assistant Vice Chief of Staff, Weapons Systems Analysis Directorate (LD 32756MA), 1971.

Staff Officer's Field Manual Nuclear Weapons Employment Doctrine and Procedures. Washington, D.C.: Department of the Army and Navy, 1968.

Steneck, N. H. (ed.), *Risk Benefit Analysis: The Microwave Case*. San Francisco, California: San Francisco Press, Inc., 1982.

Tank, Anti-Tank, and Assault Weapon Requirements Study (U). Fort Knox, Kentucky: U.S. Army Combat Development Command, Armor Agency (ACN 7356), 1968.

TEMPER, Volume II: The Theory of the Model. Bedford, Massachusetts: Raytheon (AD 471 458), 1965.

Theil, H., *Principles of Econometrics*. New York: John Wiley and Sons, Inc., 1971.

Thomas, J., *Tartarus Model Sensitivity Experiment*. Bethesda, Maryland: Army Concepts Analysis Agency (AD A018 896), 1975.

Thrall, R. M., Tsokos, C. P., Turner, J. C. (eds.), *Proceedings of the Workshop on Decision Information for Tactical Command and Control*. Houston, Texas: Robert M. Thrall (AD A052 439), 1976.

Tiede, R. V., and Leake, L. A., "A Method for Evaluating the Combat Effectiveness of a Tactical Information System in a Field Army." *Operations Research* 19:587-604, 1971.

Tinall, J. B., et al., *A Primer in Strategic Command and Control Communications*. Colorado Springs, Colorado: Air Force Academy (AD 786-034), 1973.

Tippett, *Random Sampling Numbers*. Cambridge, England: Cambridge University Press, 1927.

Tukey, J.W., *Exploratory Data Analysis*. Reading, Massachusetts: Addison-Wesley Publishing Company, 1977.

U.S. Air Force, *Guide for Air Force Writing*, AF Manual 10-4, 1 April 1960.

VECTOR-II System for Simulation of Theater-Level Combat, Part I: User's Manual. Washington, D.C.: Joint Chiefs of Staff (Computer System Manual CSM UM 244-78), 1978.

Wadsworth, G. P., and Bryan, J. G., *Introduction to Probability and Random Variables*. New York: McGraw-Hill Book Company, 1960.

Wald, A., *Statistical Decision Functions*. New York: John Wiley and Sons, Inc., 1950.

Walker, G. A., *Life Cycle/System Effectiveness Evaluation and Criteria*. Seattle, Washington: Boeing Company (AD 916 001), 1974.

Walker, L. K., et al., *Comparison and Evaluation of Four Theater-Level Models: CEM IV, IDAGAM I, LULEJIAN I, VECTOR I*. Arlington, Virginia: Weapons Systems Evaluation Group (AD A036 325), 1976.

Wallis, W. A., and Roberts, H. V., *Statistics: A New Approach*. Glencoe, Illinois: Free Press of Glencoe, 1956.

Walther, H. J., *Catalog of War Gaming and Military Simulation Models*. Washington, D.C.: Organization of the Joint Chiefs of Staff (SAGA-236-75), 1975.

Watson, G. S., and Hannan, E. J., "Serial Correlation in Regression Analysis, II," *Biometrika*, Volume 43, Parts 3 and 4, December 1956, pp. 436-448.

Weaver, W., *Lady Luck*. Garden City, New York: Doubleday, 1963.

Wetzler, E., *The Structure of the IDA Civil Defense Model*. Arlington, Virginia: Institute for Defense Analysis (AD 717 098).

White, W., et al., *DIVOPS. A Division-Level Combined Arms Engagement Model. Version 3. Volume III: Planner/User Manual*. Fort Leavenworth, Kansas: BDM Services Company (LD 36265MD), 1975.

Willard, D., *Lanchester as Force in History; An Analysis of Land Battles of the Years 1618-1905*. McLean, Virginia: Research Analysis Corporation (RAC-TP-74), 1962.

Williamson, O. E., *The Economics of Discretionary Behavior*. Englewood Cliffs, New Jersey: Prentice-Hall, 1964.

Wilson, A., *The Bomb and the Computer*. New York: Dell Publishing Company, 1970.

Yergin, D., *Shattered Peace*. Boston, Massachusetts: Houghton Mifflin Company, 1977.

Zellner, A., and Tiao, G. C., "Bayesian Analysis of the Regression Model With Autocorrelated Errors." *Journal of the American Statistical Association*, Volume 59, No. 307, September 1964, pp. 763-778

Zimmerman, R. E., "A Monte Carlo Model for Military Analyses," in *Operation Research for Management*, Volume 2, McCloskey and Coppinger (eds.). Baltimore, Maryland: Johns Hopkins University Press, 1956.

Index

Accuracy, 60–61, 70
Acquisition, 5–7, 18–21
Acronyms, 345–353
Advanced planning, 146–147
Aerospace environmental support, 69
Aerospace roles and missions, 259
Affordability, 24–26, 29–30, 49–50
Air combat maneuvering, 273–276
Alternatives, 21–28
Analysis, 126–127, 130–143, 170–171
Analysis of variance, 134–141
    Components of variance model, 134
    Linear hypothesis model, 134
Analytical modeling and simulation, 55–57, 73–95
Automatic diagnostic systems, 237–250
    Can-not duplicate, 248–250
    Fault detection, 246–247
    Measures of effectiveness, 18–21, 246–250
    Probabilistic description, 239–250
    Reliability, 237–238
Availability, 224–226
    Inherent (or intrinsic), 225, 337
    Operational, 225, 337
    Uptime ratio, 224–225
Average, 66, 321–324

Baseline, 22–28
Bathtub chart, 213–214
Bias, 63–64, 70, 166
Blue ribbon defense panel, 301
Bottom-up cost estimating, 25–26
Built-in measuring and recording, 233–235
Built-in test, 237–239, 248–250
    Can-not duplicate, 216, 241, 248
    False alarm, 248–249
    Fault detection, 248–249
    Percent fault detection, 248–249
    Percent retest OK, 248–249
Burn-in, 213–215

Calibration and alignment, 63, 69–70
Can-not duplicate, 216, 241, 248
Cause and effect relationship, 129
Chi-square, 141–143, 204–205, 327
Cinetheodolite, 67–68
Circular error average, 325, 329–333

Circular error probable, 118–119, 125–126, 329–333
Computer-aided analysis, 126–127
Concept, 99–110, 149
Conditional probabilities, 240–250
Confidence, 3–4, 36–38, 321–323
    Interval, 36–38, 321–323, 331–332
    Level, 321–323
    Mean, 321–323
    Sample population, 321–323
Confounded, 140
Correlation, 25–26
Cost analysis, 24–26
Cost and operational effectiveness analysis, 17–51, 254
Cost estimating, 25–26
Costs, 17–18, 24–27
    Acquisition, 25, 42–50
    Build-up, 24–26
    Development, 24–26
    Disposal, 24–26, 41–42
    Life cycle, 25, 41–43, 49–50
    Operation and support, 24–26, 41–42, 49–50
    Present value, 40–45
    Prior year, 25–45
    Production, 41
    R&E, 24–26, 41–45
    Risk, 36–38, 99–101
    Sunk, 25–45
    Then year, 25–45
    Thresholds, 27, 45–50
Costs and effectiveness comparisons, 29–50
Criteria, 105–106, 127–128, 262
Critical failure, 247
Critical issue/objective matrix, 154
Critical issues, 104–105, 114–116, 149–152, 261
Critical region, 122–123
Customer, 4

Data, 126–128
    Acquisition and control, 165–175
    Analysis, 106, 170–171
    Control, 167–170
    Management, 126–127, 165–176
    Reduction, 126–127, 166–170
Debugging, 180, 182–183

INDEX

Decision criteria, 23, 27–28, 106–107, 127–128, 141–143, 171
Decomposition, 76–77
Defining the problem, 54–55
Deflection Error Probable (DEP), 330–333
Degree of confidence, 113–114, 123–126, 128–141, 321–327
Degrees of freedom 134–136, 324, 333
Deming, W. Edwards, 2, 4–5
Design, 113–143
    Considerations, 113–114, 131
    Experimental T&E, 114, 128–141
    Limitations and assumptions, 128
    Procedural T&E, 114–128
    Sequential methods, 141
Design concept, 99, 103
Designed-in testing capabilities, 232–235
Deviation, 64–66, 321, 324, 329–334
Discrimination, 61
Distributed interactive simulation, 94–95
Dominant solution, 30–32, 106
Dry runs, 166–167
Dynamic error (or accuracy), 61–62
    Fidelity, 62, 271
    Speed of response, 62

Effectiveness, 53
Efficiency, 55–57
Electromagnetic environment testing, 57, 68–69
Electronic combat, 263–268
Ely, Eugene, 1
Empirical rule, 123
Engineering level T&E, 231–232
Engineering sequential photography, 69
Engineering systems theory, 76–77
Environment, 148–149
Environmental impact, 160
Errors, 62–65
    Experimental, 65–66
    Gross, 63
    Instrument, 62–64
    Measurement, 57–70
    Observational, 64
    Random, 64–65
    Residual, 64–65
    Systematic, 63–64
    Type I and Type II, 122
Estimation, 121–123
Evaluation criteria, 155–156
Experimental, 113–143
    Approach, 13–15, 54–65, 128–141, 152
    Data, 65–66
    Design, 113–114, 128–141
    Error, 59–65
    Execution, 165–171
    Phases, 8–9, 18–21
    Planning, 145–162
    Reporting, 161–162, 171–175
    Unit, 129

Experimentation, 113–143
Exponential failure distribution, 217–218, 247, 336

Face validation, 84
Factorial and fractional factorial designs, 139–140
Factors, 130–134
    Classification, 130
    Random, 130
    Treatment, 130–131
False alarm, 248–250
Fault detection, 237–240
Fault isolation, 237–240
Fault tolerant systems, 226–227
Feasibility study, 302–305
Federated systems, 230
Fidelity, 62, 271
Field/ flight test, 3–4, 54–57
Field/flight test ranges, 339–344
First order conditions, 46
Fisher's Exact Probability Test, 141–143
Force level comparisons, 32–36

Global positioning system, 68, 235
Graeco-Latin square, 138–139
Gross errors, 63

Habit interference, 272–273
Hardware-in-the-loop test facilities, 55–57
Human engineering, 192, 197–199
Human factors, 191–207
    Activity analysis, 195
    Analyzing equipment characteristics, 195–197
    Contingency analysis, 195
    Decision analysis, 195
    Fatigue and boredom, 197–199
    Flow, 195
    Functional analysis, 194
    Learning, 197
    Measuring performance, 200–205
    Operator controls, 196
    Optimum dimensions, 196–197
    Person-machine interfaces, 191
    Small sample considerations, 202–203
    System analysis, 192–199
    System evaluation, 199–201
    System functions, 193–194
    System tasks, 194–195
    Time-line analysis, 195
Hypothesis testing, 121–122, 322–324

Iconic models, 73
Infant mortality, 213–215
Installed test facilities, 55–57
Instrumentation, 59–60
Instrumentation errors, 61–65
Integrated systems, 226–227, 230

Integrated weapons systems, 229–235
Integration laboratories, 55–57
Isocost lines, 47–48
Isoquants, 47–48

Joint test and evaluation, 301–319
   Analysis and reporting, 311–312
   Examples, 313–317
   Feasibility study, 302–305
   General officer steering group, 304–305
   Lead service, 302
   Test design, 308–311
Joint test force, 13, 305–308
Joint testing, 13, 301–319
   Test organization and direction, 308

Kolmogorov-Smirnov (KS Test), 332–333
Kruskal-Wallis (H Test), 333–334

Laser time-space-position-information (TSPI), 67
Latin square, 138–139
Lead service, 302
Learning, 233–235, 272–273
Least costs, 26–28
Least significant, 134–136
Levels of models, 74–76
Limitations of experimental T&E design, 141
Line of sight, 327–328
Logistics supportability, 222

Maintainability, 211, 218–222
   Mean down time, 223–224, 336–337
   Mean time between maintenance, 225–226
   Mean time to repair, 218–219
Maintenance, 209–228
   Scheduled, 218
   Unscheduled, 218
Maintenance and support concept, 104, 219–221
   Integrated systems, 219–222
Maintenance personnel and training, 222
Marginal rate of substitution, 47
Mathematical model, 73–95, 134–141
Mature system, 209–212
Mean, 66, 321–324
Mean down time, 223–224, 336–337
Mean time between critical failure, 223, 247
Mean time between failure, 223
Mean time between maintenance, 225–226, 336–337
Mean time to repair, 225, 336–337
Measure of effectiveness, 17, 24, 77, 101–102, 118–119, 253–268, 276–300
Measure of performance, 77, 101–102, 118–119, 253–268
Measure theory, 57–58
Measurement, 57–65, 119–120, 253–270,
   Accuracy, 60–62

Discrimination, 61
Error, 62–65
Precision, 60
Resolution, 61
Sensitivity, 61
Static considerations, 60–61
Validity, 59–65
Measurement facilities, 55–57, 66–70, 92–95, 339–344
Measurement process, 59–60
Methodology, 113–114, 116–118, 154–157
Military utility, 253–255
Military worth, 253–255
Missing values, 140
Mission area analysis (or assessment), 253–255
Mission capable rate, 224
   Fully mission capable, 224
   Partially mission capable, 224
Missions, 258–261
Model code, 75–77, 81
Modeling, 73–95
Modeling and simulation, 73–95
Models, 23–24, 45–50
   Abstract, 73
   Accreditation, 23, 78–80
   Configuration control, 73
   Iconic, 73
   Level, 74–76
   Simulation, 73
   T&E, 54–57, 92–95
   Validation, 78–91
   Verification, 78–91
   Visual, 73

Non-Gaussian distributions, 331–333
Non-real time kill removal testing, 107–108
Nonparametric methods, 141–143, 326–327, 329–333
   Example problem, 141–143, 326–327, 329–333
Normal or Gaussian distribution, 122–123, 330–331

Objectives, 54–57, 104–105, 114–116, 154–157
Objectivity, 1–4, 13–15, 166
Observational error, 64
Operational concept, 21, 103–104
Operational effectiveness, 7–15, 209, 253–268
Operational level T&E, 232
Operational significance, 62, 106, 113, 123
Operational suitability, 209–212,
Operations optimization, 45–50
Operations research analysis, 1–3
Operations tasks, 129–130, 260–261
Optical time-space-position-information, 67–68
Output validation, 82–84
Overall T&E model, 54–57, 92–95

INDEX

Parametric, 25–27, 78
Parametric statistics, 121
Partial analysis, 45–49
Participating service, 302–305
Periodic evaluation and feedback, 168–170
Phase, 8–9, 18–20
Photography, 69
Planning, 145–162
    Advanced, 146–147
Precision, 60
Preliminary design, 304
Present value, 40–45
Pretest, 166–167
Probability, 38–40, 238–250
Procedural design, 114–128
Process, 4–5
Product, 5
Programming, 179–187
    Bottom-up, 184
    Top-down, 183–184
Project case file, 167–168

Qualitative measures, 57–65
Quality, 2, 4–5
Quantitative measures, 57–65
Questionnaires, 17–26, 201–205
Quick-look analyses, 127, 168–170

Radar time-space-position-information (TSPI), 67
Random, 64–66, 130
    Error, 64–65
    Test sequence, 114, 140–141
Randomization, 130, 140–141
Randomized blocks, 137–138
Range Error Probable (REP), 330–333
Range timing systems, 69
Rating scales, 201–205
    Interval, 202
    Justification, 202
    Nominal, 201
    Ordinal, 201
    Ratio, 202
Real time kill removal testing, 107–108
Real world, 3–4, 14–15, 113–114, 170–171, 231
Realistic conditions, 3–4, 14–15, 114, 170–171, 200, 231, 233, 271,306
Reconfiguration, 226–227, 233–235
Records, 167–168
Red flag measurement and debriefing system, 276
Regression analysis, 25–26
Reliability, 215–218, 237–250
    Assessment, 211, 215–218
    Burn-in, 213–214
    Bottom-up, 217–218
    Calculations, 302–303
    Combining components, 216–218
    Complex system, 217–218
    Debugging, 159
    Degradation, 215
    Design concepts and attributes, 162
    Errors, 159
    Failure, 215–216
    Fallback and recovery, 160
    Growth, 210
    Infant mortality, 218
    Maintenance, 218–222
    Maturity, 210
    Mean time between critical failure, 215–224
    Modification, 138
    Operational testing of, 232
    Parallel, 216–217
    Performance measurements, 253–270
    Probability of successful performance, 215
    Regression testing, 159
    Series, 216
    Suitability, 163
    Supportability, 163
    Test limitations, 163
    Time cycle, 212–213
Repeatability, 60, 65–66
Replication, 121–126
Reporting, 161–162, 171–175
    Activity, 161
    Final, 161, 171–177
    Interim summary, 161
    Service, 162
    Significant event, 161
    Status, 161
Reports, 171–175
    Editing, 175
    Gathering data, 172–173
    General, 171–172
    Organizing, 173
    Outlining, 173–174
    Planning, 172
    Writing, 174
Requirements, 5–9
Resolution, 61
Retest OK (RTOK), 249–250
Risk-cost modeling, 38
Risks, 28, 36–38
Roles and missions, 258–261
Roll-up/roll-down examination, 262–268
Root sum square, 61

Safety, 158
Sample, 65–66, 121–126
    Finite population, 325–326
    Infinite population, 325
    Size, 121
    Small, 202–205
Scatter, 64–66
Scenarios, 106–107
Scientific method (approach), 13–15, 166–170
Second order conditions, 46–47
Security, 158–160

Segregated systems, 230
Sensitivity analysis, 85–86, 129
Sequential methods, 141
Serviceability, 221–222
Shared databases, 101–102, 236
Significance, 113, 121–126
Software, 179–190
    Assessment, 184–187
    Attributes, 185–186
    Debugging, 180–182
    Errors, 182- 183
    Maintenance, 180
    Maturity, 186–187
    Suitability, 186–187
    Supportability, 187
    Test and evaluation, 179–190
Standard deviation, 66, 121–122, 321
    Estimation, 66, 121–122
    Units, 122
Static error, 60–61, 123–125
Statistical significance, 106, 113, 123
Structural validation, 84–85
Subjective assessments, 201–205
Suitability, 53
Sunk costs, 25
Surrogates, 108–109, 117, 235–236
Survivability assessment, 156
System acquiring, 5–15
System comparison level, 29–32
System concept level, 29
System cycles, 212–215
System degradation, 215
System design level, 29–36
System life cycle, 213–215
    Burn-in, 213–214
    Wearout, 214–215
System test and evaluation, 230–233
Systematic approach, 3–15, 113–114, 152, 166–170
Systematic error, 63–64, 69–70

t distribution, 322–323
Tactics, procedures, and training, 53–54
Tasks, 258–260
Tchebysheff's theorem, 123
Test and evaluation, 1–4, 10–15
    Analysis, 170–171
    Concept, 99–110, 170
    Conduct, 165–170
    Constraints, 113–116
    Design, 113–143
    DT&E, 10–11
    Evaluation, 4
    Facilities, 107, 339–344
    FOT&E, 12
    History, 1–3
    IOT&E, 11–12
    Issues, 101–105
    JT&E, 13, 301–319
    Management, 7–15, 157–161
    Method, 116–118, 154–157

    OT&E, 11
    PAT&E, 13
    Planning, 145–162
    Reporting, 171–175
    Test, 3–4
Test facilities, 339–344
Test variables, 128–130
    Concomitant, 130
    Dependent, 128–129
    Independent, 128–129
Testability, 226–227, 232–235
Threat, 18–19
Time and phase difference systems for time-space-position-information (TSPI), 67
Time-space-position-information (TSPI), 66, 273–276
Total error, 59–66
Tradeoff analysis, 26–28
Training, 53–54, 158, 197, 200, 271–300
    Fidelity, 271
    Measurement, 276–277
    Measures, 276–300
    Negative, 272–273
    Operators, 200
    Positive, 271–273
    Realism, 271–272
    System, 273–277
    Transfer, 197, 272–273
Transducer, 59
Transfer functions, 76–77
Transfer of training, 197, 271–273
True value, 60
Turn radius, 328
Typical range system measurement capabilities, 66–70

Uncertainty, 28, 36–38
Uptime ratio, 224–225
Utilization rate, 224

Variability, 65–67
Variables, 128–130
    Concomitant, 130
    Dependent, 128–129
    Independent, 128–129
    Random, 128–130
Variance, 66
Verification, validation, and accreditation, 23–24, 78–91
    Output, 82–83
    Plan, 88–89
    Structural, 84–85

Weapons systems, 1–16
Wright Brothers, 1

Youden square, 138–139